WEAK INTERACTIONS

Other books in this series

*

Colour Centres and Imperfections in Insulators and Semiconductors
P. D. TOWNSEND and J. C. KELLY

The Superconducting State
A. D. C. GRASSIE

Refrigeration and Thermometry below One Kelvin
D. S. BETTS

Plasma Physics
E. W. LAING

GRADUATE STUDENT SERIES IN PHYSICS

General Editor: Professor Douglas F. Brewer, M.A., D.Phil.
Professor of Experimental Physics, University of Sussex

WEAK INTERACTIONS

DAVID BAILIN

M.A., Ph.D.

School of Mathematical and Physical Sciences
University of Sussex

SUSSEX UNIVERSITY PRESS

1977

Published for
Sussex University Press
by
Chatto & Windus Ltd
40 William IV Street
London WC2N 4DF
*
Clark, Irwin & Co. Ltd
Toronto
*
Crane, Russak & Co. Inc
347 Madison Ave,
New York, N.Y. 10017

Library of Congress Catalog No. 75 34572

ISBN 0 8448 0851 2 (paper edition)

British Library Cataloguing in Publication Data
Bailin, David
Weak interactions. – (Graduate student series in physics).
Bibli. – Index.
ISBN 0-85621-023-4
1. Title 2. Series
539.7′54 QC794.8.W4
Weak interactions (Nuclear physics)

ISBN 0 85621 023 4

Printed and bound in Great Britain by
REDWOOD BURN LIMITED
Trowbridge & Esher

CONTENTS

PREFACE

The cornerstone of the modern theory of weak interactions was laid by Feynman and Gell-Mann[1] in 1958. Their theory, "the universal V-A current-current theory", was itself a development from the pioneering work of Fermi[2] twenty years earlier. For more than fifteen years the current-current theory was subjected to extensive experimental probing and survived virtually unscathed. With the passage of time, developments in other areas of particle physics led to refinement of the theory; it was formulated in terms of the strong interaction symmetry SU(3), and its richness was enhanced by the theoretical development and application of current algebra. The discovery of CP-violation in 1964 was not predicted, but the magnitude of the observed violation is so small as to be negligible in most areas of application of the theory. Thus today the theory is seen as providing an economical description of the vast majority of low energy weak phenomena.

For this reason the first five chapters of this book are devoted to an exposition of this theory and its application to experiment. Several excellent and full books on Weak Interactions have appeared in recent years[3], so I have not attempted to be exhaustive. Rather, I have sought to provide an introductory text which illustrates how one does actual calculations for processes accessible to experimentalists. The emphasis is on theory and calculation rather than experimental measurement, but naturally the theoretical predictions are compared with the current experimental data whenever possible. In doing this I have aspired to emulate Okun upon whose admirable little book[4] was reared a generation of students, including myself.

It has been known since its inception that the Feynman Gell-Mann theory could not possibly be correct at high energies, but, with only relatively low energies accessible, this problem was left for theorists to worry about when they had nothing else to do. However, in 1973 (after this book was started) an entirely new type of weak process was discovered which cannot be explained by the "old" theory: the new processes involve "neutral currents", as distinct from the "charged currents" in the Feynman Gell-Mann theory. The strength of the new interactions is comparable with that of the well-established phenomena, so a unified explanation is obviously called for.

The most natural framework in which to achieve this unification is provided by quantum field theory. In any case, it is my own prejudice that elementary particles are described by operator fields, so I have written this book using the formalism of quantum field theory. In writing it I have particularly had in mind first and second year graduate students of particle physics. Thus it seemed appropriate to include some material on those aspects of elementary relativistic free field theory which are especially important in weak interactions. This is presented in Chapter 2 as a preliminary to the definition of the "leptonic current". Space limitations precluded any coverage of Feynman diagrams or the LSZ reduction formulae. I anticipate that most readers will be following in parallel a course on these topics, if they do not already possess the minimal competence required here. In any case, all of these topics are developed extensively in other texts, as is the free field theory which I have included. In particular I have repeatedly referred to those by Bjorken and Drell[(5)]. The reader who is already familiar with this material can safely omit the first four sections of Chapter 2. Similar remarks apply to the first section of Chapter 4. This includes a modicum of SU(3) group theory prior to the statement of the Cabibbo hypothesis. This topic too is well-covered elsewhere[(6)].

The recent experimental developments, to which I have alluded above, have been matched by corresponding theoretical advances, both before and after the event. Naturally these theories are not (yet ?) grounded in experiment in the way that the Feynman Gell-Mann theory is. However their manifest power and elegance are, I hope, characteristic of the sort of theory we might eventually expect to have, and I feel that their pedagogic value outweighs their speculative nature. This material is contained in

Chapter 6.

The observation of CP-violation in weak processes has led to the tradition of this topic being discussed alongside known weak phenomena, even though the interaction responsible is not yet known to be weak. This book conforms with this tradition and the subject is dealt with in Chapter 7.

My view of this subject has of course been developed by continuous exposure to the literature and by personal contact with my colleagues in the international community of physicists. It is impossible to acknowledge individually all those who have helped me in this way. But I must at least apologise to those whose work is uncited either because of my own ignorance of it or because space limitations have prevented me from treating the subject as fully as it merits. I would like to thank Mrs. Jean Hafner for converting my almost illegible manuscript into a beautifully produced typescript. In the course of writing this book, as well as an earlier review, I have benefited from inumerable discussions with my colleagues at the University of Sussex. It is a pleasure to thank Professors J. P. Elliott, R.J. Tayler and Drs. G. Barton, J. Byrne, N. Dombey, W.D. Hamilton, D.R.T. Jones and A. Love for their continued advice. Dr. H.F. Jones of Imperial College read the final typescript and made a number of very helpful suggestions, for which I am most grateful. I also acknowledge my debt to Professor R.J. Blin-Stoyle. He urged me to write this book, encouraged me when I was wrestling with it and undertook a critical reading when it was finished. His helpful advice is much appreciated. My family, too, have participated, in a very real sense, in the production of this book. My wife's endurance of an absent and absorbed husband is repaid only in part by the dedication of this book: to Anjali.

David Bailin

Chapter 1

PRELIMINARY SURVEY

The weak interactions of elementary particles are characterised by the slowness of reactions in which they participate. Typically, the weak decay of a π^+ meson has a mean lifetime of 10^{-8} s, while the electromagnetic decay of a π^o has a lifetime of 10^{-16} s. However, within this class of slow processes there is a tremendous range of observed lifetimes; the neutron decays in 10^3 s, while the hyperons have lifetimes of the order of 10^{-10} s. Clearly, one would like to ascribe this variation to kinematical differences between the various decay modes rather than to some intrinsic difference in the dynamical interaction responsible for the decay.

Such a hope is motivated by the fantastic success of the theory of quantum electrodynamics (QED), which apparently can predict the observable properties of the electron (and muon) to arbitrary accuracy. This theory is formulated within the framework of relativistic quantum field theory, in which the electron and photon are represented by an operator field which is defined at each space-time point; the field is an "operator" in the sense that the electron's field, for example, "operates" on the space of physical states and can either destroy an electron or create a positron. The time development of any (Schrödinger picture) state is then determined by the specification of the Lagrangian or Hamiltonian for the system in terms of the field operators, just as in Classical Field Theory. The whole theory of QED is based on the assumption of a specific <u>simple</u> form for the Hamiltonian, the only input being the mass and charge of the electron.

In view of the success of field theory in QED, it is natural to hope that a similar approach to the weak interactions will yield an equally successful theory of all weak phenomena. In this book we shall see how far this approach has progressed. The theory is certainly not complete, as we shall see, though as at present formulated it has had a measure of success in explaining quite disparate weak phenomena. However, we shall see also that in its present form the theory is certainly wrong, in as much as it makes predictions which cannot be correct. It is, therefore, believed that the present version of the theory is an approximation to the 'correct' theory; the success of the present formulation is thought to be due to the fact that it has until quite recently been applied only to relatively low energy phenomena; more accurately, it has in fact been derived from low energy decay processes. In this connection, it is interesting that at the present time it is becoming feasible to perform weak scattering experiments using high energy neutrino beams. It may well be that these high energy processes will yield the clues needed to get at the next version of the theory. It is certainly true that those experiments already performed have yielded fascinating results and, perhaps, the seeds of a new theory.

We have already noted that the weak interactions are in general significantly slower than electromagnetic interactions. It is, therefore, to be expected that we may calculate weak matrix elements by using relativistic perturbation theory, just as one does in QED. In this latter case, the amplitude for any electromagnetic process is expanded as a power series in e, the electron charge; in the language of quantum field theory we may regard e as the electromagnetic 'coupling constant', which determines the strength with which the electron and photon fields are coupled in the Hamiltonian. The quantity e, as we have already noted, is input and in 'natural' units ($\hbar = c = 1$) satisfies $\alpha \equiv e^2/4\pi = 1/137$. Since the fine structure constant α is small compared with unity, it is believed that the perturbation series converges in QED, and consequently that any matrix element can be calculated to arbitrary accuracy. The slowness of weak interactions compared with electromagnetic processes leads one to hope that the weak coupling of the fields in the Hamiltonian has a coupling constant which is even smaller than the fine structure constant α. However, as we shall see, it turns out that the weak coupling constant G is not dimensionless. Its magnitude is about 10^{-49} erg cm^3. In order to compare with α we must scale G with some dimensional quantity. In natural units, if we scale with the square of the nucleon mass, then $G m_n^2$ is about 10^{-5}. Thus, if the nucleon mass is a suitable unit with which to scale G, then one could hope that the perturbation approach is applicable to weak processes. Further, if this is so, then it is an excellent approximation to retain only the first term in the perturbation series, since hardly any weak processes are measured accurately to one part in 10^5. This is what is usually done, and it is the approach we shall adopt for most of this book.

However, it is important to appreciate from the outset that one does not have to scale G with the nucleon mass. It may well appear to be an appropriate unit for the relatively low energy decay phenomena from which the present form of the theory has been obtained by induction. At higher energies though, which are becoming available in the weak scattering experiments for example, it is not so clear what the scaling unit should be, although it would evidently have to be fairly massive, if it were to lead us to retain even the second term in the perturbation series. At very high energies, where it is anticipated that all masses are negligible, the only mass unit available is the centre-of-mass energy W. If W is sufficiently large, it is clear that eventually GW^2 is of the order of unity. This occurs when W is around 350 GeV, and at this enormous energy all terms in the perturbation series have the same order of magnitude. Thus at this energy, and beyond, the perturbation approach is useless, and we must solve the problem by 'other means', which have yet to be devised.

The study of weak interactions has been considerably simplified by the experimental observation that some of the elementary particles which occur in nature do not participate in strong interactions. This is of considerable importance, because, while it is possible to treat the weak interactions to lowest order in the perturbation series, as we have argued, it is clear that this is not possible for the strong interactions, which are characterized by the pion-nucleon coupling constant g satisfying $g^2/4\pi = 13.6$. The particles which do not have strong interactions are the photon γ, which has only electromagnetic interactions, the neutrinos and antineutrinos, ν_e, ν_μ, $\bar{\nu}_e$, $\bar{\nu}_\mu$, which only have weak interactions, and the electron and muon and their antiparticles, e^-, μ^-, e^+, μ^+, which have both electromagnetic and weak interactions. Excluding the photon, these particles are referred to collectively as the leptons. The other known particles, which do participate in strong interactions are defined as hadrons.

If, as a first approximation, we neglect the electromagnetic interactions of the charged leptons, then we may infer the precise way in which the leptonic field operators enter the weak Hamiltonian; an experimental study of the weak processes in which the leptons participate in principle enables one to determine the matrix element of the process being studied. The form of this matrix element precisely mirrors the way in which the lepton field operators enter the weak Hamiltonian, since there are no renormalization effects for the leptons, so long as we ignore electromagnetic and higher order weak effects which we assume to be small. As a result of such experimental studies, it is now believed that the leptons' field operators enter the weak Hamiltonian in a combination known as the "leptonic current". We shall define this notion precisely in the following chapters. For the moment we confine ourselves to some more general observations. The word "current" is used because, like the electromagnetic current in QED, it is a bilinear combination of the field operators which transforms as a vector under restricted Lorentz transformations. However, the analogy ends there; the use of the word "current" should not be taken as implying (necessarily) that the leptonic current is generated by a gauge transformation of some underlying field (i.e. a Noether current). And in certain respects the leptonic current is quite different from the electromagnetic current. We shall see that under orthochronous Lorentz transformations (i.e. including the discrete operation of space inversion or parity reversal P) the leptonic current is found to be the sum of vector and axial vector pieces. (The space components of a vector change sign under parity reversal, like the electric field $\underline{E}$, while those of an axial vector are unaltered, like the magnetic field $\underline{B}$). In contrast the electromagnetic current transforms just as a vector.

What can be said about the way in which the hadrons enter the weak Hamiltonian ? Let us restrict ourselves for the moment to that part of the Hamiltonian which is responsible for semileptonic processes, i.e. weak processes involving both hadrons and leptons. Since angular momentum is found to be conserved in all known processes, it follows that the weak Hamiltonian, or more precisely the Hamiltonian density $\mathcal{H}_w$, must be invariant under restricted Lorentz transformations. Thus, since we know that the leptons' fields enter $\mathcal{H}_w$ in the form of the leptonic current, it follows that the hadrons' fields enter $\mathcal{H}_w$ in some form which also transforms as a vector under restricted Lorentz transformations; only a vector can be coupled to another vector via their scalar product to give a scalar or invariant product. This vector combination of the hadronic fields is called the "hadronic current", the word "current" being used in the same sense as before. Like the leptonic current, the hadronic current is the sum of vector and axial vector pieces. Under orthochronous Lorentz transformations the resultant Hamiltonian is the sum of scalar and pseudoscalar pieces; the scalar part is invariant under parity reversal, while the pseudoscalar part changes sign. Thus $\mathcal{H}_w$ is not invariant under parity reversal. This implies that the transition probability for a semileptonic process contains observable pseudoscalars, and is therefore not invariant under parity reversal. The existence of parity violation in nuclear beta decays was established in 1957. As we shall see, the observations are most elegantly explained by a theory in which parity invariance is maximally violated. Since experiments have put very stringent limits on possible parity violation in strong and electromagnetic interactions, it has come to be believed that the only interactions which violate parity invariance are the weak interactions. Thus to some extent the existence of parity violation has now become one of the defining features of the weak processes.

One might hope that it would be possible to infer the precise form of the hadronic current in terms of the hadronic field operators, in just the same way that the form of the leptonic current has been derived. The semileptonic processes certainly determine, in principle, the matrix elements of the hadronic current between hadronic states. Unfortunately, however, this does not enable us to determine the form of the hadronic current itself, because there is no reliable way of calculating the effects of the strong interactions. The best that can be done is to characterise the hadronic current in terms of the quantities which are believed to be conserved by the strong interactions. For example, the observation of the beta decay of a neutron, $n \rightarrow p\, e^- \bar{\nu}_e$, indicates that the hadronic current contains a piece which has a non-vanishing matrix element between the initial neutron state and the final proton state. Thus this piece of the hadronic current must have zero hypercharge Y, since the neutron and proton have the <u>same</u> hypercharge and Y is conserved by the strong interactions. Likewise, we may infer that this piece of the hadronic current has third component of isospin I_3 equal to unity. In the same way, we may classify the pieces of the hadronic current according to their total isospin I, since isospin is believed to be an exact symmetry of the strong interactions. It is known, however, that the electromagnetic interactions do <u>not</u> conserve isospin. Thus, in order to test any hypothesis we may make about the isospin properties of the hadronic current, it is important that we are able to calculate the electromagnetic effects in any process in which the experimental accuracy is of the same order as the fine structure constant α.

In addition to the semileptonic decays of the type already discussed, there is another type, in which the hypercharge of the initial and final hadron states differ by one unit; for example, the beta decay of a lambda hyperon, $\Lambda \rightarrow p e^- \bar{\nu}_e$, must proceed by a piece of the hadronic current which has $Y = 1$, $I_3 = \frac{1}{2}$. No semileptonic decays have ever been observed in which the hypercharges of the initial and final hadron states differ by two or more units; for example, the decay $\Xi^0 \rightarrow p e^- \bar{\nu}_e$ has not been observed. Thus until very recently it was believed that all semileptonic processes were accounted for by a Hamiltonian involving only the leptonic current and the two pieces of the hadronic current which we have mentioned. However, it is now clear that a different type of semileptonic process exists, which cannot be described by such a Hamiltonian. In the "common" semileptonic processes the leptons invariably carry away one unit of charge. This is apparent from the examples which we have given, and it can also be deduced from the Gell-Mann-Nishijima formula $Q = I_3 + \frac{1}{2} Y$; evidently $Q = 1$ if $(Y, I_3) = (0, 1)$ or $(1, \frac{1}{2})$. So the initial and final hadron states have charges differing by one unit which, by charge conservation, must be carried away by the leptons, $e^- \bar{\nu}_e$. In the newly discovered semileptonic processes the leptons carry away no charge, so plainly cannot be described by a Hamiltonian involving the hadronic current pieces we have mentioned. For this reason the new processes are called "neutral current" processes, although it certainly has not been established that the leptons or hadrons enter this weak Hamiltonian in the form of a current, even in the restricted sense in which we have used the term.

The observation that the hadronic current has two pieces, one with $(Y, I, I_3) = (0, 1, 1)$ and the other with $(Y, I, I_3) = (1, \frac{1}{2}, \frac{1}{2})$, naturally leads one to wonder whether these two pieces may be related somehow by the use of a "higher" symmetry of the strong interactions. Such a group would have to include the isospin group, SU(2), and the hypercharge group, U(1), as subgroups. The simplest group which has these properties is the group SU(3), which appears to be an approximate symmetry of the strong interactions. In this classification the eight pseudoscalar mesons $(\pi, K, \bar{K}, \eta)$ have the correct quantum numbers for inclusion in the eight dimensional, or octet, representation of the group. It is plain, therefore, that SU(3) is not an exact symmetry of the strong interactions; if it were, the eight pseudoscalar mesons would have the same mass, whereas experimentally it is observed that their masses are very different. However, unlike the electromagnetic interaction which violates the isospin invariance of the strong interactions, the nature of the interaction which breaks the unitary or SU(3) symmetry is not well understood. Some tentative hypotheses on the form of this symmetry violation have been advanced, but it is certainly true that its form is considerably less certain than that of the electromagnetic interaction. Thus, although we shall postulate that the two pieces of the hadronic current which we have discussed have a definite character with respect to the SU(3) group, it is not clear how accurate we may expect the resulting predictions to be. A problem of principle is then presented: in the event that our only partially accurate predictions are only approximately satisfied by experiment, how is one to decide whether the original hypothesis is "correct" or only "nearly correct" ? In this, as in other areas of physics, we are guided by essentially aesthetic considerations. It is generally believed, or more accurately hoped, that any "correct" theory in physics will not only be consistent with the experimental data, but also that it will be, in essence, "simple" in some immediately recognizable way. In particle physics this belief is reinforced by the outstanding success of QED, which, as we have already remarked, is specified by an outstandingly simple Hamiltonian.

We have seen that the common semileptonic weak interactions are believed to be described by a Hamiltonian in which the leptonic and hadronic currents are coupled at the same space-time point. In addition to the semileptonic processes there are two other types of weak interaction. Firstly, there are the purely leptonic processes, in which only the leptons participate. Secondly, there are the purely hadronic processes involving only the hadrons.

The only purely leptonic process which has been extensively studied in the laboratory is the decay of the muon into an electron, a neutrino and an antineutrino. The absence of any blurring of the point-like weak interaction, by unknown strong interactions, makes this process an ideal one in which to study the precise way in which the leptons' fields enter the weak Hamiltonian, particularly since any electromagnetic effects can, in principle, be calculated reliably using the known electromagnetic interactions of the leptons. It turns out that a Hamiltonian for this process, which satisfies our requirements of experimental consistency and theoretical simplicity, can be written down by coupling the leptonic current to itself, or rather its hermitian conjugate, at the same space-time point. We emphasize that the reason this simple form of the Hamiltonian is actually believed is based on aesthetic criteria, since there is no doubt, as we shall see, that the experimental data can tolerate quite substantial departures from this preferred form.

The Hamiltonian responsible for the purely hadronic processes is much more difficult to tie down. The absence of any leptons denies us the probe with which to investigate the precise form of the weak Hamiltonian. We may certainly characterize the Hamiltonian in terms of the quantities which are conserved, or approximately conserved, by the strong interactions, in just the same way as the hadronic current may be so specified. Of course, one would really like to go much further than this, and actually say something more specific about the actual form of the weak Hamiltonian. The most readily observed weak hadronic processes are the nonleptonic hyperon decays, in which, for example, a lambda hyperon decays into a pion and a nucleon, and the decay of the kaons into two or three pions. Such processes evidently do not conserve hypercharge, since the hypercharge of the initial and final states differs by one unit. On the other hand, no hadronic decays have been observed in which the hypercharge changes by more than one unit; the decay of the cascade (xi) hyperon into a pion and a nucleon has not been seen, while its decay into a lambda and a pion is seen. We have already seen how the semileptonic processes may be described by coupling the leptonic current to the hadronic current, and similarly that muon decay is described by coupling the lepton current to its hermitian conjugate. It is, therefore, natural to <u>hope</u> that the hadronic processes can be described by coupling the hadronic current to its hermitian conjugate. The resulting coupling of the piece of the hadronic current which conserves hypercharge to that piece which does not conserve hypercharge would certainly imply the existence of hadronic processes in which the hypercharge changes by just one unit. Equally, by coupling the hadronic current to its hermitian conjugate we ensure the absence of processes in which the hypercharge changes by two or more units. On the other hand, this also implies the existence of hadronic processes in which hypercharge is <u>conserved</u>. Since the strong interactions conserve hypercharge, one might suppose that such weak processes would constitute an unobservable correction to a dominantly strong interaction, which in any case cannot yet be calculated. However, as we have already observed, the currents we are considering are the sum of vector and axial vector pieces, so that the Hamiltonian which results from the current-current coupling does not conserve parity. Thus our hypothesis implies the existence of parity violation in what might otherwise be considered as strong processes. In particular, we expect there to be parity violating nuclear forces as a first order effect. Such effects have indeed been observed, and we may therefore suppose that the current-current hypothesis for the hadronic processes merits quantitative investigation. We shall see that this hypothesis does make predictions which are reasonably consistent with the experimental data. On the other hand, it does not predict some of the salient features of the data. Unfortunately, it is not at all clear whether this lack of predictive power results from our hypothesis being too "weak", or whether it is due to our inability to use all of the information contained in the hypothesis. Or perhaps other "currents", such as the newly discovered "neutral currents", participate in addition to the well-known currents. This is a continuing problem for the hadronic processes and one which we feel it is wise not to prejudge. We shall, therefore, study alternative "simple" hypotheses to see whether they are richer in accurate predictions than the current-current hypothesis.

We have already remarked that none of the Hamiltonians we have been discussing is invariant under the discrete operation of space inversion P. The same is true of their behaviour under the charge conjugation operation C, which changes the field operator of a particle into that of its anti-particle. However, all of the Hamiltonians are invariant under the product, CP, of these two operations, and most of the weak processes observed experimentally are consistent with this hypothesis. But there is, by now, no doubt that CP is not in fact an exact symmetry of nature. Both the semileptonic and hadronic decays of the long-lived neutral kaon provide unmistakable evidence of a small admixture from some interaction which violates CP-invariance. The trouble is that it is not at all clear which interaction is responsible for the violation, and it could be that it is not even a weak interaction. Since the only manifestation of the violation so far encountered has been in weak processes, it has become traditional to include this topic in discussions on the nature of the weak interaction. We shall maintain this tradition in this book, and a later chapter will be devoted to a consideration of some of the models which have been advanced to explain the observed violation.

However, our task in the main body of the book is to formulate the CP-invariant current-current hypothesis in a more precise way. We shall then be able to make theoretical predictions for all three types of weak processes and to compare these predictions with the experimental facts. After that we shall turn to the speculative attempts which have been made to give the whole subject a more acceptable theoretical basis. These unified field theories of the weak and electromagnetic interactions are designed to reproduce QED and the current-current theory, at least at low energies. But we shall see also how some of them naturally contain neutral currents of the type recently observed. It remains to be seen whether or not any of these new candidates is able to withstand the rigours of time and experiment as sturdily as its predecessors. Any survivor will surely have to be accorded a more central role in the next book on Weak Interactions.

Chapter 2

DIRAC FIELD THEORY AND THE LEPTONIC CURRENT

The leptonic current is defined as a certain bilinear combination of the leptons' fields, as we shall see. Since all of the leptons have spin $\frac{1}{2}$, it is therefore useful to summarize some of the properties of such fields before proceeding to discuss the leptonic current itself. One of the primary requirements of <u>any</u> theory in elementary particle physics is that it be relativistically invariant. That is to say, we require that the laws of motion have the same form in different inertial frames. To ensure that this is the case in weak interaction theory we must know the behaviour of the fields and currents under Lorentz transformations. We therefore start this chapter by defining our relativistic notation. Throughout the book we use natural units, i.e. $\hbar = c = 1$.

2.1 RELATIVISTIC NOTATION

We denote the coordinates of a space-time point x by x^μ ($\mu = 0,1,2,3$), where $x^o = t$ is the time coordinate and x^i ($i = 1,2,3$) are the spatial coordinates. In general, Greek indices are to be understood to refer to the Lorentz coordinates and to take the values 0,1,2,3, while Latin indices refer to the Euclidean coordinates and take the values 1,2,3. We shall on different occasions write

$$x = (x^o, x^i) = (x^o, \underline{x}) . \tag{2.1}$$

We also use the metric tensor $g_{\mu\nu}$ with components satisfying

$$g_{oo} = -g_{11} = -g_{22} = -g_{33} = +1 \tag{2.2a}$$

$$g_{\mu\nu} = 0 \quad (\mu \neq \nu) \quad . \tag{2.2b}$$

We then define

$$x^2 \equiv x.x = g_{\mu\nu} x^\mu x^\nu \quad , \tag{2.3}$$

where we are using the convention that repeated suffixes are to be summed over all permissible values. Thus (2. 3) together with (2. 1) and (2. 2) imply

$$x^2 = (x^o)^2 - (x^1)^2 - (x^2)^2 - (x^3)^2 = t^2 - \underline{x}^2 . \tag{2.4}$$

A Lorentz transformation Λ is defined as a real linear transformation of the coordinates which leaves the quantity x^2 invariant. Thus if

$$x'^{\mu} = \Lambda^{\mu}{}_{\rho}\, x^{\rho} \tag{2.5}$$

is a Lorentz transformation, we require

$$x'^2 = x^2 . \tag{2.6}$$

Thus the matrix Λ must satisfy

$$g_{\rho\sigma} = \Lambda^{\mu}{}_{\rho}\, g_{\mu\nu}\, \Lambda^{\nu}{}_{\sigma} , \tag{2.7a}$$

or

$$g = \Lambda^T g \Lambda , \tag{2.7b}$$

using matrix notation; the superfix T indicates that the transpose is to be taken. In general, we define any quantity A as a contravariant vector, if under the Lorentz transformation (2. 5) its coordinates A' transform like the space-time coordinates; that is

$$A'^{\nu} = \Lambda^{\nu}{}_{\sigma}\, A^{\sigma} . \tag{2.8}$$

Similarly, we may define the scalar product of any two vectors A, B by

$$A.B = g_{\mu\nu}\, A^{\mu} B^{\nu} = A^o B^o - \underline{A}.\underline{B} . \tag{2.9}$$

It is then clear from (2. 7) that this product is invariant under Lorentz transformations. We may use the tensor $g_{\mu\nu}$ to define a covariant vector A_{μ} from any contravariant vector A^{ν}. Thus

$$A_{\mu} \equiv g_{\mu\nu} A^{\nu} = (A^o , -\underline{A}) . \tag{2.10}$$

Then it follows that the scalar product of any two vectors A. B may be written

$$A.B = A_{\mu} B^{\mu} = A^{\mu} B_{\mu} . \tag{2.11}$$

Now if A^{μ} is a contravariant vector, its behaviour under a Lorentz transformation is given by (2.8). Thus using (2.7) we see that

$$\Lambda^{\mu}{}_{\rho}\, g_{\mu\nu}\, A'^{\nu} = g_{\rho\sigma}\, A^{\sigma} , \tag{2.12}$$

and from (2.10)

$$\Lambda^{\mu}{}_{\rho}\, A'_{\mu} = A_{\rho} \tag{2.13}$$

This equation specifies the transformation law for covariant vectors. We may define a tensor $g^{\mu\nu}$ by the requirement

$$g^{\mu\nu}\, g_{\nu\lambda} = \delta^{\mu}{}_{\lambda} , \tag{2.14}$$

where $\delta^{\mu}{}_{\lambda}$ is the Kronecker delta: $\delta^{\mu}{}_{\lambda} = 1$, if $\mu = \lambda$, $\delta^{\mu}{}_{\lambda} = 0$, if $\mu \neq \lambda$. Then, in fact,

$$g^{\mu\nu} = g_{\mu\nu} . \tag{2.15}$$

Thus we may invert (2.10), so that

$$A^{\mu} = g^{\mu\nu}\, A_{\nu} . \tag{2.16}$$

In general we may use $g^{\mu\nu}$ and $g_{\mu\nu}$ to raise and lower suffixes. In particular we may invert (2.8) and (2.13):

$$A'_{\nu} = \Lambda_{\nu}{}^{\sigma}\, A_{\sigma} \tag{2.17a}$$

$$A^{\rho} = \Lambda_{\mu}{}^{\rho}\, A'^{\mu} . \tag{2.17b}$$

It follows from (2.7b) that

$$\det \Lambda = \pm 1 , \tag{2.18}$$

and from (2.7a), taking $\rho = \sigma = 0$

$$| \Lambda^{o}{}_{o} | \geq 1 . \tag{2.19}$$

Thus the Lorentz group may be divided into four sets characterized by the sign of the determinant and the sign of $\Lambda^{o}{}_{o}$. Those with $\Lambda^{o}{}_{o} \geq +1$ are called <u>orthochronous</u>; they have the property of mapping the forward light cone $\{x : x^{o} > 0,\ x^{2} > 0\}$ into itself. The orthochronous transformations which also have $\det \Lambda = +1$ are called <u>restricted</u>. In particular it is clear that the identity transformation $(x' = x)$ belongs to the restricted set, which is in fact a group.

2.1.1 INFINITESIMAL LORENTZ TRANSFORMATIONS

Let us consider an infinitesimal Lorentz transformation, that is, one which differs from this identity by an infinitesimal amount; such a transformation evidently has

$$\Lambda^{\mu}{}_{\nu} = \delta^{\mu}{}_{\nu} + \epsilon\lambda^{\mu}{}_{\nu} \,, \tag{2.20}$$

where ϵ is infinitesimal. Then it follows from (2.7a), dropping the ϵ^2 terms, that

$$\lambda_{\rho\sigma} = -\lambda_{\sigma\rho} \,. \tag{2.21}$$

Thus it is clear that the general infinitesimal transformation can be expressed in terms of a linear combination of six independent "generators", since a general matrix satisfying (2.21) has six independent elements. This may be understood physically as well; three of the generators are responsible for infinitesimal rotations about the three spatial axes (i.e. they have $x'^{o} = x^{o}$ and $x'^{i} = x^{i}$ for a rotation about the x^{i} axis); the other three generate infinitesimal boosts in the three spatial directions (i.e. they have $x'^{j} = x^{j}$ if $j \neq i$, for a boost in the x^{i} direction). For an infinitesimal rotation about the 3-axis for example, we have

$$x'^{o} = x^{o} \tag{2.22a}$$

$$x'^{1} = x^{1} + \epsilon x^{2} \tag{2.22b}$$

$$x'^{2} = -\epsilon x^{1} + x^{2} \tag{2.22c}$$

$$x'^{3} = x^{3} \tag{2.22d}$$

and from (2.20) the corresponding $\lambda^{\mu}{}_{\nu}$ satisfies

$$-i\lambda_{\rho\sigma} = \begin{pmatrix} 0 & 0 & 0 & 0 \\ 0 & 0 & i & 0 \\ 0 & -i & 0 & 0 \\ 0 & 0 & 0 & 0 \end{pmatrix} \equiv (J^{3})_{\rho\sigma} \,. \tag{2.23}$$

On the other hand an infinitesimal boost in the 3-direction, for example, has

$$x'^{o} = x^{o} - \epsilon x^{3} \tag{2.24a}$$

$$x'^{1} = x^{1} \tag{2.24b}$$

$$x'^2 = x^2 \tag{2.24c}$$

$$x'^3 = -\epsilon x^o + x^3 \quad . \tag{2.24d}$$

Thus in this case the corresponding $\lambda^\mu{}_\nu$ satisfies

$$-i\lambda_{\rho\sigma} = \begin{pmatrix} 0 & 0 & 0 & i \\ 0 & 0 & 0 & 0 \\ 0 & 0 & 0 & 0 \\ -i & 0 & 0 & 0 \end{pmatrix} \equiv (K^3)_{\rho\sigma} \; . \tag{2.25}$$

Note that both (2.23) and (2.25) satisfy (2.20) and multiplication by $-i$ ensures that they are hermitian. It is apparent from (2.23) and (2.25) that we may write the rotation generators J^i and the boost generators K^i as follows

$$J^i = \tfrac{1}{2}\epsilon^{ijk} M^{jk} \tag{2.26a}$$

$$K^i = M^{oi} \quad , \tag{2.26b}$$

where

$$(M^{\mu\nu})_{\rho\sigma} = i(\delta^\mu{}_\rho \delta^\nu{}_\sigma - \delta^\mu{}_\sigma \delta^\nu{}_\rho) = -(M^{\nu\mu})_{\rho\sigma} \; . \tag{2.26c}$$

Thus an arbitrary infinitesimal Lorentz transformation may be written

$$\Lambda^\rho{}_\sigma = \delta^\rho{}_\sigma + \tfrac{1}{2} i\,\omega_{\mu\nu} (M^{\mu\nu})^\rho{}_\sigma \quad , \tag{2.27a}$$

where

$$\omega_{\mu\nu} = -\omega_{\nu\mu} \; . \tag{2.27b}$$

It is easy to verify that the generators $M^{\mu\nu}$ satisfy the following Lie algebra

$$[M^{\mu\nu}, M^{\rho\sigma}] = -i(g^{\mu\rho} M^{\nu\sigma} + g^{\nu\sigma} M^{\mu\rho} - g^{\nu\rho} M^{\mu\sigma} - g^{\mu\sigma} M^{\nu\rho}), \tag{2.28}$$

where the invariant matrix product is understood, i.e.

$$(M^{\mu\nu} M^{\rho\sigma})_{\alpha\beta} \equiv (M^{\mu\nu})_{\alpha\gamma} (M^{\rho\sigma})^\gamma{}_\beta \; .$$

In fact, any finite restricted Lorentz transformation may be expressed in terms of these six infinitesimal generators. This is fairly clear from (2.7), since the 16 components of a general 4 x 4 matrix must satisfy the 10 conditions imposed by the symmetric condition (2.7) in order to be a Lorentz transformation. Thus the behaviour of any entity under infinitesimal Lorentz transformations enables us to determine its behaviour under a general restricted transformation.

2.1.2 SPACE INVERSION AND TIME REVERSAL

A general Lorentz transformation may be obtained by multiplying one of the restricted transformations by one of the following discrete transformations :

(i) Space inversion or parity transformation

$$x'^{o} = x^{o}, \quad x'^{r} = - x^{r} , \tag{2.29a}$$

i. e.

$$\Lambda(i_s) = \begin{pmatrix} 1 & 0 & 0 & 0 \\ 0 & -1 & 0 & 0 \\ 0 & 0 & -1 & 0 \\ 0 & 0 & 0 & -1 \end{pmatrix} \tag{2.29b}$$

(ii) Time reversal

$$x'^{o} = - x^{o}, \quad x'^{r} = x^{r} \tag{2.30a}$$

$$\Lambda(i_t) = - \Lambda(i_s) . \tag{2.30b}$$

(iii) Space-time reversal

$$x' = - x \tag{2.31a}$$

$$\Lambda(i_{st}) = \Lambda(i_s) \Lambda(i_t) = - I_4 . \tag{2.31b}$$

2.1.3 THE POINCARE GROUP

To establish the full relativistic invariance of our theory we shall require that it be invariant under Poincaré transformations. The Poincaré transformation $P = \{\Lambda, a\}$ is defined as a restricted Lorentz transformation Λ followed by a space-time translation a. Thus under such a transformation

$$x'^{\mu} = \Lambda^{\mu}_{\ \nu} x^{\nu} + a^{\mu} . \tag{2.32}$$

It is easy to see from the definition that the Poincaré transformations form a group with the product defined as

$$\{\Lambda_1, a_1\} \{\Lambda_2, a_2\} = \{\Lambda_1 \Lambda_2, \Lambda_1 a_2 + a_1\} . \tag{2.33}$$

As before, we may ask what the generators of this group are. Clearly, they include the six generators $M^{\mu\nu}$, of the restricted Lorentz group. To find the others we consider an infinitesimal translation $P = \{I, a\}$ for which

$$x'_\rho = x_\rho + a_\rho . \tag{2.34}$$

There are evidently four independent translation generators corresponding to the four possible directions of translation. We choose these to be P^λ, where

$$(P^\lambda)_\rho = -i\delta^\lambda_\rho . \tag{2.35}$$

Then the general infinitesimal translation is given by

$$x'_\rho = x_\rho + i a_\lambda (P^\lambda)_\rho . \tag{2.36}$$

The Poincaré group thus has ten generators in all, and their Lie algebra, which follows from (2.26c), (2.33) and (2.35), is seen to be

$$[M^{\mu\nu}, M^{\rho\sigma}] = -i(g^{\mu\rho} M^{\nu\sigma} + g^{\nu\sigma} M^{\mu\rho} - g^{\nu\rho} M^{\mu\sigma} - g^{\mu\sigma} M^{\nu\rho}) \tag{2.37a}$$

$$[M^{\mu\nu}, P^\lambda] = i(g^{\nu\lambda} P^\mu - g^{\mu\lambda} P^\nu) \tag{2.37b}$$

$$[P^\lambda, P^\rho] = 0 . \tag{2.37c}$$

From (2.37b, c) we see that the scalar quantity $P^2 \equiv P_\lambda P^\lambda$ commutes with all ten generators of the Poincaré group, and there is one other scalar which has this property. Let us define

$$W_\sigma = \tfrac{1}{2}\epsilon_{\sigma\mu\nu\lambda} M^{\mu\nu} P^\lambda , \tag{2.38}$$

where $\epsilon_{\sigma\mu\nu\lambda}$ is the totally antisymmetric invariant tensor with $\epsilon_{0123} = +1$, then it follows from (2.37) that

$$[M^{\mu\nu}, W^\sigma] = i(g^{\nu\sigma} W^\mu - g^{\mu\sigma} W^\nu) , \tag{2.39a}$$

$$[P^\lambda, W^\sigma] = 0 , \tag{2.39b}$$

$$[W^\rho, W^\sigma] = -i\epsilon^{\rho\sigma\mu\nu} W_\mu P_\nu . \tag{2.39c}$$

Then from (2.39a, b) it follows that the quantity $W^2 \equiv W_\sigma W^\sigma$ also commutes with all ten generators. Thus in any irreducible representation of the Poincaré group it follows from Schur's Lemma that the quantities P^2, W^2 are multiples of the identity and that their eigenvalue spectra may be used to classify the irreducible representations.

Now the fundamental hypothesis which underlies the whole of particle physics is that every physical state is represented by a unique vector in a linear vector space, and that this vector space is a representation space of the Poincaré group.

The Poincaré transformation, and therefore the infinitesimal generators, are represented by operators within this space. The fact that the quantities P^2 and W^2 commute with all of the generators means that their values are unaltered by a Poincaré transformation, since any finite transformation can be built up out of infinitesimal transformations. The physical interpretation of this is well known; the operator P_μ represents the energy-momentum vector, so that P^2 is the invariant mass squared or the centre-of-mass energy squared. The representations of physical interest are those with $P^2 = m^2 > 0$ and those with $P^2 = 0$.

2.1.4 MASSIVE REPRESENTATIONS OF THE POINCARE GROUP

We consider first those with $P^2 = m^2 > 0$. In this case it is easy to see that the sign of the energy, that is $\epsilon(P^0)$, is also unaffected by a Poincaré transformation; this follows because, if $P^2 = m^2 > 0$, then $|P^0| \geq m$, and this remains true after a Poincaré transformation, since P^2 is invariant. Thus $\epsilon(P^0)$ is also an invariant, since P^0 must vary continuously under the continuous transformations we are discussing. The physical interpretation of W follows from its definition (2.38);

$$W^0 = \underline{J} \cdot \underline{P} \tag{2.40a}$$

$$\underline{W} = \underline{J}\, P^0 + \underline{K} \times \underline{P} \tag{2.40b}$$

using (2.26). Then in the centre-of-mass frame, where $P = (m, \underline{0})$, we see that $W = (0, m\underline{J})$. Thus since W^2 is an invariant quantity,

$$W^2 = -m^2 \underline{J}^2 = -m^2 s(s+1), \tag{2.41}$$

where s is an integer or half an integer. This follows because of the well known identification of $\underline{J}$ with the angular momentum vector. Thus the irreducible representations with $m \neq 0$ are specified by the three quantities $\{m, s, \epsilon(P^0)\}$. To specify a basis within an irreducible representation we must select a complete set of commuting generators, and the basis is then the set of all eigenvectors of these operators. It is conventional to choose the complete set to be P_μ and one component of W, since it is apparent from (2.39c) that any two components do not commute. We may choose the component of W which is to be diagonalized to be W . N, where N is a unit space-like vector whose rest frame form is $N = (0, \underline{n})$, where $\underline{n}$ is a unit vector. Thus the general form for N is easily seen to be

$$N^o = (\underline{P} \cdot \underline{n}) \, (P^2)^{-\frac{1}{2}} \tag{2.42a}$$

$$\underline{N} = \epsilon(P_o) \left\{ \underline{n} + (\underline{P} \cdot \underline{n}) \, \underline{P} \, (P^2)^{-\frac{1}{2}} \left[(P^2)^{\frac{1}{2}} + |P^o| \right]^{-1} \right\} . \tag{2.42b}$$

Then in the centre-of-mass frame, where $P = (\pm m, \underline{0})$, we see that

$$W.N = -m \, \underline{J} \cdot \underline{n} = -m\zeta \; , \tag{2.43a}$$

where

$$\zeta = s, s-1, \ldots, -s, \text{ if } \underline{J}^2 = s(s+1) \; . \tag{2.43b}$$

The possible eigenvalues of ζ again follow from the angular momentum algebra satisfied by $\underline{J}$. Thus, in summary, we have chosen a basis of states within the irreducible representation $\{m, s, \epsilon(p_o)\}$ to be $|\underline{p}, \zeta\rangle$, where

$$P^2 \, | \underline{p}, \zeta \rangle = m^2 | \underline{p}, \zeta \rangle \tag{2.44a}$$

$$W^2 \, | \underline{p}, \zeta \rangle = -m^2 s(s+1) | \underline{p}, \zeta \rangle \tag{2.44b}$$

$$P^i \, | \underline{p}, \zeta \rangle = p^i \, | \underline{p}, \zeta \rangle \tag{2.44c}$$

$$P^o \, | \underline{p}, \zeta \rangle = p^o \, | \underline{p}, \zeta \rangle \; , \tag{2.44d}$$

where

$$p^o = \epsilon(p_o) \, (\underline{p}^2 + m^2)^{\frac{1}{2}}$$

$$W.N \, | \underline{p}, \zeta \rangle = -m\zeta \, | \underline{p}, \zeta \rangle \; , \tag{2.44e}$$

where

$$\zeta = s, s-1, \ldots -s \; .$$

2.1.5 MASSLESS REPRESENTATIONS OF THE POINCARE GROUP

We turn next to the case $P^2 = 0$, corresponding to massless particles like the photon or neutrino. As before, we may choose a basis of states which are eigenvectors of the operators P_μ. The only states of physical interest are those with non-zero eigevalues p_μ. Since these eigenvalues satisfy $p^2 = 0$, it is no longer possible to find a rest frame and the quantity $\epsilon(P_o)$ is still unaffected by a Poincaré transformation. We have already observed that W^2 is an invariant, and it turns out that the only case of physical interest is when the eigenvalue of W^2 is zero.

We define

$$\underline{W}_{\|} = (\underline{W} \cdot \underline{P}) \, \underline{P} \, (\underline{P}^2)^{-1} \tag{2.45a}$$

$$\underline{W}_{\perp} = \underline{W} - \underline{W}_{\|} \, . \tag{2.45b}$$

Now it follows from (2.38) that

$$W \cdot P = W_o P_o - \underline{W} \cdot \underline{P} = 0 \, . \tag{2.46}$$

Using the notation of (2.45) this implies that

$$W^2 = W_o^2 - \underline{W}_{\|}^2 - \underline{W}_{\perp}^2 = (\underline{W} \cdot \underline{P})^2 P_o^{-2} - (\underline{W} \cdot \underline{P})^2 (\underline{P}^2)^{-1} - \underline{W}_{\perp}^2 \, . \tag{2.47}$$

Then since $P^2 = 0$ and $W^2 = 0$, we have

$$\underline{W}_{\perp} = 0 \, . \tag{2.48}$$

It then follows from (2.40) that in this case

$$W_\mu = \lambda P_\mu \, , \tag{2.49a}$$

where

$$\lambda = (\underline{J} \cdot \underline{P}) \, P_o^{-1} \, . \tag{2.49b}$$

It is easy to see from (2.37) and (2.39) that the quantity λ commutes with all of the generators of the Poincaré group and is therefore an invariant in the same way as P^2, W^2 and $\epsilon(P^o)$ are. Since $P^2 = 0$, we see that for positive energy representations $P_o = |\underline{P}|$, so that λ is just the component of angular momentum along the momentum. λ is called the <u>helicity</u>. Its eigenvalues are evidently integers or half odd integers. Thus, in summary, the irreducible representations of the Poincaré group with $P^2 = W^2 = 0$ are labelled by the two quantities $\{\epsilon(p^o), \lambda\}$ and the basis states within this representation may be chosen to be $|\underline{p}\rangle$, where

$$P^2 |\underline{p}\rangle = 0 \tag{2.50a}$$

$$W^2 \,|\, \underline{p} > \; = \; 0 \qquad (2.50b)$$

$$P^i \,|\, \underline{p} > \; = \; p^i |\underline{p} > \qquad (2.50c)$$

$$P^o \,|\underline{p} > = p^o \,|\underline{p} > \;, \quad \text{where} \quad p^o = \epsilon(p^o)\,|\underline{p}| \qquad (2.50d)$$

$$W_\mu \,|\, \underline{p} > \; = \; \lambda\, P_\mu |\underline{p} > \quad . \qquad (2.50e)$$

2.2 THE DIRAC EQUATION

The time development of the Schrödinger state describing a single free electron was derived originally by Dirac[1]. The wave function for such a state is $\psi_\alpha(x)$, where $\alpha = 1,2,3,4$ is a 'spinor index', and it satisfies the Dirac equation :

$$(i\gamma^\mu_{\alpha\beta}\,\partial_\mu - m\delta_{\alpha\beta})\,\psi_\beta(x) = 0 \;, \qquad (2.51a)$$

where

$$\partial_\mu \equiv \frac{\partial}{\partial x^\mu} \quad . \qquad (2.51b)$$

The matrices γ^μ $(\mu = 0,1,2,3)$ satisfy the Clifford Algebra :

$$\gamma^\mu\gamma^\nu + \gamma^\nu\gamma^\mu = 2\,g^{\mu\nu}\,I_4 \;. \qquad (2.52)$$

2.2.1 POINCARE INVARIANCE OF THE DIRAC EQUATION

To establish the Poincaré invariance of the Dirac equation we must show that it satisfies two requirements. The first is that, given the wave function $\psi(x)$, which describes the motion of the state for an observer O in an inertial frame, we can find a prescription which enables us to calculate the wave function $\psi'(x')$, which describes the same state for an observer O' in another inertial frame. The second requirement is that the equation of motion has the same form in the two inertial frames. Thus we are required to show that the wave function $\psi'(x')$ satisfies the equation

$$(i\gamma^\mu\partial'_\mu - m)\,\psi'(x') = 0 \quad , \qquad (2.53)$$

where we leave the summation over spinor indices understood, and $\partial'_\mu \equiv \partial/\partial x^{\mu'}$. If a point P has coordinates x in the frame of reference of O and x' in the frame of reference of O', then since the two frames are inertial frames the coordinates x and x' are related to each other by a Poincaré transformation (2.32). So

$$x'^{\mu} = \Lambda^{\mu}_{\ \nu} x^{\nu} + a^{\mu} , \tag{2.54}$$

and

$$\partial_{\nu} = \frac{\partial x'^{\mu}}{\partial x^{\nu}} \partial'_{\mu} = \Lambda^{\mu}_{\ \nu} \partial'_{\mu} . \tag{2.55}$$

Thus from (2.51) we find

$$(i \gamma^{\nu} \Lambda^{\mu}_{\ \nu} \partial'_{\mu} - m) \psi(\Lambda^{-1}(x' - a)) = 0 . \tag{2.56}$$

Now the 4 x 4 matrices

$$\gamma'^{\mu} \equiv \Lambda^{\mu}_{\ \nu} \gamma^{\nu} \tag{2.57}$$

satisfy the Clifford Algebra

$$\gamma'^{\mu} \gamma'^{\nu} + \gamma'^{\nu} \gamma'^{\mu} = 2 g^{\mu\nu} I_4 .$$

This follows from the definition (2.57) and equations (2.7) and (2.52). Now it can be shown that any two representations of the Clifford Algebra by 4 x 4 matrices are equivalent, and it follows that there exists a non-singular matrix $S(\Lambda)$ having the property

$$\gamma'^{\mu} = \Lambda^{\mu}_{\ \nu} \gamma^{\nu} = S(\Lambda)^{-1} \gamma^{\mu} S(\Lambda) , \tag{2.58a}$$

where

$$\det S(\Lambda) = 1 \tag{2.58b}$$

fixes the normalization convention for the matrices S. Substituting into (2.56) then gives

$$(i \gamma^{\mu} \partial'_{\mu} - m) S(\Lambda) \psi(\Lambda^{-1}(x' - a)) = 0 . \tag{2.59}$$

Thus if we write

$$\psi'(x') = S(\Lambda) \psi(\Lambda^{-1}(x' - a)) , \tag{2.60}$$

we see that $\psi'(x')$ satisfies (2.53), and that (2.60) gives the prescription for calculating ψ' in terms of ψ. Thus we have established the Poincaré invariance of the Dirac equation.

Let us now find the precise form of the Poincaré group generators.

For a pure translation $\Lambda^{\mu}{}_{\nu} = \delta^{\mu}{}_{\nu}$ and it is clear that $S(\Lambda) = I_4$. Thus for an infinitesimal translation a

$$\psi'(x') = \psi(x' - a) \tag{2.61a}$$

$$= \psi(x') - a^{\lambda} \partial'_{\lambda} \psi(x') + O(a^2) \tag{2.61b}$$

$$= (1 + i\, a^{\lambda} P_{\lambda}) \psi(x') + O(a^2) , \tag{2.61c}$$

and so

$$P_{\lambda} = i\, \partial_{\lambda} . \tag{2.62}$$

As required P_{μ} satisfies (2.37c) and we may rewrite (2.51)

$$(\not{P} - m)\, \psi(x) = 0 ; \tag{2.63}$$

here and elsewhere we use the Feynman notation

$$\not{Q} \equiv \gamma^{\mu} Q_{\mu} . \tag{2.64}$$

Now

$$(\not{P})^2 = \gamma^{\mu} P_{\mu} \gamma^{\nu} P_{\nu} = \tfrac{1}{2} (\gamma^{\mu} \gamma^{\nu} + \gamma^{\nu} \gamma^{\mu})\, P^{\mu} P^{\nu} = P^2 . \tag{2.65}$$

Thus it follows from (2.63) that

$$P^2 \psi(x) = (\not{P})^2 \psi(x) = m^2 \psi(x) , \tag{2.66}$$

and we see that $\psi(x)$ is a representation of the Poincaré group with mass m.

Let us now consider an infinitesimal Lorentz transformation. We take $a = 0$ and Λ given by (2.27)

$$\Lambda^{\mu}{}_{\nu} = \delta^{\mu}{}_{\nu} - \epsilon\, (g^{\rho\mu} \delta^{\sigma}{}_{\nu} - g^{\sigma\mu} \delta^{\rho}{}_{\nu}) . \tag{2.67}$$

If we write in this case

$$S(\Lambda) = I_4 + i\, \epsilon\, s^{\rho\sigma} , \tag{2.68}$$

where the trace of $s^{\rho\sigma}$ is zero so as to ensure that (2.58b) is satisfied, it follows from (2.58a) that

$$[\gamma^{\mu}, s^{\rho\sigma}] = i\, (g^{\rho\mu} \gamma^{\sigma} - g^{\sigma\mu} \gamma^{\rho}) . \tag{2.69}$$

This may be solved using (2.52), with the result that

$$s^{\rho\sigma} = \frac{i}{4} [\gamma^{\rho}, \gamma^{\sigma}] \equiv \tfrac{1}{2} \sigma^{\rho\sigma} . \tag{2.70}$$

Further, with Λ given by (2.67)

$$\psi(\Lambda^{-1}x) = \psi[x^\mu + \epsilon(g^{\rho\mu}x^\sigma - g^{\sigma\mu}x^\rho)] \tag{2.71a}$$

$$= \psi(x) + \epsilon(g^{\rho\mu}x^\sigma - g^{\sigma\mu}x^\rho)\partial_\mu\psi(x) \tag{2.71b}$$

$$= [1 - i\epsilon(x^\sigma P^\rho - x^\rho P^\sigma)]\psi(x) \quad . \tag{2.71c}$$

Thus it follows from (2.60) that if we write

$$\psi'(x') = (1 + i\epsilon M^{\rho\sigma})\psi(x') \quad , \tag{2.72}$$

then

$$M^{\rho\sigma} = x^\rho P^\sigma - x^\sigma P^\rho + \frac{i}{4}[\gamma^\rho, \gamma^\sigma] \ . \tag{2.73}$$

It is easy to verify that $M^{\rho\sigma}$ and P^λ satisfy the commutation relations (2.37). Using (2.73) and (2.62) we may calculate W defined in (2.38)

$$W_\mu = \frac{i}{8}\epsilon_{\mu\rho\sigma\lambda}[\gamma^\rho, \gamma^\sigma]P^\lambda \quad . \tag{2.74}$$

Then using the defining relation (2.52) we may calculate W^2.

$$W^2 = -\tfrac{3}{4}P^2 \quad . \tag{2.75}$$

Thus

$$W^2\psi(x) = -m^2\tfrac{1}{2}(1 + \tfrac{1}{2})\psi(x) \tag{2.76}$$

using (2.66), and we see that ψ is a representation of the Poincaré group having spin $s = \frac{1}{2}$.

Now let us take the hermitian conjugate of the Dirac equation (2.51):

$$\psi^\dagger(x)(-i\gamma^{\mu\dagger}\overleftarrow{\partial}_\mu - m) = 0 \ . \tag{2.77}$$

Now it is clear from (2.52) that the matrices $\gamma^{\mu\dagger}$ also satisfy the Clifford Algebra. Thus there is a non-singular matrix A having the property

$$\gamma^{\mu\dagger} = A\gamma^\mu A^{-1} \ . \tag{2.78}$$

If we define

$$\overline{\psi}(x) = \psi^\dagger(x)A \quad , \tag{2.79a}$$

then

$$\overline{\psi}(x)(-i\gamma^\mu\overleftarrow{\partial}_\mu - m) = 0 \quad . \tag{2.79b}$$

Now, if $\psi(x)$ and $\varphi(x)$ are both solutions of the Dirac equation (2.51a), then using (2.79b) it follows that

$$\partial_\mu \{\bar{\psi}(x)\, \gamma^\mu \varphi(x)\} = 0 \ . \tag{2.80}$$

Hence, if we define, for fixed x^o,

$$(\psi, \varphi)_{x_o} \equiv \int d^3 x\, \bar{\psi}(x)\, \gamma^o \varphi(x) \ , \tag{2.81}$$

(2.80) shows that (ψ, φ) is in fact independent of x^o. Thus (ψ, φ) is an <u>invariant</u> scalar product, provided the matrix $A\gamma^o$ is hermitian and positive definite. We shall now show that both of these properties are true.

It follows from (2.78) that

$$[A^{\dagger -1} A, \gamma^\mu] = 0 \ , \tag{2.82a}$$

so that

$$A = a A^\dagger \ , \tag{2.82b}$$

where a is a constant. It is easy to see that the number a has the same value in all representations of the Clifford Algebra. For, suppose we consider another representation $\tilde{\gamma}_\mu$, then $\tilde{\gamma}_\mu$ must be equivalent to γ_μ; so there exists a non-singular matrix S such that

$$\tilde{\gamma}_\mu = S^{-1} \gamma_\mu S \ . \tag{2.83}$$

The matrix $\tilde{A}$ having the property corresponding to A satisfies

$$\tilde{\gamma}^\dagger_\mu = \tilde{A}\, \tilde{\gamma}_\mu\, \tilde{A}^{-1} \ . \tag{2.84}$$

It follows from (2.78) and (2.83) that we may take $\tilde{A}$ to be the matrix

$$\tilde{A} = S^\dagger A S \ . \tag{2.85}$$

With this choice, using (2.82b), we find

$$\tilde{A} = a\, \tilde{A}^\dagger \ ,$$

so that the constant a is the same in all representations.

Now consider the following representation of the Clifford Algebra

$$\gamma^o = \begin{pmatrix} I_2 & 0 \\ 0 & -I_2 \end{pmatrix} \qquad \gamma^i = \begin{pmatrix} 0 & \sigma^i \\ -\sigma^i & 0 \end{pmatrix} \ , \tag{2.86}$$

where σ^i are the 2 x 2 Pauli matrices, which are, of course, hermitian. Plainly in this representation γ^o is hermitian and γ^i are anti-hermitian. Thus we may take A to be the matrix γ^o. Hence, in this, and therefore in all representations, we may take a to be unity, so that

$$A = A^\dagger .$$

This ensures that $A\gamma^o$ is hermitian in all representations. Further, it follows from (2.83), (2.85) that

$$\tilde{A}\,\tilde{\gamma}^o = S^\dagger (A\gamma^o)\, S .$$

Thus since $A\gamma^o$ is positive definite in the representation (2.86), it is positive definite in all representations.

Let us consider the matrix $s^{\rho\sigma}$; from (2.70) and (2.78) we find

$$s^{\rho\sigma\dagger} = \frac{i}{4}[\gamma^{\rho\dagger}, \gamma^{\sigma\dagger}] = A\, s^{\rho\sigma} A^{-1} . \tag{2.87}$$

Thus in general $s^{\rho\sigma}$ is not hermitian, and correspondingly the matrix $S(\Lambda)$ for finite transformations is <u>not</u> in general unitary. So it follows that the solutions of the Dirac equation define a non-unitary representation of the Poincaré group. In fact, this might have been anticipated; the 4 x 4 matrices $S(\Lambda)$, as indicated, depend only upon the Lorentz transformation, and they therefore define a representation of the Lorentz group. The Lie Algebra of this group is given in (2.28), and it is easy to see that all of its finite dimensional irreducible representations are non-unitary (we ignore, of course, the trivial one-dimensional representation in which all of the generators are zero).

As an example, let us consider the 2-dimensional representations of the Lie Algebra (2.28). We may take J^3 and K^3 given in (2.26) to be the complete set of commuting generators with which we label our basis. Thus, there are just two representations of the Lie Algebra (2.28). Either

$$J^i = \tfrac{1}{2}\sigma^i \quad \text{and} \quad K^i = \tfrac{1}{2} i\sigma^i \tag{2.88a}$$

or

$$J^i = \tfrac{1}{2}\sigma^i \quad \text{and} \quad K^i = -\tfrac{1}{2} i\sigma^i . \tag{2.88b}$$

Thus in both of these representations the infinitesimal generators K^i are anti-hermitian, and as a result it follows that the finite pure Lorentz transformations will not have a unitary representation.

2.2.2 PARITY INVARIANCE OF THE DIRAC EQUATION

Returning now to the Dirac equation, it is apparent that our proof of its Poincaré invariance did not utilize the fact that the Lorentz transformation Λ is in fact a restricted transformation; the only property required of Λ was that it satisfied (2.7), and this is satisfied by all Lorentz transformations including those with det $\Lambda = -1$ and/or $\Lambda^o{}_o \leq -1$.

Consider first the parity transformation $\Lambda(i_s)$ defined in (2.29). Then the Dirac equation is parity invariant provided we can find a matrix $S(\Lambda(i_s))$ having the property (2.58) :

$$S^{-1} \gamma^o S = \gamma^o \tag{2.89a}$$

$$S^{-1} \gamma^r S = -\gamma^r \quad . \tag{2.89b}$$

Clearly the matrix S having the required property is γ^o which satisfies (2.89) by virtue of (2.52); it also ensures that

$$\det \gamma^o = \pm 1 \quad .$$

Thus, it follows that the wave function ψ^p in the parity reversed system is given by

$$\psi^p(x^o, \underline{x}) = \epsilon \gamma^o \psi(x^o, -x) \quad , \tag{2.90}$$

where ϵ is a phase factor. Thus the Dirac equation furnishes us with a 4-dimensional representation of the orthochronous Lorentz Group. In fact, it is easy to see that the 2-dimensional representations of the restricted Lorentz Group cannot constitute a representation of the orthochronous group. The matrix Π representing the parity transformation must satisfy

$$[\Pi, J^i] = 0 \tag{2.91a}$$

$$\{\Pi, K^i\} \equiv \Pi K^i + K^i \Pi = 0 \quad . \tag{2.91b}$$

These follow in precisely the way in which we deduced (2.28). Now any matrix which commutes with the J^i given in (2.88) evidently also commutes with the matrices K^i in (2.88). Thus the parity transformation cannot be accommodated within this representation. If we wish to incorporate parity within the representation we may do so by taking a 4-dimensional representation which is the direct sum of the two 2-dimensional representations given in (2.88). Thus we may take

$$2J^i = \begin{pmatrix} \sigma^i & 0 \\ 0 & \sigma^i \end{pmatrix} \qquad 2K^i = \begin{pmatrix} i\sigma^i & 0 \\ 0 & -i\sigma^i \end{pmatrix} . \tag{2.92}$$

Then if we choose Π to be the matrix which interchanges the two representations it will clearly satisfy (2.92). Thus we may take

$$\Pi = \begin{pmatrix} 0 & I_2 \\ I_2 & 0 \end{pmatrix} \equiv \gamma^o , \tag{2.93}$$

where we have chosen the normalization of Π so that it satisfies $\Pi^2 = 1$, which is a defining property of γ^o. It then follows from (2.92) and (2.70) that

$$\gamma^i = \begin{pmatrix} 0 & -\sigma^i \\ \sigma^i & 0 \end{pmatrix} . \tag{2.94}$$

Note that in <u>this</u> representation also γ^o is hermitian, γ^i are anti-hermitian and the matrix A in (2.78) can be taken to be γ^o; then, as required, A is hermitian and $A\gamma^o$ is positive definite.

2.2.3 BILINEAR COVARIANTS

Now it follows from (2.87) that the matrix $S(\Lambda)$ in (2.68) which represents the infinitesimal Lorentz transformation (2.67) satisfies

$$S(\Lambda)^\dagger = A\,S(\Lambda)^{-1} A^{-1} . \tag{2.95}$$

Likewise the matrix

$$S(\Lambda(i_s)) = \epsilon\gamma^o , \tag{2.96}$$

representing the parity reversal operation given in (2.90), satisfies a similar equation

$$S(\Lambda(i_s)) = A\,S(\Lambda(i_s))^{-1} A^{-1} \tag{2.97}$$

by virtue of (2.78). Thus for any <u>orthochronous</u> Lorentz transformation Λ, we have from (2.60)

$$\overline{\psi'(x')} \equiv \psi'(x')^{\dagger} A$$
$$= \psi(\Lambda^{-1}(x'-a))^{\dagger} S(\Lambda)^{\dagger} A$$
$$= \psi(\Lambda^{-1}(x'-a))^{\dagger} A S(\Lambda)^{-1}$$
$$= \overline{\psi}(\Lambda^{-1}(x'-a)) S(\Lambda)^{-1} . \tag{2.98}$$

Hence,

$$\overline{\psi'(x')}\, \psi'(x') = \overline{\psi}(x)\, \psi(x) , \tag{2.99}$$

and we see that this combination is invariant under <u>any</u> orthochronous Lorentz transformation. It is therefore called a <u>scalar</u> combination. If we define

$$\gamma_5 \equiv \frac{i}{4!} \epsilon_{\lambda\mu\rho\sigma} \gamma^{\lambda} \gamma^{\mu} \gamma^{\rho} \gamma^{\sigma} = i \gamma^{o} \gamma^{1} \gamma^{2} \gamma^{3} , \tag{2.100}$$

it follows using (2. 58a) that

$$\overline{\psi'(x')}\, \gamma_5 \psi'(x') = \frac{i}{4!} \epsilon_{\mu_o\mu_1\mu_2\mu_3} \Lambda^{\mu_o}{}_{\nu_o} \Lambda^{\mu_1}{}_{\nu_1} \Lambda^{\mu_2}{}_{\nu_2} \Lambda^{\mu_3}{}_{\nu_3}$$

$$\overline{\psi}(x) \gamma^{\nu_o} \gamma^{\nu_1} \gamma^{\nu_2} \gamma^{\nu_3} \psi(x) = (\det \Lambda)\, \overline{\psi}(x) \gamma_5 \psi(x) . \tag{2.101}$$

Thus the combination $\overline{\psi} \gamma_5 \psi$ is invariant under restricted Lorentz transformations, but changes sign if the orthochronous transformation includes parity reversal. For this reason the combination is called <u>pseudoscalar.</u> Similarly, we may show that

$$\overline{\psi}'(x') \gamma^{\lambda} \psi'(x') = \Lambda^{\lambda}{}_{\mu} \overline{\psi}(x) \gamma^{\mu} \psi(x) \tag{2.102}$$

$$\overline{\psi}'(x') \gamma^{\lambda} \gamma_5 \psi'(x') = (\det \Lambda) \Lambda^{\lambda}{}_{\mu} \overline{\psi}(x) \gamma^{\mu} \gamma_5 \psi(x) , \tag{2.103}$$

and these combinations are called respectively the <u>vector</u> and <u>axial vector</u> covariants. Finally, we have

$$\overline{\psi}'(x') s^{\lambda\mu} \psi'(x') = \Lambda^{\lambda}{}_{\rho} \Lambda^{\mu}{}_{\sigma} \overline{\psi}(x) s^{\rho\sigma} \psi(x) , \tag{2.104}$$

which is called the <u>tensor</u> covariant. Now, it follows from (2.100) and (2. 52) that

$$\epsilon_{\mu\rho\sigma\lambda} [\gamma^{\rho}, \gamma^{\sigma}] = 2 i [\gamma_{\mu}, \gamma_{\lambda}] \gamma_5 . \tag{2.105}$$

Hence, the <u>pseudotensor</u> covariant $\overline{\psi} s^{\lambda\mu} \gamma_5 \psi$ is expressible in terms of the tensor covariant :

$$\bar{\psi}\, s_{\mu\lambda}\, \gamma_5\, \psi \;=\; -\frac{i}{2}\, \epsilon_{\mu\rho\sigma\lambda}\, \bar{\psi}\, s^{\rho\sigma}\, \psi \;. \tag{2.106}$$

2.2.4 TIME REVERSAL INVARIANCE OF THE DIRAC EQUATION

Let us now consider the behaviour of the Dirac equation under the time reversal transformation $\Lambda(i_t)$ given in (2.30). To orient ourselves let us consider first the free Schrödinger equation

$$i\,\dot{\psi}\,(t, \underline{x}) \;=\; -\frac{1}{2m}\;\nabla^2\,\psi(t, \underline{x}) \;. \tag{2.107}$$

Its plane wave solutions are given by

$$\psi(t, \underline{x}) \;=\; \exp[\,-i\,(E\,t - \underline{p}.\underline{x})\,] \tag{2.108a}$$

with

$$E \;=\; \frac{p^2}{2m} \;. \tag{2.108b}$$

We see that the wave function

$$\psi(-\,t, \underline{x}) \;=\; \exp[\,i\,(Et + \underline{p}\,.\,\underline{x})\,] \;, \tag{2.109}$$

which one might have expected to be the wave function describing the plane wave state in a time reversed frame, does not satisfy the equation of motion (2.107). Further the wave function $\psi(t, \underline{x})$ evidently describes a particle with momentum $\underline{p}$, as does $\psi(-t, \underline{x})$. Clearly, we would require that the wave function in the time-reversed frame would describe a particle of momentum $-\underline{p}$. In these two respects the wave function $\psi(-t, \underline{x})$ is unsatisfactory for describing the time-reversed state. On the other hand, the wave function

$$\psi^*(-t, \underline{x}) \;=\; \exp[\,-i\,(Et + \underline{p}.\underline{x})\,] \tag{2.110}$$

has neither of these defects. Further, since quantum mechanics is concerned with probabilities, which are just the modulus squared of relevant scalar products, it is easy to see that the connection (2.110) preserves the required probabilities. We are therefore led to the conclusion that the time-reversal transformation is anti-linear; that is to say that we are to take the complex conjugate of all complex numbers in the theory.

Motivated by these considerations we expect the solutions to the Dirac equation in a time-reversed frame to be linearly related to the complex conjugate of the original solution. We start from the transpose of (2.79b)

$$(-i\gamma^{\mu T}\partial_\mu - m)\,\bar{\psi}^T(x^o, \underline{x}) = 0 \quad , \tag{2.111}$$

and rewrite it in terms of the time reversed coordinates x' which are related to x by (2.30); namely $x'^o = -x^o$, $\underline{x}' = \underline{x}$. Then we have

$$(i\gamma^{oT}\partial'^o + i\gamma^{rT}\partial'^r - m)\,\bar{\psi}^T(-x'^o, \underline{x}') = 0 \ . \tag{2.112}$$

Premultiplying by γ^{oT} we have

$$(i\gamma^{\mu T}\partial'_\mu - m)\,\gamma^{oT}\,\bar{\psi}^T(-x'^o, \underline{x}') = 0 \ . \tag{2.113}$$

The matrices $\gamma^{\mu T}$ also satisfy the Clifford Algebra (2.52) and must therefore be equivalent to γ^μ; thus there is a non-singular matrix B having the property

$$\gamma^{\mu T} = B\gamma^\mu B^{-1} \ . \tag{2.114}$$

Hence, it follows that

$$(i\gamma^\mu\partial'_\mu - m)\,B^{-1}\gamma^{oT}\,\bar{\psi}^T(-x'^o, \underline{x}') = 0 \ , \tag{2.115}$$

and that the wave function $\psi^t(x')$ representing the same state in the time-reversed frame of reference is given by

$$\psi^t(x'^o, \underline{x}') = \eta\, B^{-1}\gamma^{oT}\,\bar{\psi}^T(-x'^o, \underline{x}') \quad , \tag{2.116}$$

where η is a complex number. Now, the defining properties of A, B given in (2.78) and (2.114) imply that

$$[B^{T-1}B, \gamma^\mu] = 0 \tag{2.117}$$

$$[B^{-1}A^* B^{-1\dagger}A, \gamma^\mu] = 0 \ . \tag{2.118}$$

Then proceeding as we did following (2.82) we can show that in <u>all</u> representations we may choose B so that

$$B = -B^T \tag{2.119}$$

$$B^{-1\dagger}A^* = A^{*-1}B \ . \tag{2.120}$$

Note that in the representation given in (2.93) and (2.94) we may take

$$A = \gamma^o \tag{2.121}$$

$$B = \gamma^1 \gamma^3 , \qquad (2.122)$$

and that these choices satisfy (2.78), (2.114), (2.119) and (2.120). It follows from (2.116) that

$$\overline{\psi^t} (x'^o, \underline{x}') = \eta^* \psi^T (-x'^o, \underline{x}') A^* \gamma^{o*} B^{-1\dagger} A . \qquad (2.123)$$

Hence, by virtue of (2.78) and (2.120)

$$\overline{\psi}^t (x'^o, \underline{x}') = \eta^* \psi^T (-x'^o, \underline{x}') \gamma^{oT} B . \qquad (2.124)$$

Thus taking η to be unimodular ensures that

$$\overline{\psi}^t (x') \psi^t (x') = \overline{\psi} (x) \psi (x) \qquad (2.125a)$$

$$\overline{\psi}^t (x') i\gamma_5 \psi^t (x') = - \psi (x) i\gamma_5 \psi (x) \qquad (2.125b)$$

$$\overline{\psi}^t (x') \gamma^\lambda \psi^t (x') = - \Lambda (i_t)^\lambda{}_\mu \overline{\psi} (x) \gamma^\mu \psi (x) \qquad (2.125c)$$

$$\overline{\psi}^t (x') \gamma^\lambda \gamma_5 \psi^t (x') = - \Lambda (i_t)^\lambda{}_\mu \overline{\psi} (x) \gamma^\mu \gamma_5 \psi (x) \qquad (2.125d)$$

$$\overline{\psi}^t (x') s^{\lambda\mu} \psi^t (x') = - \Lambda (i_t)^\lambda{}_\rho \Lambda (i_t)^\mu{}_\sigma \overline{\psi} (x) s^{\rho\sigma} \psi (x) \qquad (2.125e)$$

$$\overline{\psi}^t (x') s^{\lambda\mu} \gamma_5 \psi^t (x') = \Lambda (i_t)^\lambda{}_\rho \Lambda (i_t)^\mu{}_\sigma \overline{\psi} (x) s^{\rho\sigma} \gamma_5 \psi (x) . \qquad (2.125f)$$

Thus the scalar and pseudoscalar bilinear covariants have opposite behaviour under time reversal, as do the tensor and pseudotensor quantities. This should be contrasted with the vector and axial vector covariants which transform in the same way. This observation is of crucial significance for the theory of weak interactions, as we shall see subsequently.

2.2.5 NEGATIVE ENERGY AND CHARGE CONJUGATE SOLUTIONS OF THE DIRAC EQUATION

We have already observed that the Dirac Equation is Poincaré Invariant. However, the solutions of the equation admit of both positive and negative eigenvalues for P_o; this follows from (2.66). Physically these negative energy solutions can be related to the wave functions describing anti-particle states, as we shall now show.

The interaction of a charged particle with an external electromagnetic field is most simply obtained by making the "minimal" substitution

$$P_\mu \rightarrow P_\mu - e A_\mu \quad , \tag{2.126}$$

which is made in classical electrodynamics to describe the interaction of a point charge $+ e$ with an external vector potential A_μ. Thus the wave function of an electron having charge $- e$ in an external field satisfies

$$(i \gamma_\mu \partial^\mu + e \gamma_\mu A^\mu - m) \psi (x) = 0 \quad . \tag{2.127}$$

Since the external field is real we have, as in (2.111),

$$(- i \gamma_\mu^T \partial^\mu + e \gamma_\mu^T A^\mu - m) \overline{\psi}^T (x) = 0 \; . \tag{2.128}$$

The matrices $- \gamma_\mu^T$ satisfy the Clifford Algebra, so there is a non-singular matrix C satisfying

$$- \gamma_\mu^T = C^{-1} \gamma_\mu C \quad . \tag{2.129}$$

Thus, defining

$$\psi^c (x) = \omega C \overline{\psi}^T (x) \quad , \tag{2.130}$$

where ω is a phase factor, we see that ψ^c satisfies

$$(i \gamma_\mu \partial^\mu - e \gamma_\mu A^\mu - m) \psi^c (x) = 0 \; . \tag{2.131}$$

Clearly, therefore, ψ^c is the wave function of a state having charge $+ e$. Further, if $\psi(x)$ is a negative energy solution of the Dirac equation, $\psi^c (x)$ is a positive energy solution because of the complex conjugation involved in $\overline{\psi}^T$ in (2.130). In the absence of an external potential (2.131) is just the Dirac equation, so ψ^c describes a state of mass m and spin $\frac{1}{2}$. For these reasons we interpret $\psi^c(x)$ as being the wave function describing a positron state when $\psi(x)$ is a negative energy solution. The defining properties of A and C given in (2.78) and (2.129) imply that

$$[C^T C^{-1} , \gamma_\mu] = 0 \tag{2.132a}$$

$$[C A^* C^\dagger A, \gamma_\mu] = 0 \; . \tag{2.132b}$$

Proceeding as we did following (2.82), we can show in <u>all</u> representations we may choose C so that

$$C = - C^T \tag{2.133a}$$

$$A^* C^\dagger A = - C^{-1} . \qquad (2.133b)$$

Note that in the representation given in (2.93) and (2.94) we may take

$$A = \gamma^o , \qquad (2.134a)$$

$$C = \gamma^o \gamma^2 , \qquad (2.134b)$$

and that these choices satisfy (2.78), (2.129) and (2.133).

It follows from (2.130) that

$$\overline{\psi}^c(x) = \omega^* \psi^T(x) A^* C^\dagger A . \qquad (2.135)$$

Hence, by virtue of (2.133b)

$$\overline{\psi}^c(x) = - \omega^* \psi^T(x) C^{-1} . \qquad (2.136)$$

The following behaviour of the bilinear covariants may then be verified immediately :

$$\overline{\psi}^c \psi^c = - \overline{\psi} \psi \qquad (2.137a)$$

$$\overline{\psi}^c i \gamma_5 \psi^c = - \overline{\psi} i \gamma_5 \psi \qquad (2.137b)$$

$$\overline{\psi}^c \gamma^\lambda \psi^c = \overline{\psi} \gamma^\lambda \psi \qquad (2.137c)$$

$$\overline{\psi}^c \gamma^\lambda \gamma_5 \psi^c = - \overline{\psi} \gamma^\lambda \gamma_5 \psi \qquad (2.137d)$$

$$\overline{\psi}^c s^{\lambda\mu} \psi^c = \overline{\psi} s^{\lambda\mu} \psi \qquad (2.137e)$$

$$\overline{\psi}^c s^{\lambda\mu} \gamma_5 \psi^c = \overline{\psi} s^{\lambda\mu} \gamma_5 \psi . \qquad (2.137f)$$

2.2.6 PLANE WAVE SOLUTIONS OF THE DIRAC EQUATION

Let us first find the solutions of the Dirac equation which constitute a basis of the irreducible and non-unitary representation of the Poincaré group $\{m, \frac{1}{2}, +\}$ and which diagonalize the operators P_μ and W.N, as in (2.44). Clearly, the positive energy solutions which diagonalize P_μ are given by

$$\psi_p(x) = e^{-ipx} u(p) , \qquad (2.138a)$$

where

$$(\not{p} - m) u(p) = 0 \qquad (2.138b)$$

and

$$p_o = +(\underline{p}^2 + m^2)^{\frac{1}{2}} . \qquad (2.138c)$$

Then, from (2.62),

$$P_\lambda \psi_p(x) = p_\lambda \psi_p(x) , \qquad (2.139)$$

as required. To diagonalize W.N we note from (2.105) and (2.74) that

$$W.N = -\tfrac{1}{4}[\not{N}, \not{P}]\, \gamma_5 . \qquad (2.140)$$

Now it follows from the definition of γ_5 (2.100) that

$$\gamma_5 \gamma_\mu + \gamma_\mu \gamma_5 = 0 , \; (\gamma_5)^2 = 1 . \qquad (2.141)$$

Hence, from the definition of N given in (2.42),

$$(W.N)^2 = \tfrac{1}{4} P^2 .$$

Thus the eigenvalues of W.N are $-m(\pm\frac{1}{2})$, as stated in (2.43). The required eigenvectors are given by

$$\psi_{p,\pm\frac{1}{2}}(x) = u(p, \pm s)\, e^{-ipx} , \qquad (2.142a)$$

where, from (2.138), (2.139), (2.140),

$$\gamma_5 \not{s}\, u(p,\pm s) = \pm u(p, \pm s) . \qquad (2.142b)$$

s_μ is the eigenvalue of N_μ and, therefore, satisfies

$$s_o = (\underline{p} . \underline{n})\, m^{-1} , \; \underline{s} = \underline{n} + (\underline{p} . \underline{n})\, \underline{p}\, m^{-1} (m + p_o)^{-1} , \qquad (2.142c)$$

so that

$$s^2 = -1 , \quad s.p = 0 . \qquad (2.142d)$$

It then follows that, if, as in (2.79), we define

$$\bar{u}(p, \pm s) \equiv u(p, \pm s)^\dagger A ,$$

where A is specified in (2.78), then $\bar{u}(p, s)$ satisfies

$$\bar{u}(p, \pm s)(\not{p} - m) = 0 \qquad (2.143a)$$

$$\bar{u}(p, \pm s)\, \gamma_5 \not{s} = \pm \bar{u}(p, s) . \qquad (2.143b)$$

Hence

$$\bar{u}(p, \pm s)\, u(p, \mp s) = 0 . \qquad (2.143c)$$

Finally, we fix the normalization of our states by requiring

$$\bar{u}(p, \pm s)\; u(p, \pm s) = 2m \quad . \tag{2.144}$$

Then the Dirac equation and (2. 52) imply that

$$\bar{u}(p,s)\gamma_\mu u(p,s) = \frac{1}{2m}\bar{u}(p,s)\{\gamma_\mu \not{p} + \not{p}\gamma_\mu\} u(p,s) = 2p_\mu \tag{2.145a}$$

$$\bar{u}(p,s)\gamma_\mu\; u(p,-s) = 0 \quad . \tag{2.145b}$$

Likewise

$$\bar{u}(p,s)\, i\gamma_5\, u(p, \pm s) = 0 \tag{2.145c}$$

and

$$\bar{u}(p,s)\;\gamma_\mu\;\gamma_5\, u(p,s) = 2m s_\mu \quad . \tag{2.145d}$$

The operators which connect the different spin eigenstates within the same basis are plainly the ladder operators $W.(N^{(1)} \pm i N^{(2)})$, where $N^{(1),(2)}$ are the two unit space-like vectors perpendicular to each other as well as to P and $N \equiv N^{(3)}$. In the rest frame it follows from (2. 40) that

$$W.N^{(i)} = -m\underline{J}.\underline{n}^{(i)} \qquad (i = 1,2,3) \quad .$$

Plainly for a fixed value of p the spinors $u(p, \alpha)$ $(\alpha = \pm s)$ constitute a 2-dimensional representation of the rotation group. Hence

$$W.N^{(i)}\;\psi_{p,\alpha}(x) = -\frac{m}{2}\;\sigma^i_{\beta\alpha}\;\;\psi_{p,\beta}(x) \qquad (i = 1,2,3) \quad .$$

Then using (2. 140) we find that the spinors $u(p, \alpha)$ satisfy

$$\gamma_5\,\not{n}^{(i)}\; u(p,\alpha) = \sigma^i_{\beta\alpha}\;\; u(p,\beta) \qquad (i = 1,2,3) \quad ,$$

where $n^{(i)}$ are the eigenvalues of the vectors $N^{(i)}$. In the special case $i = 3$ we obtain (2. 142b). The vectors $n^{(i)}$, p form a complete set, so

$$m^{-2}\, p_\mu\, p_\nu \;-\; n^{(i)}_\mu\, n^{(i)}_\nu = g_{\mu\nu} \quad .$$

Thus

$$n^{(i)}_\mu\, n^{(i)}_\nu\, \gamma_5\, \gamma^\nu\; u(p,\alpha) = (p_\mu\, \gamma_5\, m^{-1} - \gamma_5\, \gamma_\mu)\, u(p,\alpha) \quad ,$$

and using (2. 145c), (2. 143c) and (2. 144) we find

$$\bar{u}(p,\beta)\,\gamma_\mu\gamma_5\,u(p,\alpha) = n_\mu^{(i)}\,\bar{u}(\beta)\,\sigma^i_{\gamma\alpha}\,u(\gamma)$$
$$= 2m\,\sigma^i_{\beta\alpha}\,n^{(i)}_\mu\,. \tag{2.145e}$$

We see that this reproduces (2.145d) in the case $\alpha = \beta$, since only σ^3 has non-zero diagonal matrix elements. In the same way we may show that

$$\bar{u}(p,\beta)\,s_{\rho\sigma}\,u(p,\alpha) = 2\,\epsilon_{\rho\sigma\mu\nu}\,p^\mu n^{(i)\nu}\sigma^i_{\beta\alpha}\,. \tag{2.145f}$$

Similarly, we may consider the plane wave negative energy solutions of the Dirac equation. We see from (2.131) that these may be constructed from the positive energy solutions already discussed by using the prescription (2.130). Thus,

$$\psi^c_{p,\pm\frac{1}{2}}(x) = v(p,\pm s)\,e^{ipx}\,, \tag{2.146a}$$

where

$$v(p,\pm s) \equiv \omega(p,\pm s)\,C\,\bar{u}^T(p,\pm s)\,. \tag{2.146b}$$

Evidently,

$$P_\mu\,\psi^c_{p,\pm\frac{1}{2}} = -p_\mu\,\psi^c_{p,\pm\frac{1}{2}}\,,$$

so that ψ^c is indeed a negative energy solution of the Dirac equation, and the negative energy spinor v, defined in (2.146), satisfies

$$(\not{p} + m)\,v(p,\pm s) = 0\,. \tag{2.147}$$

Likewise, it follows from the defining property of C in (2.129), and equation (2.143b), that

$$\gamma_5\,\not{s}\,v(p,\pm s) = \omega(p,\pm s)\,C[\bar{u}(p,\pm s)\,\gamma_5\,\not{s}]^T$$
$$= \pm\,v(p,\pm s)\,. \tag{2.148}$$

Now, the eigenvalue of N_μ given in (2.42) for the solution $\psi^c_{p,\pm\frac{1}{2}}$ is $-s_\mu$. Hence, from (2.140) and (2.148)

$$W.N\,\psi^c_{p,\pm\frac{1}{2}} = -\tfrac{1}{2}m\,(-\gamma_5\,\not{s})\,\psi^c_{p,\pm\frac{1}{2}} = -m(\mp\tfrac{1}{2})\,\psi^c_{p,\pm\frac{1}{2}}\,.$$

Thus the negative energy solution $\psi^c_{p,+\frac{1}{2}}$ has its spin anti-parallel with $\underline{n}$ in the frame where $\underline{p} = 0$, whereas $\psi_{p,+\frac{1}{2}}$ has its spin parallel to $\underline{n}$ in this frame. The adjoint negative energy spinor is given by

(2.136) to be

$$\bar{v}(p, \pm s) = -\omega(p,s)^* u^T(p, \pm s) C^{-1} . \tag{2.149}$$

These equations may be inverted with the aid of (2.133a) so that

$$u(p, s) = \omega(p, s) C \bar{v}^T(p, s) \tag{2.150a}$$

$$\bar{u}(p, s) = -\omega(p, s)^* v^T(p, s) C^{-1} . \tag{2.150b}$$

The normalization properties of the positive energy spinors and (2.137) then yield the following properties for the negative energy spinors :

$$\bar{v}(p, s) v(p, s) = -\bar{u}(p, s) u(p, s) = -2m \tag{2.151a}$$

$$\bar{v}(p, s) \gamma_5 v(p, s) = -\bar{u}(p, s) \gamma_5 u(p, s) = 0 \tag{2.151b}$$

$$\bar{v}(p, s) \gamma_\mu v(p, s) = \bar{u}(p, s) \gamma_\mu u(p, s) = 2p_\mu \tag{2.151c}$$

$$\bar{v}(p, s) \gamma_\mu \gamma_5 v(p, s) = -\bar{u}(p, s) \gamma_\mu \gamma_5 u(p, s) = -2m s_\mu \tag{2.151d}$$

$$\bar{v}(p, s) \sigma_{\rho\sigma} v(p, s) = \bar{u}(p, s) \sigma_{\rho\sigma} u(p, s) = 4\epsilon_{\rho\sigma\mu\nu} p^\mu s^\nu . \tag{2.151e}$$

As before we can also show that

$$\bar{v}(p, s) v(p, -s) = 0 = v(p, s) \gamma_5 v(p, -s) \tag{2.151f}$$

and the Dirac equation yields

$$\bar{v}(p, s) u(p, \pm s) = 0 . \tag{2.151g}$$

Using our previous results it is easy to see how the plane wave spinors transform under Lorentz transformations. Consider first the orthochronous Lorentz transformations. Clearly, the positive energy spinors transform into positive energy spinors (see (2.60) and (2.90)). Thus we may write

$$u(p', s') = \epsilon(p, s) S(\Lambda) u(p, s) , \tag{2.152}$$

where $\epsilon(p, s)$ is a phase factor. Using (2.145) and (2.102)

$$2 p'^\lambda = \bar{u}(p',s')\gamma^\lambda u(p',s') = \Lambda^\lambda{}_\mu \bar{u}(p,s)\gamma^\mu u(p,s) = 2 \Lambda^\lambda{}_\mu p^\mu$$

So

$$p'^\lambda = \Lambda^\lambda{}_\mu p^\mu \tag{2.153a}$$

as might have been anticipated. Likewise, using (2.103)

$$s'^{\lambda} = (\det \Lambda)\ \Lambda^{\lambda}_{\ \mu}\ s^{\mu} . \tag{2.153b}$$

Thus the covariant spin-vector transforms as an axial vector. In particular in the rest frame, where $p = (m, \underline{0})$, the spin vector $s = (0, \underline{n})$, and under the parity reversal transformation $s' = (0, \underline{n})$. Thus the direction of the spin in the rest frame is unaltered when viewed in a parity reversed frame of reference; this too might have been anticipated, because spin transforms like angular momentum $\underline{r} \times \underline{p}$, which is invariant in a frame where $\underline{r}' = -\underline{r}$ and $\underline{p}' = -\underline{p}$.

Likewise, if we consider the time-reversal transformation, we may write

$$u(p^t, s^t) = \eta(p,s)\, B^{-1}\, \gamma^{oT}\, \bar{u}^T(p,s) , \tag{2.154a}$$

and (2.125) shows that

$$p^t = (p^o, -\underline{p}) \quad s^t = (s^o, -\underline{s}) . \tag{2.154b}$$

Again, these should have been expected, since in the time reversed frame

$$\underline{r}^t = \underline{r} \quad \text{and} \quad \underline{p}^t = -\underline{p} .$$

2.2.7 PLANE WAVE PROJECTION OPERATORS

We have now specified up to a phase factor four linearly independent plane wave solutions of the Dirac equation: $u(p, \pm s)$, $v(p, \pm s)$. If we were to choose a particular representation of the Clifford Algebra it would be quite straightforward to determine these four solutions explicitly. In principle (and in practice) this is not necessary, because any physical results must be independent of the choice of representation for the γ-matrices. The physical consequences of any theory involving spin $\frac{1}{2}$ particles can only be tested by the observation of the transition probabilities predicted by the theory. Since these involve multiplying a wave function by its complex conjugate, we shall often be confronted by the matrix:

$$\lambda_{\alpha\beta}(p,s) \equiv u_{\alpha}(p,s)\, \bar{u}_{\beta}(p,s) . \tag{2.155}$$

This is essentially a projection operator for $u(p,s)$, since the normalization and orthogonality properties discussed previously ensure that

$$\lambda(p,s)\ \lambda(p,s) = 2m\,\lambda(p,s) \tag{2.156a}$$

$$\lambda(p,s)\ u(p,s) = 2m\,u(p,s) \tag{2.156b}$$

$$\lambda(p,s)\ u(p,-s) = 0 \tag{2.156c}$$

$$\lambda(p,s)\ v(p,\pm s) = 0\ . \tag{2.156d}$$

Note that the unspecified spinor phase does not affect the matrix $\lambda_{\alpha\beta}$, which must therefore be expressible in terms of the γ-matrices and the vectors p and s. It follows from (2.152) that

$$\lambda(p', s') = S(\Lambda)\ \lambda(p,s)\ S(\Lambda)^{-1}\ , \tag{2.157}$$

where p', s' are given in (2.153). The only matrices having this property are 1, $\not{p}$, $\gamma_5 \not{s}$. Further, since the squares of all of these matrices are proportional to the unit matrix, λ can be at most linear in $\not{p}$ and $\gamma_5 \not{s}$. Without loss of generality, then, we may take γ to be a linear combination of the four matrices

$$(\not{p} \pm m)\ (1 \pm \gamma_5 \not{s})\ . \tag{2.158}$$

Equations (2.138b) and (2.142b) imply that

$$(\not{p} - m)\ \lambda(p,s) = 0$$

$$(1 - \gamma_5 \not{s})\ \lambda(p,s) = 0\ .$$

Then using (2.158) it follows that

$$\lambda(p,s) = k\,(\not{p} + m)\ (1 + \gamma_5 \not{s})\ ,$$

and (2.156b) enables us to determine k

$$\lambda(p,s)\ u(p,s) = 4mk\,u(p,s) = 2m\,u(p,s)\ .$$

Hence

$$u(p,s)\,\bar{u}(p,s) = (\not{p} + m)\,\tfrac{1}{2}(1 + \gamma_5 \not{s})\ . \tag{2.159}$$

In the same way we find

$$v(p,s)\,\bar{v}(p,s) = (\not{p} - m)\,\tfrac{1}{2}(1 + \gamma_5 \not{s})\ . \tag{2.160}$$

Thus without any further specification of the γ-matrices we are able to write down these projection operators. In order to compute any required transition probabilities, these matrices will be multiplied by other combinations of γ-matrices and the trace of the product will be taken, as we shall see later. Thus to verify that physically observable quantities are indeed independent of any choice of representation of the

Clifford Algebra we must show how these traces can be evaluated independently of the representation using only the defining property (2. 52).

2.2.8 TRACE PROPERTIES OF THE γ-MATRICES

Recall first the additive and cyclic properties of traces

$$\mathrm{Tr}\,(A+B) = \mathrm{Tr}(A) + \mathrm{Tr}(B)$$
$$\mathrm{Tr}\,(AB) = \mathrm{Tr}(BA) \quad .$$

Then from (2. 52), since

$$\mathrm{Tr}\,(I_4) = 4 \quad ,$$

we deduce

$$\mathrm{Tr}\,(\gamma_\mu \gamma_\nu) = 4\, g_{\mu\nu} \quad . \tag{2.161}$$

It follows from the definition of γ_5 that

$$\gamma_5 \gamma_\mu = -\gamma_\mu \gamma_5 \quad \text{and} \quad (\gamma_5)^2 = +1 \quad .$$

So

$$\gamma_{\mu_1} \gamma_{\mu_2} \cdots \gamma_{\mu_n} = \gamma_{\mu_1} \gamma_{\mu_2} \cdots \gamma_{\mu_n} \gamma_5 \gamma_5 = (-)^n \gamma_5 \gamma_{\mu_1} \cdots \gamma_{\mu_n} \gamma_5 \quad .$$

Taking the trace and using the cyclic property we have

$$\mathrm{Tr}\left(\gamma_{\mu_1} \cdots \gamma_{\mu_n}\right) = (-)^n\, \mathrm{Tr}\left(\gamma_{\mu_1} \cdots \gamma_{\mu_n}\right) \quad . \tag{2.162}$$

Thus the trace of an odd number of γ-matrices is zero. In particular (since γ_5 has four γ-matrices)

$$\mathrm{Tr}(\gamma_\mu) = \mathrm{Tr}(\gamma_\lambda \gamma_\mu \gamma_\nu) = \mathrm{Tr}(\gamma_\mu \gamma_5) = 0 \quad . \tag{2.163}$$

Likewise, since

$$\gamma_5 = i\, \gamma^o \gamma^1 \gamma^2 \gamma^3 = -\, i\, \gamma^1 \gamma^2 \gamma^3 \gamma^o \quad , \tag{2.164}$$

the cyclic property yields

$$\mathrm{Tr}(\gamma_5) = 0 \quad . \tag{2.165}$$

From (2.105) and (2. 52) we have

$$2\,\gamma_\mu \gamma_\nu \gamma_5 = [\gamma_\mu, \gamma_\nu]\, \gamma_5 + \{\gamma_\mu, \gamma_\nu\}\, \gamma_5$$
$$= -\frac{i}{2}\, \epsilon_{\mu\rho\sigma\nu} [\gamma^\rho, \gamma^\sigma] + 2\, g_{\mu\nu}\, \gamma_5 \quad .$$

Thus

$$\mathrm{Tr}\,(\gamma_\mu \gamma_\nu \gamma_5) = 0 \tag{2.166}$$

using the cyclic property and (2.165). Finally, we note that a product containing any more than three γ-matrices can always be reduced using the identity

$$\gamma_\lambda \gamma_\mu \gamma_\nu = g_{\lambda\mu}\gamma_\nu + g_{\mu\nu}\gamma_\lambda - g_{\lambda\nu}\gamma_\mu + i\epsilon_{\lambda\mu\nu\rho}\gamma_5\gamma^\rho , \qquad (2.167)$$

which is most easily proved by considering the various cases. In particular we may deduce immediately for any four vectors A, B, C, D

$$\mathrm{Tr}\,(\not{A}\not{B}\not{C}\not{D}) = 4\,(A.B\ C.D + B.C\ A.D - A.C\ B.D) \qquad (2.168a)$$

$$\mathrm{Tr}\,(\not{A}\not{B}\not{C}\not{D}\,\gamma_5) = 4\,i\,\epsilon_{\alpha\beta\gamma\delta}\,A^\alpha B^\beta C^\gamma D^\delta . \qquad (2.168b)$$

2.3 THE FIELD OPERATOR OF THE ELECTRON

We are now in a position to make the transition to quantum field theory. This is dealt with fully in many standard text books so we shall present only a fairly abbreviated treatment. Since this second quantization is effected most elegantly in terms of canonical variables, we must first develop a Lagrangian formulation which yields the wave equation we have been discussing. Consider the Lagrangian

$$L = \int d^3x\ \mathcal{L}\ (\psi, \partial_\mu \psi) , \qquad (2.169a)$$

where $\mathcal{L}$ is the Lagrangian density given by

$$\mathcal{L} = \bar{\psi}(x)\ (i\not{\partial} - m)\ \psi(x) . \qquad (2.169b)$$

Regarding $\psi(x)$ as an independent generalized coordinate for each value of $\underline{x}$ we may minimize the Action Integral in a way analogous to that used in classical mechanics with a finite number of degrees of freedom. The action S is given by

$$S = \int_{t_1}^{t_2} dt\ L = \int_\Omega d^4x\ \mathcal{L} , \qquad (2.170)$$

where Ω is the space-time volume between the hyper-planes $t = t_1, t_2$. Thus

$$\delta S = \int_\Omega d^4x \left[\frac{\partial \mathcal{L}}{\partial \psi}\,\delta\psi + \frac{\partial \mathcal{L}}{\partial(\partial_\mu \psi)}\,\delta\,\partial_\mu \psi \right] . \qquad (2.171)$$

Then since $\delta\,\partial_\mu \psi = \partial_\mu \delta\psi$, we may integrate by parts and find

$$\delta S = \int_\Omega d^4x \left[\frac{\partial \mathcal{L}}{\partial \psi} - \partial_\mu \frac{\partial \mathcal{L}}{\partial(\partial_\mu \psi)} \right] \delta\psi(x) + \int_\Sigma d\sigma\, n_\mu \frac{\partial \mathcal{L}}{\partial(\partial_\mu \psi)} \delta\psi(x), \quad (2.172)$$

where Σ is the hyper-surface of Ω, and n_μ is the unit normal to it. The least action principle requires that $\delta S = 0$ for arbitrary $\delta\psi$ which vanishes on Σ. Hence, we obtain the Euler-Lagrange equations

$$\partial_\mu \frac{\partial \mathcal{L}}{\partial(\partial_\mu \psi)} = \frac{\partial \mathcal{L}}{\partial \psi} \quad . \qquad (2.173)$$

With the particular choice (2.169) for $\mathcal{L}$ we have

$$\frac{\partial \mathcal{L}}{\partial(\partial_\mu \psi)} = \bar{\psi}\, i\gamma_\mu \qquad (2.174a)$$

$$\frac{\partial \mathcal{L}}{\partial \psi} = - m\bar{\psi} \quad . \qquad (2.174b)$$

Thus the Euler-Lagrangian equation (2.173) yields

$$\bar{\psi}(x)\;(i\overleftarrow{\not\partial} + m) = 0 \quad . \qquad (2.175)$$

This is simply the adjoint form of the Dirac equation which is given in (2.79). The dynamical variable conjugate to $\psi(x)$ is defined analogously to classical mechanics. We define

$$\pi_\alpha(x) \equiv \frac{\partial \mathcal{L}}{\partial(\partial_o \psi_\alpha)} = \bar{\psi}_\beta(x)\, i\gamma^o_{\beta\alpha} \quad . \qquad (2.176)$$

The canonical quantization postulate can now be stated. The classical field $\psi(x)$ and its canonical conjugate $\pi(x)$ are to be operators in a Hilbert space of possible physical states, and are to satisfy certain quantum conditions analogous to the classical Poisson brackets. The quantum conditions imposed are

$$\{ \psi_\alpha(\underline{x}, t),\ \psi_\beta(\underline{y}, t) \} = 0 \qquad (2.177a)$$

$$\{ \pi_\alpha(\underline{x}, t),\ \pi_\beta(\underline{y}, t) \} = 0 \qquad (2.177b)$$

$$\{ \psi_\alpha(\underline{x}, t),\ \pi_\beta(\underline{y}, t) \} = i\delta(\underline{x}-\underline{y})\,\delta_{\alpha\beta} \quad . \qquad (2.177c)$$

As before, the notation $\{\ ,\ \}$ indicates that we are to take the anti-commutators of the operators enclosed. The use of anti-commutators for spin $\frac{1}{2}$ fields is necessitated by the Fermi statistics possessed by the particles so described, as we shall see. We may rewrite (2.177) in terms of the fields ψ, $\bar{\psi}$:

$$\{\psi_\alpha(\underline{x}, t), \psi_\beta(\underline{y}, t)\} = 0 = \{\bar{\psi}_\alpha(\underline{x}, t), \bar{\psi}_\beta(\underline{y}, t)\} \tag{2.178a}$$

$$\{\psi_\alpha(\underline{x}, t), \bar{\psi}_\beta(\underline{y}, t)\} = \gamma^o_{\alpha\beta}\, \delta(\underline{x} - \underline{y}) \quad . \tag{2.178b}$$

The character of the field operators ψ, $\bar{\psi}$ is best appreciated by considering the significance of (2.178) for the Fourier transforms of the fields. Since ψ, $\bar{\psi}$ satisfy the Dirac equation we may write

$$\psi(x) = (2\pi)^{-3} \int d^3p\, (2p_o)^{-1} \sum_{\pm s} \{a(p,s)\, u(p,s)\, e^{-ipx} + b^\dagger(p,s)\, v(p,s)\, e^{ipx}\} \tag{2.179a}$$

$$\bar{\psi}(x) = (2\pi)^{-3} \int d^3p\, (2p_o)^{-1} \sum_{\pm s} \{b(p,s)\, \bar{v}(p,s)\, e^{-ipx} + a^\dagger(p,s)\, \bar{u}(p,s)\, e^{ipx}\} , \tag{2.179b}$$

where $p_o = +(\underline{p}^2 + m^2)^{\frac{1}{2}}$. $u(p,s)$ are the spinors defined in (2.138), (2.142), while the $v(p,s)$ are the negative energy spinors defined in (2.146), (2.148). Inverting the Fourier transforms in (2.179) gives

$$a(p,s) = \int d^3x\, \bar{u}(p,s)\, \gamma_o\, \psi(x)\, e^{ipx} \tag{2.180a}$$

$$a^\dagger(p,s) = \int d^3x\, \bar{\psi}(x)\, \gamma_o\, u(p,s)\, e^{-ipx} \tag{2.180b}$$

$$b(p,s) = \int d^3x\, \bar{\psi}(x)\, \gamma_o\, v(p,s)\, e^{ipx} \tag{2.180c}$$

$$b^\dagger(p,s) = \int d^3x\, \bar{v}(p,s)\, \gamma_o\, \psi(x)\, e^{-ipx} , \tag{2.180d}$$

where we have used the orthogonality condition

$$\bar{u}(p_o, \underline{p})\, \gamma^o\, v(p_o, -\underline{p}) = 0 , \tag{2.181}$$

which follows from the defining equations of the positive and negative energy spinors. Using the anti-commutation relations (2.178) and the orthogonality condition (2.181) we may verify that

$$\{a(p,s), a(p',s')\} = \{b(p,s), b(p',s')\} = 0 \tag{2.182a}$$

$$\{a(p,s), b(p',s')\} = \{a(p,s), b^\dagger(p',s')\} = 0 \tag{2.182b}$$

$$\{ a(p,s),\ a^{\dagger}(p',s') \} = \{ b(p,s),\ b^{\dagger}(p',s') \}$$

$$= (2\pi)^3\ 2p_o\ \delta(\underline{p}-\underline{p}')\ \delta_{ss'} \quad . \tag{2.182c}$$

The Hamiltonian H for the system we are discussing is constructed, as before, in strict analogy to what is done in classical mechanics.

We write H in the form

$$H = \int d^3x\ \mathcal{H} \ , \tag{2.183a}$$

where the Hamiltonian density $\mathcal{H}$ is given by

$$\mathcal{H} = \pi(x)\ \dot{\psi}(x) - \mathcal{L} \tag{2.183b}$$

$$= \overline{\psi}\ i\gamma^o \partial_o \psi - \overline{\psi}(i\not\partial - m)\psi \ . \tag{2.183c}$$

Substituting (2.179) and using the various orthonormality properties of the spinors, we find

$$H = \int d^3p\ (2\pi)^{-3}\ (2p_o)^{-1} \sum_{\pm s} \{ a^{\dagger}(p,s)\ a(p,s)\ p_o - b(p,s)\ b^{\dagger}(p,s)\ p_o \} \tag{2.184a}$$

$$= \int d^3p\ (2\pi)^{-3}\ (2p_o)^{-1} \sum_{\pm s} \{ a^{\dagger}(p,s)\ a(p,s) + b^{\dagger}(p,s)\ b(p,s) \}\ p_o$$

$$- \int d^3p\ (2\pi)^{-3}\ (2p_o)^{-1} \sum_{\pm s} \{ b(p,s),\ b^{\dagger}(p,s) \}\ p_o \ . \tag{2.184b}$$

Ignoring, for the moment, the last term of (2.184b), which according to (2.182c) is infinite, this decomposition of the Hamiltonian leads to the interpretation of

$$N(e^-) \equiv \int d^3p\ (2\pi)^{-3}\ (2p_o)^{-1} \sum_{\pm s} a^{\dagger}(p,s)\ a(p,s) \tag{2.185}$$

as the number operator for electrons, and to the identification of $a(p,s)$, $a^{\dagger}(p,s)$ as operators which respectively destroy and create an electron of momentum p and spin s. The anti-commutation relations (2.182) show that

$$[N(e^-),\ a(p,s)] = \int d^3q\ (2\pi)^{-3}\ (2q_o)^{-1} \sum_{\pm r} [a^{\dagger}(q,r)\ \{ a(q,r),\ a(p,s) \}$$

$$- \{ a^{\dagger}(q,r),\ a(p,s) \}\ a(q,r)] \tag{2.186}$$

$$= -\ a(p,s) \ .$$

Thus if $|n>$ is a state containing n electrons, so that

$$N(e^-) | n> = n | n> , \tag{2.187}$$

it follows from (2.186) that

$$N(e^-) a(p,s) | n> = (n-1) a(p,s) | n> . \tag{2.188}$$

Thus either $a(p,s) | n>$ is zero, or it contains (n-1) electrons. This is consistent with our interpretation of $a(p,s)$, since either the state $|n>$ has no electron with momentum p and spin s, in which case $a(p,s)|n>$ vanishes, or, if it has, $a(p,s)$ destroys it, in which case $a(p,s)|n>$ is a state with (n-1) electrons. In the same way we can show that

$$[N(e^-), a^\dagger(p,s)] = a^\dagger(p,s) , \tag{2.189}$$

and verify the interpretation of $a^\dagger(p,s)$ as a creation operator. Likewise

$$N(e^+) = \int d^3p (2\pi)^{-3} (2p_o)^{-1} \sum_{\pm s} b^\dagger(p,s) b(p,s) \tag{2.190}$$

is identified as the positron number operator, and $b(p,s)$, $b^\dagger(p,s)$ as operators which destroy and create positrons of momentum p and spin s. The vacuum state $| 0>$ is defined as the state with no particles, so that

$$< 0 | N(e^-) | 0> = < 0 | N(e^+) | 0> = 0 , \tag{2.191a}$$

and hence

$$a(p,s) | 0> = b(p,s) | 0> = 0 \text{ for all } p,s. \tag{2.191b}$$

We may define the single electron state of momentum p and spin s as

$$| e, p, s> = a^\dagger(p,s) | 0> .$$

Its normalization is determined using (2.191b) and (2.182c) :

$$\begin{aligned} < e,p,s | e,p',s'> &= < 0| a(p,s) a^\dagger(p',s') |0> \\ &= < 0| \{ a(p,s), a^\dagger(p',s') \} |0> \\ &= (2\pi)^3 2p_o \delta(\underline{p}-\underline{p}') \delta_{ss'} , \end{aligned} \tag{2.192a}$$

where the normalization of the vacuum is fixed by taking

$$< 0 | 0> = 1 . \tag{2.192b}$$

Finally, we note that (2.182) implies that

$$a^\dagger(p,s)\, a^\dagger(p,s) = 0 \quad . \tag{2.193}$$

Thus it is impossible to have a state with more than one electron having momentum p and spin s. Thus we recognise that the exclusion principle is directly connected to our use of anti-commutators in (2.177). Likewise, the anti-commutator implies for any state $|\psi\rangle$

$$a^\dagger(p,s)\, a^\dagger(p',s')\,|\psi\rangle = -\, a^\dagger(p',s')\, a^\dagger(p,s)\,|\psi\rangle \tag{2.194}$$

thereby exhibiting the connection between our use of anti-commutators and the Fermi statistics of the electron. Precisely similar conclusions apply to the positron states.

Let us now consider the infinite constant c-number appearing in the last term of (2.184b). It follows from (2.191b) and (2.192b) that the energy of the vacuum is

$$\langle 0|H|0\rangle = -\,2\delta\,(\underline{0}) \int d^3 p\, p_o \,. \tag{2.195}$$

This (minus) infinite value for the energy of the vacuum is plainly unsatisfactory, since we would obviously like the state with no particles to have zero energy. It is seen to arise from the fact that H (and L) contains terms in which a creation operator appears to the <u>left</u> of an annihilation operator; plainly such terms permit the creation and subsequent annihilation of a particle independently of what state the operators are acting upon. This constant addition to the energy of every state necessitates a redefinition of H (and of L). We replace the definitions by their <u>normal product</u> form. This consists of reordering all operator products so that all annihilation operators stand to the <u>right</u> of all creation operators, and then multiplying by a factor $(-1)^\epsilon$, where ϵ is the number of interchanges of fermion operators required to perform the reordering. We use colons to denote normal ordering

$$:b(p,s)\, b^\dagger(p,s): \;=\; -\; b^\dagger(p,s)\, b(p,s) \quad . \tag{2.196}$$

Thus defining

$$\mathcal{H} = \; :\pi(x)\,\dot{\psi}(x) - \mathcal{L}: \;, \tag{2.197a}$$

we obtain the form for H with the desired properties

$$H = \int d^3 p\, (2\pi)^{-3}\, (2p_o)^{-1} \sum_{\pm s} \{a^\dagger(p,s)\, a(p,s) + b^\dagger(p,s)\, b(p,s)\}\, p_o \,. \tag{2.197b}$$

2.3.1 RELATIVISTIC INVARIANCE OF THE ELECTRON'S FIELD

We have already considered the behaviour of the wave function $\psi(x)$ under the Poincaré group and the inversions (in this theory ψ is a complex number or c-number). The Poincaré invariance was demonstrated by showing that the wave function $\psi'(x')$ describing the electron state with respect to an observer O' satisfies an equation of the same form as that satisfied by the wave function $\psi(x)$ describing the same state with respect to an observer O. In quantum field theory $\psi(x)$ becomes an operator (or q-number) acting in a Hilbert space, and the requirements of Poincaré invariance are applied to the matrix elements of the operators. It is these matrix elements which play the role of the c-number wave functions. Thus equation (2.60), which relates the c-number wave functions describing the same state with respect to observers in different inertial frames, is replaced by

$$<\alpha'|\,\psi(x')\,|\,\beta'> \;=\; S(\Lambda) <\alpha\,|\,\psi(\Lambda^{-1}(x'-a))\,|\,\beta> \quad . \tag{2.198}$$

$|\alpha>$, $|\beta>$ are arbitrary physical states as described by the observer O, and $|\alpha'>$, $|\beta'>$ are respectively the same states as described by the observer O'. We assume that the operator relating corresponding state vectors is unitary, so that the state normalization is preserved. Thus we write

$$|\,\alpha'> \;=\; U(\Lambda, a)\,|\,\alpha> \;, \tag{2.199a}$$

$$|\,\beta'> \;=\; U(\Lambda, a)\,|\,\beta> \;, \tag{2.199b}$$

where

$$U(\Lambda, a)\; U(\Lambda, a)^{\dagger} = U^{\dagger}(\Lambda, a)\, U(\Lambda, a) = 1 \quad . \tag{2.199c}$$

Since (2.198) is true for arbitrary state vectors we may deduce the operator (q-number) identity

$$U(\Lambda, a)^{-1}\,\psi(x')\,U(\Lambda, a) = S(\Lambda)\,\psi(\Lambda^{-1}(x'-a)) \quad . \tag{2.200}$$

Thus to establish the Poincaré covariance of the quantized Dirac field theory we are required to show that there exists a unitary operator $U(\Lambda, a)$ satisfying (2.200). As before, it suffices to consider an infinitesimal Poincaré transformation. Then for an infinitesimal translation a, we may write

$$U(1,a) = I + i a^{\lambda} P_{\lambda} + O(a^2) . \tag{2.201}$$

Since U is unitary, P_λ is an hermitian or self-adjoint operator. Substituting in (2.200), and using (2.61), we find that P_λ must satisfy

$$[P_\lambda , \psi(x)] = - i \partial_\lambda \psi(x) . \tag{2.202}$$

It is easy to show that an operator with the required properties does exist for the free field case we have considered so far. Using the Fourier decomposition (2.179a) we may rewrite (2.202) in terms of the creation and annihilation operators. In this way we find that P_λ is required to satisfy

$$[P_\lambda , a(p,s)] = - p_\lambda a(p,s) \tag{2.203a}$$

$$[P_\lambda , b^\dagger(p,s)] = p_\lambda b^\dagger(p,s) . \tag{2.203b}$$

Inspection of (2.186) shows us immediately that the required operator is given by

$$P_\lambda = \int d^3p \, (2\pi)^{-3} (2p_0)^{-1} \sum_{\pm s} \{ a^\dagger(p,s) \, a(p,s) \, p_\lambda + b^\dagger(p,s) \, b(p,s) \, p_\lambda \} . \tag{2.204}$$

Thus the operator P_λ is just the total momentum operator for the field theory. In the same way, if we consider the infinitesimal Lorentz transformation (2.67), we may write

$$U(\Lambda, 0) = I + i \epsilon M^{\rho\sigma} , \tag{2.205}$$

where $M^{\rho\sigma}$ is an hermitian operator. Then $M^{\rho\sigma}$ is required to satisfy

$$[M_{\rho\sigma} , \psi(x)] = - (s_{\rho\sigma} + i x_\rho \partial_\sigma - i x_\sigma \partial_\rho) \psi(x) . \tag{2.206}$$

The required operator $M_{\rho\sigma}$ can be shown to be just the angular momentum tensor operator of the field theory.

In classical field theory we demonstrated the Poincaré invariance of the theory by showing that the field on the right-hand side of (2.200) or (2.60) satisfies the same equation as does $\psi(x)$; equivalently we could equally have shown that replacement of $\psi(x)$ in (2.169b) by the right hand side of (2.200) leaves the Lagrangian density (and hence the equation of motion) invariant. In quantum field theory this requirement is necessary but not sufficient. To establish the Poincaré invariance of the field theory we must also demonstrate the existence of the unitary operator $U(\Lambda, a)$ connecting the two fields. For all of the field theories we shall be discussing, the unitary operator $U(\Lambda, a)$ is generated by the momentum vector and angular momentum tensor operators, and the Poincaré invariance can be established, as in the classical case, by verifying that the Lagrangian density is invariant.

2.3.2 P-, C- AND T-INVARIANCE

Similarly, to demonstrate the parity invariance of the field theory we must show that there is a unitary operator $\mathscr{P}$ having the property

$$\mathscr{P}^{-1} \psi(x^o, \underline{x}) \mathscr{P} = \epsilon \gamma^o \psi(x^o, -\underline{x}) , \qquad (2.207)$$

where the right hand side is given in (2.90). Equivalently, $\mathscr{P}$ satisfies

$$\mathscr{P} \psi(x^o, \underline{x}) \mathscr{P}^{-1} = \epsilon^* \gamma^o \psi(x^o, -\underline{x}) \qquad (2.208a)$$

$$\mathscr{P} \overline{\psi}(x^o, \underline{x}) \mathscr{P}^{-1} = \epsilon \overline{\psi}(x^o, -\underline{x}) \gamma^o . \qquad (2.208b)$$

For the free field theory we have been discussing an explicit form for can be written down[(2)]. For interacting field theories we can similarly construct an operator $\mathscr{P}$ satisfying (2.208), and we must then see whether the Lagrangian density is invariant. If so

$$\mathscr{P} \mathscr{L}(x^o, \underline{x}) \mathscr{P}^{-1} = \mathscr{L}(x^o, -\underline{x}) , \qquad (2.209a)$$

and

$$[\mathscr{P}, H] = 0 , \qquad (2.209b)$$

so that $\mathscr{P}$ is a symmetry of the theory.

To demonstrate the charge conjugation invariance of a field theory it follows from (2.130), (2.133), (2.136) that we must show that there is a unitary operator $\mathscr{C}$, having the properties

$$\mathscr{C} \psi(x) \mathscr{C}^{-1} = \omega^* C \overline{\psi}^T(x) \qquad (2.210a)$$

$$\mathscr{C} \overline{\psi}(x) \mathscr{C}^{-1} = -\omega \psi^T(x) C^{-1} , \qquad (2.210b)$$

which leaves the Lagrangian invariant:

$$\mathscr{C} \mathscr{L}(x) \mathscr{C}^{-1} = \mathscr{L}(x) . \qquad (2.211)$$

In the free Dirac field theory (2.210) implies

$$\mathcal{C} a(p,s) \mathcal{C}^{-1} = \epsilon(p,s)\, b(p,s); \quad \mathcal{C} b(p,s) \mathcal{C}^{-1} = \epsilon^*(p,s)\, a(p,s) \tag{2.212a}$$

$$\mathcal{C} a^\dagger(p,s) \mathcal{C}^{-1} = \epsilon^*(p,s)\, b^\dagger(p,s); \quad \mathcal{C} b^\dagger(p,s) \mathcal{C}^{-1} = \epsilon(p,s)\, a^\dagger(p,s)\,, \tag{2.212b}$$

if we substitute (2.179) and use (2.150), $\epsilon(p,s)$ is a phase factor. Applying these operator equations to the vacuum state, we see that the operator $\mathcal{C}$ has the effect of changing an electron state into a positron state, and vice versa, as might have been anticipated. The explicit form for $\mathcal{C}$ is given in Bjorken and Drell[(2)].

We have already written down the transformation law for the classical Dirac field under the time reversal transformation (see (2.116), (2.124)). We recall that this operation is anti-linear so we must be a little careful when deriving the corresponding law for the q-number field. The classical transformation law (2.116) is replaced by the following relation between the operator expectation values

$$< \alpha^t | \psi(x^o, \underline{x}) | \beta^t > = \eta\, B^{-1} \gamma^{oT} A^T < \alpha | \psi(-x^o, \underline{x}) | \beta >^* \,, \tag{2.213}$$

where $|\alpha^t>$, $|\beta^t>$ are the time reversed states corresponding to $|\alpha>$, $|\beta>$. Notice that the complex conjugate of the c-number wave function in (2.116) is replaced by the complex conjugate of the matrix element of this q-number field. If we write

$$|\beta^t> = \mathcal{T} | \beta > \tag{2.214a}$$

and

$$|\alpha^t> = \mathcal{T} | \alpha > \,, \tag{2.214b}$$

then to derive any operator transformation we must demand that $\mathcal{T}$ be anti-unitary; that is, it is a product of a unitary operator and the operation of complex conjugating c-numbers. It has the property

$$< \mathcal{T}\alpha | \mathcal{T}\beta > = < \alpha | \beta >^* \tag{2.215}$$

for all states $|\alpha>$, $|\beta>$. Thus, although $\mathcal{T}$ is not an isometric operator, it does leave the modulus of the scalar product invariant, and hence the probability. Taking $\mathcal{T}$ to be anti-unitary then gives

$$\begin{aligned} < \alpha^t | \psi(x^o, \underline{x}) | \beta^t > &= < \mathcal{T}\alpha | \mathcal{T}\mathcal{T}^{-1} \psi(x^o, \underline{x}) | \beta^t > \\ &= < \alpha | \mathcal{T}^{-1} \psi(x^o, \underline{x})\, \mathcal{T} | \beta >^* \,, \end{aligned} \tag{2.216}$$

and it follows thus from (2.213) that

$$\mathcal{T}^{-1}\, \psi(x^o, \underline{x})\, \mathcal{T} = \eta^* B^{-1*} \gamma^{o\dagger} A^\dagger \psi(-x^o, \underline{x}) \; . \tag{2.217}$$

Or equivalently, remembering that $\mathcal{T}$ is anti-linear,

$$\mathcal{T}\, \psi(x^o, \underline{x})\, \mathcal{T}^{-1} = \eta^* A^{T-1} \gamma^{oT} B\, \psi(-x^o, \underline{x}) \tag{2.218a}$$

$$\mathcal{T}\, \bar{\psi}(x^o, \underline{x})\, \mathcal{T}^{-1} = \eta\, \bar{\psi}(-x^o, \underline{x})\, B^{-1} \gamma^{oT} A^* \; . \tag{2.218b}$$

To demonstrate that the field theory is time-reversal invariant it must then be shown that

$$\mathcal{T}\, \mathcal{L}(x^o, \underline{x})\, \mathcal{T}^{-1} = \mathcal{L}(-x^o, \underline{x}) \; . \tag{2.219}$$

2.3.3. ILLUSTRATION: THE P-, C-, T-INVARIANCE OF QUANTUM ELECTRODYNAMICS

The interaction Lagrangian which generates all known electrodynamic phenomena can be written succinctly in terms of the electron's field operator $\psi(x)$ and the electromagnetic field operator $A_\mu(x)$

$$\mathcal{L}_I(x) = e{:}\,\bar{\psi}(x)\, \gamma_\mu \psi(x)\, A^\mu(x){:} \qquad (e > 0) \; . \tag{2.220}$$

The vector potential field A_μ transforms as a vector, rather than an axial vector, so

$$\mathcal{P} A^\mu(x^o, \underline{x})\, \mathcal{P}^{-1} = \Lambda(i_s)^\mu{}_\nu\, A^\nu(x^o, -\underline{x}) \; . \tag{2.221}$$

Likewise, it follows from (2.208) that

$$\mathcal{P} j^\mu(x^o, \underline{x})\, \mathcal{P}^{-1} = \Lambda(i_s)^\mu{}_\rho\, j^\rho(x^o, -\underline{x}) \; , \tag{2.222a}$$

where

$$j^\mu(x) = {:}\,\bar{\psi}(x)\, \gamma^\mu \psi(x){:} \; , \tag{2.222b}$$

just as in the c-number case (see (2.102)). Thus

$$\mathcal{P} \mathcal{L}_I(x^o, \underline{x})\, \mathcal{P}^{-1} = \mathcal{L}_I(x^o, -\underline{x}) \; , \tag{2.223}$$

and the interaction Lagrangian is parity invariant. In the same way we can show that the free field part of the Lagrangian is also invariant and hence that QED is a parity conserving theory.

Under charge conjugation the electromagnetic field of course reverses, so that

$$\mathcal{C} A_\mu \mathcal{C}^{-1} = - A_\mu , \qquad (2.224a)$$

and from (2.210)

$$\mathcal{C} : \bar{\psi} \gamma_\mu \psi : \mathcal{C}^{-1} = : \psi^T \gamma_\mu^T \bar{\psi}^T : , \qquad (2.224b)$$

so that

$$\mathcal{C} j_\mu \mathcal{C}^{-1} = - j_\mu . \qquad (2.224c)$$

As compared with the c-number theory, see (2.137c), the field theory has an extra minus sign, which arises from the interchange of fermion field operators when we take the transpose. Thus the electromagnetic current reverses under charge conjugation, as we would expect, and QED is a charge conjugation invariant theory.

Using (2.218) and the anti-linearity of $\mathcal{T}$ we find

$$\mathcal{T} j^\mu (x^o, \underline{x}) \mathcal{T}^{-1} = - \Lambda (i_t)^\mu{}_\nu \, j^\nu (-x^o, \underline{x}) , \qquad (2.225)$$

just as in the c-number case (see (2.125c)). The vector potential A_μ transforms in the same way, thus

$$\mathcal{T} A^\mu (x^o, \underline{x}) \mathcal{T}^{-1} = - \Lambda (i_t)^\mu{}_\nu \, A^\nu (-x^o, \underline{x}) , \qquad (2.226)$$

and it follows that QED is invariant under time reversal.

2.3.4 GENERAL CONSEQUENCES OF P- AND T-INVARIANCE

In a theory which is parity-reversal invariant it is easy to see that (2.209b) implies that

$$\mathcal{P}^{-1} S \mathcal{P} = S , \qquad (2.227)$$

where S is the S-matrix whose elements determine the transition probabilities of all processes generated by a theory. It then follows that for arbitrary states $|\alpha>$, $|\beta>$

$$< \alpha^p | S | \beta^p > = < \alpha | S | \beta > , \qquad (2.228a)$$

where

$$| a^p > = \mathcal{P} | \alpha > , \; | \beta^p > = \mathcal{P} | \beta > , \qquad (2.228b)$$

since $\mathcal{P}$ is a unitary operator. Thus any transition probability W calculated with this theory, which may be a function of various momenta, collectively called $\underline{p}$, and spins, collectively called $\underline{s}$, must satisfy

$$W(\underline{p}, \underline{s}) = W(-\underline{p}, \underline{s}) \quad , \qquad (2.229)$$

since $\underline{p}$ reverses under parity and $\underline{s}$ does not.

The analogous statement does not hold for the time-reversal transformation whose operator $\mathcal{T}$ is anti-linear. In fact, the S-matrix operator in a time reversal invariant theory satisfies

$$\mathcal{T} S \mathcal{T}^{-1} = S^{\dagger} \ . \qquad (2.230)$$

Hence, for arbitrary states $|\alpha>$, $|\beta>$

$$<\alpha|S|\beta> = <\mathcal{T}\alpha|\mathcal{T} S|\beta>^{*} = <\alpha^{t}|S^{\dagger}|\beta^{t}>^{*} = <\beta^{t}|S|\alpha^{t}> , \qquad (2.231a)$$

where

$$|\alpha^{t}> = \mathcal{T}|\alpha> , \quad |\beta^{t}> = \mathcal{T}|\beta> \ . \qquad (2.231b)$$

Thus, in general

$$<\alpha|S|\beta> \neq <\alpha^{t}|S|\beta^{t}> \ , \qquad (2.232)$$

and the analogue of (2.229) is not, in general, true.

2.4 WAVE FUNCTION AND FIELD THEORY FOR THE NEUTRINO

2.4.1 DIRAC EQUATION FOR THE NEUTRINO

The wave function and field operator satisfying (2.51) have been used successfully in QED to describe the electromagnetic interaction of the electron. It is, therefore, natural to suppose that any formulation of the theory of the weak interaction of the electron will also use this form for the wave function and field operator. It is an experimental fact that the weak interactions of the electron frequently involve the participation of a neutrino or antineutrino. The success of the Dirac equation for the electron then suggests that we should formulate the theory for the massless spin $\frac{1}{2}$ neutrino field in the same way. We, therefore, start with the Dirac equation (2.51) with $m = 0$

$$i \not{\partial} \psi(x) = 0 \ . \qquad (2.233)$$

The Poincaré invariance of this equation follows precisely as before. Likewise from (2.66)

$$P^{2} \psi(x) = 0 \ , \qquad (2.234)$$

and $\psi(x)$ does indeed describe a massless particle; similarly from (2.75) we find

$$W^2 \psi(x) = 0 , \tag{2.235}$$

as anticipated in section 2.1.5. Using (2.74), (2.105) and the Dirac equation we have

$$\begin{aligned} W_\mu \psi(x) &= \frac{i}{8} \epsilon_{\mu\sigma\lambda\rho} [\gamma^\rho, \gamma^\sigma] P^\lambda \psi(x) \\ &= -\tfrac{1}{4} [\gamma_\mu, \not{P}] \gamma_5 \psi(x) \\ &= \tfrac{1}{2} P_\mu \gamma_5 \psi(x) , \end{aligned} \tag{2.236a}$$

and likewise

$$W_\mu \gamma_5 \psi(x) = \tfrac{1}{2} P_\mu \psi(x) . \tag{2.236b}$$

Hence

$$W_\mu (1 \pm \gamma_5) \psi(x) = \pm \tfrac{1}{2} P_\mu (1 \pm \gamma_5) \psi(x) . \tag{2.237}$$

Thus defining

$$a = \tfrac{1}{2}(1 - \gamma_5), \; a' = \tfrac{1}{2}(1 + \gamma_5) \tag{2.238a}$$

$$\psi_L(x) = a\,\psi(x), \; \psi_R(x) = a'\psi(x) , \tag{2.238b}$$

we can see that

$$\psi(x) = \psi_L(x) + \psi_R(x) . \tag{2.239}$$

It is easy to see that $\psi_{L,R}$ separately satisfy the Dirac equation as well as (2.234) and (2.235). Also, from (2.237) we have

$$W_\mu \psi_L(x) = -\tfrac{1}{2} P_\mu \psi_L(x) \tag{2.240a}$$

$$W_\mu \psi_R(x) = +\tfrac{1}{2} P_\mu \psi_R(x) . \tag{2.240b}$$

Hence, by virtue of (2.49), we see that the positive energy solutions of the Dirac equation (2.233) are the sum of <u>two</u> irreducible representations of the Poincaré group, one with helicity $\lambda = -\frac{1}{2}$ and the other with $\lambda = +\frac{1}{2}$. Positive helicity occurs (for positive energy) when $\underline{J}$ is parallel to $\underline{P}$, so this representation is called right-handed and the negative helicity representation left-handed.

2.4.2 PLANE WAVE SOLUTIONS

As before, we may write down the plane wave solutions of the Dirac equation which form a basis of the irreducible representations of the

Poincare group. Plainly the bases for each of the representations $\{+, +\frac{1}{2}\}$ or $\{+, -\frac{1}{2}\}$, specified in (2.50), which diagonalize P_μ are given by

$$\psi_p(x) = u(p)\, e^{-ipx} , \qquad (2.241a)$$

where

$$\not{p}\, u(p) = 0 \qquad (2.241b)$$

$$p_o = +\,(\underline{p}^2 + m^2)^{\frac{1}{2}} , \qquad (2.241c)$$

so that

$$P_\mu \psi_p(x) = p_\mu \psi_p(x) .$$

The helicity operator is, from (2.49),

$$\lambda = \frac{W_o}{P_o} .$$

Thus from (2.236), if we define $u(p,h)$ by

$$\tfrac{1}{2}\gamma_5 u(p,h) = h\, u(p,h), \quad h = \pm\tfrac{1}{2} , \qquad (2.242)$$

$$\bar{u}(p,h)\, \tfrac{1}{2}\gamma_5 = -\, h\, \bar{u}(p,h) ,$$

the required bases of $\{+, h\}$ are given by

$$\psi_p^{(h)}(x) = u(p,h)\, e^{-ipx} . \qquad (2.243)$$

We define

$$\bar{u}(p,h) = u^\dagger(p,h)\, A ,$$

so $\quad \bar{u}(p,h)\not{p} = 0 .$

Thus

$$2p_\mu \bar{u}(p,h)\, u(p,\pm h) = \bar{u}(p,h)\, \{\not{p}, \gamma_\mu\}\, u(p,\pm h) = 0 ,$$

so

$$\bar{u}(p,h)\, u(p,\pm h) = 0 , \qquad (2.244a)$$

and as before

$$\bar{u}(p,h)\, \gamma_5\, u(p,\pm h) = 0 . \qquad (2.244b)$$

We define the normalization of our spinors by the requirement

$$\bar{u}(p,h)\ \gamma_\mu\, u(p,h)\ =\ 2p_\mu\ , \tag{2.245}$$

which parallels the result (2.145a) for the massive spinors. It then follows from (2.242) that

$$\bar{u}(p,h)\ \gamma_\mu\gamma_5\, u(p,h)\ =\ 4h\,p_\mu \tag{2.246a}$$

and

$$\bar{u}(p,h)\,\gamma_\mu\, u(p,-h)\ =\ 0\ =\ \bar{u}(p,h)\,\gamma_\mu\gamma_5\, u(p,-h)\ . \tag{2.246b}$$

Likewise

$$\bar{u}(p,h)\, s^{\rho\sigma}\, u(p,\pm h)\ =\ 0\ =\ \bar{u}(p,h)\, s^{\rho\sigma}\,\gamma_5\, u(p,h)\ . \tag{2.246c}$$

Proceeding as before, we then find that the matrix

$$u(p,h)\,\bar{u}(p,h)\ =\ \tfrac{1}{2}(1+2h\,\gamma_5)\,\not{p}\ , \tag{2.247}$$

using the normalization (2.245). Note that this matrix is <u>not</u> a projection operator in this case, and the two factors in (2.247) do not commute.

We may also consider the negative energy spinors $v(p,h)$ defined as in (2.146b) by

$$v(p,h)\ \equiv\ \omega(p,h)\, C\bar{u}^T(p,h)\ . \tag{2.248}$$

It then follows, from the defining property of C and (2.242), that

$$\tfrac{1}{2}\gamma_5\, v(p,h)\ =\ -\,hv(p,h) \tag{2.249}$$

$$\bar{v}(p,h)\,\tfrac{1}{2}\gamma_5 = +\,h\,\bar{v}(p,h)\ .$$

Hence, if, as before,

$$\psi_p^{(h)c}(x)\ =\ C\,\bar{\psi}_p^{(h)T}(x)\ , \tag{2.250a}$$

then

$$\lambda\,\psi_p^{(h)c}(x) = -h\,\psi_p^{(h)c}(x)\ , \tag{2.250b}$$

and we see that the wave functions $\psi_p^{(h)c}(x)$ form a basis of the irreducible representations $\{-,\ -h\}$ of the Poincaré group. The only non-zero bilinear combinations of the $v(p,h)$ are

$$\bar{v}(p,h)\,\gamma_\mu\, v(p,h)\ =\ 2p_\mu \tag{2.251a}$$

$$\bar{v}(p,h)\,\gamma_\mu\gamma_5\, v(p,h)\ =\ 4h\,p_\mu\ . \tag{2.251b}$$

The analogue of (2.247) for the negative energy spinors is

$$v(p,h)\,\overline{v}(p,h) = \tfrac{1}{2}(1-2h\,\gamma_5)\,\not{p} \;. \tag{2.252}$$

2.4.3 FIELD THEORY OF THE NEUTRINO

We may now quantize the neutrino theory using the canonical techniques applied in the electron field theory. As before, we write

$$\psi(x) = (2\pi)^{-3}\int d^3p\,(2p_0)^{-1} \sum_{h=\pm\frac{1}{2}} \{c(p,h)\,u(p,h)\,e^{-ipx} + d^\dagger(p,h)\,v(p,h)\,e^{ipx}\} \tag{2.253a}$$

$$\overline{\psi}(x) = (2\pi)^{-3}\int d^3p\,(2p_0)^{-1} \sum_{h=\pm\frac{1}{2}} \{d(p,h)\,\overline{v}(p,h)\,e^{-ipx} + c^\dagger(p,h)\,\overline{u}(p,h)\,e^{ipx}\}\,, \tag{2.253b}$$

where the operators c, $c^\dagger$, d, $d^\dagger$ satisfy anti-commutation relations analogous to (2.182). Thus

$$\{c(p,h),\; c(p',h')\} = \{d(p,h),\; d(p',h')\} = 0 \tag{2.254a}$$

$$\{c(p,h),\; d(p',h')\} = \{c(p,h),\; d^\dagger(p',h')\} = 0 \tag{2.254b}$$

$$\{c(p,h),\; c^\dagger(p',h')\} = \{d(p,h),\; d^\dagger(p',h')\} = (2\pi)^3\,2p_0\,\delta(\underline{p}-\underline{p}')\,\delta_{hh'}\;. \tag{2.254c}$$

The Poincaré invariance of this field theory is established just as for the electron field theory, and as before we interpret $c(p,h)$ and $d(p,h)$ as operators which respectively destroy a particle and anti-particle with momentum p and helicity h; $c^\dagger(p,h)$ and $d^\dagger(p,h)$ are likewise the corresponding creation operators.

Now it is an experimental fact that <u>only</u> left-handed neutrinos and right-handed anti-neutrinos have been observed in nature. It follows from (2.242) and (2.249) that

$$(1-\gamma_5)\,u(p,h) = (1-2h)\,u(p,h) \tag{2.255a}$$

$$(1-\gamma_5)\,v(p,h) = (1+2h)\,v(p,h)\;. \tag{2.255b}$$

Thus, multiplying (2.253a) on the left by $\frac{1}{2}(1-\gamma_5)$, we find

$$\tfrac{1}{2}(1-\gamma_5)\psi(x) = (2\pi)^{-3}\int d^3p(2p_o)^{-1}\ \{c(p,-\tfrac{1}{2})\ u(p,-\tfrac{1}{2})\,e^{-ipx}$$

$$+ d^{\dagger}(p,+\tfrac{1}{2})\ v(p,+\tfrac{1}{2})\ e^{ipx}\}\ , \qquad (2.256)$$

and the experimental fact may be accommodated within the theory by requiring that the neutrino field $\psi(x)$ occurs only when premultiplied by the matrix $\tfrac{1}{2}(1-\gamma_5) \equiv a$. Likewise

$$\overline{\psi}(x)\tfrac{1}{2}(1+\gamma_5) = (2\pi)^{-3}\int d^3p(2p_o)^{-1}\ \{d(p,\tfrac{1}{2})\overline{v}(p,+\tfrac{1}{2})e^{-ipx} + c^{\dagger}(p,-\tfrac{1}{2})\overline{u}(p,-\tfrac{1}{2})e^{ipx}\},$$

$$(2.257)$$

and $\overline{\psi}$ occurs only when postmultiplied by the matrix $\tfrac{1}{2}(1+\gamma_5) \equiv a'$. Plainly there is no chance of any such theory being parity invariant. In a parity reversed frame of reference the helicity of a particle is reversed, and we know that right-handed neutrinos do not occur in nature. This simple observation is of course contained within the formalism we have developed. Suppose there is a unitary operator $\mathscr{P}$, representing the parity transformation. As in (2.208) it has the property

$$\mathscr{P}\psi(x^o,\underline{x})\ \mathscr{P}^{-1} = \epsilon^*\ \gamma^o\ \psi(x^o,-\underline{x})\ . \qquad (2.258)$$

Thus,

$$\mathscr{P}a\,\psi(x^o,\underline{x})\ \mathscr{P}^{-1} = \epsilon^*\ \gamma^o\ a'\ \psi(x^o,-\underline{x})\ , \qquad (2.259)$$

since γ^o and γ_5 anti-commute. So there is no way we can choose the phase factor ϵ so that $\psi(x^o,-\underline{x})$ is premultiplied by the matrix a; consequently any Hamiltonian containing the neutrino field <u>only</u> when it is premultiplied by a cannot satisfy (2.209), and such a theory cannot, therefore, be parity invariant.

In the same way, it is apparent that such theories cannot be invariant under charge conjugation; the charge conjugate of a neutrino (with negative helicity) is evidently an anti-neutrino also with negative helicity, which, as we have said, is not seen in nature. Again using the formalism we see from (2.210) that

$$\mathscr{C}\ a\,\psi(x)\ \mathscr{C}^{-1} = \omega^*\ a\ C\ \overline{\psi}^{T}(x)$$

$$= \omega^*\ C\,[\overline{\psi}(x)\ a]^{T}\ . \qquad (2.260)$$

Thus there is no way we can choose the phase factor ω so that $\overline{\psi}$ occurs only when multiplied by a', and the theory cannot be charge conjugation invariant.

On the other hand, if we consider the combined transformation we find

$$(\mathcal{C}\mathcal{P})\, a\,\psi(x^o, \underline{x})\, (\mathcal{C}\mathcal{P})^{-1} = \mathcal{C}\,\epsilon^*\gamma^o a'\,\psi(x^o - \underline{x})\,\mathcal{C}^{-1}$$

$$= \epsilon^*\omega^*\gamma^o C[\overline{\psi}(x^o, -\underline{x})\, a']^T . \qquad (2.261)$$

Thus we cannot exclude the possibility that such a theory may be CP-invariant. As before this can be understood without the formalism, since under the combined transformation a neutrino (with negative helicity) is transformed into an anti-neutrino with positive helicity, and this does not contradict our fundamental physical input. In fact, as we shall see, almost but not quite all of the experimental data are consistent with the weak interaction being CP-invariant.

Again using our formalism and (2.218) we have

$$\mathcal{T} a\,\psi(x^o, \underline{x})\,\mathcal{T}^{-1} = a^*\eta^*A^{T-1}\gamma^{0T} B\,\psi(-x^o, \underline{x})$$

$$= \eta^* A^{T-1}\gamma^{0T} B\, a\psi(-x^o, \underline{x}) , \qquad (2.262)$$

and we see that such a theory <u>may</u> be time reversal invariant. The experimental data are also consistent with the weak interactions being T-invariant.

In fact these two invariances, CP- and T- , plainly imply each other in any theory which is invariant under the combined transformation $\mathcal{T}\mathcal{C}\mathcal{P}$. Now there is a theorem, which is fundamental to the whole of particle physics, due to Lüders[3]. It states that any local Lagrangian field theory which is invariant under <u>restricted</u> Lorentz transformations is also invariant under the combined transformation, provided that the Lagrangian density $\mathcal{L}(x)$ has been correctly symmetrized; the symmetrization requirement is simply that $\mathcal{L}$ must be symmetrized with respect to Bose fields and anti-symmetrized with respect to Fermi fields. The requirements of a local field theory and of restricted Lorentz invariance are conditions which are almost taken for granted in the formulation of <u>any</u> field theory, and consequently the resulting theory is necessarily invariant under $\mathcal{T}\mathcal{C}\mathcal{P}$. Certainly, all of the field theories of weak interactions presented in this book fulfil the conditions. As a result it follows that we may choose the arbitrary phase factors of the fields so that

$$(\mathcal{T}\mathcal{C}\mathcal{P})\mathcal{L}(x)(\mathcal{T}\mathcal{C}\mathcal{P})^{-1} = \mathcal{L}(-x) . \tag{2.263}$$

2.5 THE LEPTONIC CURRENT

We have noted in the previous chapter that the electron e^-, its neutrino ν_e, the muon μ^-, its neutrino ν_μ and their anti-particles $e^+, \bar{\nu}_e, \mu^+, \bar{\nu}_\mu$, do not participate in the strong interactions. The theory of QED tells us precisely how the electron interacts with the electromagnetic field, and, so far as is known, it appears that precisely the same theory describes the interaction of the muon with the electromagnetic field; the only quantitative differences which arise are those which stem from the fact that the mass of the muon is some two hundred times larger than that of the electron. Aside from the universal gravitational attraction, which is in any case some thirty orders of magnitude weaker than any other interaction, the only other known interactions of the leptons are their weak interactions; for the neutrinos the weak interactions are the only direct interactions. We are now in a position to state precisely how the leptons participate in those weak interactions in which they do participate.

Until very recently it was believed that the lepton field operators enter the Hamiltonian responsible for weak interactions only in a certain bilinear combination known as the leptonic "current" L_α and its hermitian conjugate $L_\alpha^\dagger$. Using the notation defined previously

$$L_\alpha(x) = \bar{\psi}_{(e)}(x)\,\gamma_\alpha(1-\gamma_5)\,\psi_{(\nu_e)}(x) + \bar{\psi}_{(\mu)}(x)\,\gamma_\alpha(1-\gamma_5)\,\psi_{(\nu_\mu)}(x) \tag{2.264a}$$

$$L_\alpha^\dagger(x) = \bar{\psi}_{(\nu_e)}(x)\,\gamma_\alpha(1-\gamma_5)\,\psi_{(e)}(x) + \bar{\psi}_{(\nu_\mu)}(x)\,\gamma_\alpha(1-\gamma_5)\,\psi_{(\mu)}(x) . \tag{2.264b}$$

We shall see shortly how recent experiments have shown that this leptonic current hypothesis cannot be the whole truth. It now appears that the leptons' fields also enter the weak Hamiltonian in certain other combination(s) whose form is yet to be established. Nevertheless, it remains true that the leptonic current alone is believed to describe the role of the leptons in very many weak processes; for reasons which will become apparent, the processes to which the hypothesis is now thought to apply are called "charged leptonic current phenomena".

Of course the form (2.264) for the leptonic current is not by any means self-evident. It emerged as the fruit of more than twenty years research starting from the original theory proposed by Fermi[(4)]. In this book we shall start from the hypothesis contained in (2.264). In this, and the following chapters, we shall explore the theoretical consequences of this postulate and investigate the extent to which these are consistent with the experimental data. Broadly speaking we shall see that this simple assumption describes consistently the leptonic behaviour in all known processes to which it is applicable, and in addition forbids those conceivable processes which have not been seen. This power of our hypothesis to explain and predict experimental phenomena is of course the root cause for its general acceptance by practising physicists. They (we) are supported in this belief by the essential simplicity of the assumption. Statements like (2.264) can of course never be "proven" by the experimental data (only disproven). We shall see, in fact, that individual processes can tolerate quite substantial deviations from the preferred form. Notwithstanding this, our faith that "Nature is Beautiful" leads us to prefer a simple hypothesis which hopefully explains a large class of leptonic weak phenomena. However, despite the success of the theory in describing the relatively low energy phenomena so far observed, the theory itself cannot be complete, as we have already mentioned. Its hitherto widespread success is thus construed as indicating that the theory contains at least a germ of 'The Truth'.

Let us now explore some of the immediate consequences of our "current" hypothesis (2.264) using the formalism developed earlier in this chapter. Firstly, it is apparent that the leptonic current L_α is the sum of two pieces which transform respectively like the vector and axial vector covariants, defined in (2.102) and (2.103), under orthochronous Lorentz transformations. The different behaviour of these quantitites under transformations which include parity reversal ensures that no theory involving the leptonic current as defined in (2.264) can be parity reversal invariant. Indeed, we see from (2.264) that the neutrino fields $\psi_{(\nu_e)}$ and $\psi_{(\nu_\mu)}$ are both premultiplied by the matrix $(1-\gamma_5)$. We have already observed that this ensures that only left-handed neutrinos and right-handed anti-neutrinos occur in nature, and that such a situation necessarily implies parity violation. As a result, we shall not in general expect weak transition probabilities to satisfy equations of the type shown in (2.229). In particular, we shall see that the transition probability for polarized muon decay, for example, has the form

$$W(E) = A(E) + B(E)\ \underline{t}\cdot\underline{p} \quad , \tag{2.265}$$

where $\underline{t}$ is the spin vector of the decaying muon and $\underline{p}$, E the momentum and energy of the observed final state electron. Plainly, (2.265) does not satisfy (2.229); physically it indicates that at a given energy the distribution of the observed electrons is asymmetric with respect to the plane perpendicular to the spin of the polarized muon.

2.5.1 ELECTRON NUMBER AND MUON NUMBER

Secondly, the form of the leptonic current predicts the conservation of two quantum numbers, analogous to electric charge, in all weak processes. The first of these, electron-number N_e, is assigned as follows:

$$e^- \text{ and } \nu_e \quad \text{have} \quad N_e = 1 \tag{2.266a}$$

$$e^+ \text{ and } \bar{\nu}_e \quad \text{have} \quad N_e = -1 \ . \tag{2.266b}$$

$$\text{All other particles have} \quad N_e = 0 \ . \tag{2.266c}$$

The field operator $\bar{\psi}_{(e)}$, as we have said, creates an electron $(N_e = 1)$ or destroys a positron $(N_e = -1)$; it thus has the net effect of increasing the electron-number by one unit. Likewise, $\psi_{(\nu_e)}$ destroys an electronic neutrino $(N_e = 1)$ or creates an electronic anti-neutrino $(N_e = -1)$; its net effect is therefore to increase the electron number by minus one unit, that is to decrease it by one unit. Thus overall the first term of L_α conserves electron-number. The same is true of the second term, which involves only field operators of particles all of which have zero electron-number. Thus the leptonic current L_α, and equally its hermitian conjugate $L^\dagger_\alpha$, both conserve the quantum number N_e. Thus if, as our hypothesis states, the lepton fields enter the weak Hamiltonian in the combinations L_α and $L_\alpha^\dagger$, then it follows that these weak processes conserve N_e; that is to say, the total electron-number of the initial state (computed by summing the electron-number of all particles in that state) is equal to the total electron-number of the final state. In precisely the same way, the conservation of the second quantum number N_μ is predicted by the leptonic current hypothesis. The assignments parallel those in (2.266) :

$$\mu^- \text{ and } \nu_\mu \quad \text{have} \quad N_\mu = +1 \tag{2.267a}$$

$$\mu^+ \text{ and } \bar{\nu}_\mu \quad \text{have} \quad N_\mu = -1 \ . \tag{2.267b}$$

$$\text{All other particles have } N_\mu = 0 \ . \tag{2.267c}$$

We may restate these conservation laws using the field theory formalism developed earlier in this chapter. The electron-number of a state is then the eigenvalue of the electron-number operator $\mathbb{N}_e$. The assignments made in (2.266) then plainly demand that $\mathbb{N}_e$ satisfies

$$[\mathbb{N}_e , \psi_{(e)}(x)] = -\psi_{(e)}(x) \tag{2.268a}$$

$$[\mathbb{N}_e , \bar{\psi}_{(e)}(x)] = +\bar{\psi}_{(e)}(x) \tag{2.268b}$$

$$[\mathbb{N}_e , \psi_{(\nu_e)}(x)] = -\psi_{(\nu_e)}(x) \tag{2.268c}$$

$$[\mathbb{N}_e , \bar{\psi}_{(\nu_e)}(x)] = +\bar{\psi}_{(\nu_e)}(x) \ , \tag{2.268d}$$

and that the commutator of $\mathbb{N}_e$ with all other field operators is zero. In the case that the fields involved are all free fields the required operator is plainly just

$$\mathbb{N}_e^{\text{free}} = \int d^3p\,(2\pi)^{-3}(2p_o)^{-1}\Big\{\sum_{\pm s}[a^\dagger(p,s)\,a(p,s) - b^\dagger(p,s)\,b(p,s)]$$

$$+ \sum_{h=\pm\frac{1}{2}} [c^\dagger(p,h)\,c(p,h) - d^\dagger(p,h)\,d(p,h)]\Big\} \tag{2.269a}$$

$$= \int d^3x : \{\bar{\psi}_{(e)}(x)\,\gamma_o\,\psi_{(e)}(x) + \bar{\psi}_{(\nu_e)}(x)\,\gamma_o\,\psi_{(\nu_e)}(x)\} : , \tag{2.269b}$$

where the operators a,b,c,d respectively destroy e^-, e^+, ν_e, $\bar{\nu}_e$. It is easy to see that (2.269b) and the free field commutation relations (2.178) imply the desired commutators (2.268).

Since the commutators of $\mathbb{N}_e$ with $\bar{\psi}_{(\mu)}$ and $\psi_{(\nu_\mu)}$ are both zero,

$$[\,\mathbb{N}_e, L_\alpha(x)\,] = [\,\mathbb{N}_e, \bar{\psi}_{(e)}(x)\,]\,\gamma_\alpha\,(1-\gamma_5)\,\psi_{(\nu_e)}(x)$$

$$+\ \bar{\psi}_{(e)}(x)\,\gamma_\alpha\,(1-\gamma_5)\,[\,\mathbb{N}_e, \psi_{(\nu_e)}(x)\,]\ , \tag{2.270}$$

and using (2.268b, c) we deduce that

$$[\,\mathbb{N}_e,\ L_\alpha(x)\,] = 0\ . \tag{2.271}$$

It then follows from the leptonic current hypothesis that $\mathbb{N}_e$ commutes with the weak interaction Hamiltonian. In the same way it is easily seen that $\mathbb{N}_e$ commutes with the free field Hamiltonian; this is given for the electron field in (2.183), and analogously for the neutrino field by setting $m = 0$. $\mathbb{N}_e$ also commutes with the electromagnetic current j_μ given in (2.222b), and therefore with the electromagnetic interaction Hamiltonian. If the free Hamiltonians and the electromagnetic interaction Hamiltonians are the only other terms containing the lepton field operators, it follows that $\mathbb{N}_e$ commutes with the full Hamiltonian and therefore with the scattering matrix S. Thus for arbitrary physical states $|\alpha\rangle$, $|\beta\rangle$

$$\langle\alpha|[\,\mathbb{N}_e, S\,]|\beta\rangle = 0\ . \tag{2.272}$$

The physical states are eigenstates of $\mathbb{N}_e$, so that

$$\mathbb{N}_e\,|\alpha\rangle = N_e^{(\alpha)}\,|\alpha\rangle \tag{2.273a}$$

$$\mathbb{N}_e\,|\beta\rangle = N_e^{(\beta)}\,|\beta\rangle\ , \tag{2.273b}$$

where $N_e^{(\alpha)}$, $N_e^{(\beta)}$ are the electron-numbers of the states $|\alpha\rangle$, $|\beta\rangle$. It thus follows from (2.272) that

$$(\,N_e^{(\alpha)} - N_e^{(\beta)}\,)\,\langle\alpha|S|\beta\rangle = 0\ . \tag{2.274}$$

Any process $|\alpha\rangle \to |\beta\rangle$ which can occur physically has, by definition, a non-zero matrix element $\langle\alpha|S|\beta\rangle$, so we deduce that

$$N_e^{(\alpha)} = N_e^{(\beta)} \tag{2.275}$$

for all physical processes. In the same way we may of course show that the leptonic current hypothesis predicts that muon-number is conserved in all physical processes.

The strongest experimental evidence that electron-number is conserved, as predicted by the leptonic current hypothesis, comes from the apparent absence of neutrinoless double beta decay

$$(Z,A) \rightarrow (Z+2,\ A) + 2\,e^- \ . \qquad (2.276)$$

By definition all nuclear states (Z, A) have electron-number $N_e = 0$, and using the assignments (2.266), we see that the final state in (2.276) has $N_e = 2$. Thus plainly N_e is not conserved if any of the processes (2.276) are found to occur in nature. Of course ordinary double beta decay

$$(Z,A) \rightarrow (Z+2,\ A) + 2\,e^- + 2\bar{\nu}_e \qquad (2.277)$$

does conserve electron-number. However, in the absence of the electron-number conservation law, it can be shown that the neutrinoless process (2.276) would be favoured by a large kinematic factor (typically in the range $10^5 - 10^6$) over the neutrino-acoompanied process (2.277). Thus the search for these processes can provide a very sensitive test of electron-number conservation.

The most rigorous bound so far derived has arisen from the search for the decay

$$^{130}Te \rightarrow {}^{130}Xe + 2e^- \ . \qquad (2.278)$$

The experiment consists of looking at tellurium ores of known age and searching for any excess of xenon over that occurring naturally in the atmosphere. Any observed excess must have been created by a combination of the neutrinoless process (2.278) and the corresponding neutrino accompanied process (2.277). A priori there is no way of knowing which of the two decay modes is responsible for any observed excess, since only an observation of the electron spectrum can distinguish them; this is of course impossible in such experiments. In fact, a small xenon excess has been observed, and similar experiments in selenium ores also reveal a small excess of krypton(5). There is thus unambiguous experimental evidence that double beta decay of one sort or another does occur. The theory which will be developed in the succeeding chapters permits the theoretical calculation of the decay amplitudes for the neutrino accompanied decays. It is then possible to compare these predictions with the observed excesses of xenon and krypton and, in a reasonably model independent manner, obtain information on the magnitude of the neutrinoless decay amplitude[6]. The experiments referred to show that the fraction ϵ_e of the decay amplitude which does not conserve electron-number satisfies

$$\epsilon_e \lesssim 3 \times 10^{-3} \quad . \tag{2.279}$$

Tests such as these of course provide no evidence in support of the second conservation law predicted: conservation of muon-number. The processes analogous to (2.276,7) involving emission of muons are not allowed kinematically. The best evidence for the conservation of N_μ comes from the apparent absence of the scattering processes

$$\nu_\mu + (Z, A) \to (Z-1, A) + \mu^+ \quad . \tag{2.280}$$

Using the assignments (2.267), we see that the initial state has $N_\mu = 1$, while the final state has $N_\mu = -1$. On the other hand, the process

$$\nu_\mu + (Z, A) \to (Z+1, A) + \mu^- \tag{2.281}$$

is allowed by muon number conservation, since both initial and final states have $N_\mu = 1$.

The muonic neutrinos are produced by the decay in flight of π^+ mesons which have been produced by strong interactions in an accelerator. The decay products are then directed into a large mass of shielding which absorbs all particles except the neutrinos. The required beam of muonic-neutrinos, together with a small contamination of electronic-neutrinos, emerges from the shielding and is directed into a bubble chamber or spark chamber in which the charged muons μ^+ and μ^- may be produced and detected. In fact, none of the processes (2.280) has ever been positively identified. The significance of this result may be quantified by comparing it with the counting rate for the observation of μ^-'s produced in the allowed process (2.281). This leads to the conclusion[7] that the ratio ϵ_μ of the amplitudes for the N_μ nonconserving process (2.280) to the allowed process (2.281) satisfies

$$\epsilon_\mu \lesssim 4.6 \times 10^{-3} \quad . \tag{2.282}$$

Thus the predicted conservation laws for N_e and N_μ are consistent with the experimental data to about the same degree of accuracy.

In addition to the tests we have discussed, there are a number of other processes which are forbidden if N_e or N_μ is conserved. The forbidden processes include

$$\nu_\mu + (Z, A) \to (Z - 1, A) + e^+ \tag{2.283a}$$

$$\nu_\mu + (Z, A) \to (Z + 1, A) + e^- \quad . \tag{2.283b}$$

Neither of these has ever been observed. The absence of the latter[(8)] led to the belief, implicit in our hypothesis, that the electronic and muonic neutrinos are distinct. Hitherto the data had been consistent with the hypothesis that they were identical and that there was just one lepton number conservation law. In our terminology this is equivalent to conserving $N_e + N_\mu$. While this conservation law is sufficient to forbid (2.283a), it plainly allows the unobserved process (2.283b).

Conservation of N_e or N_μ also forbids the processes

$\mu^+ \to e^+ \gamma$	(B.R. $\lesssim 2.9 \times 10^{-8}$)
$\mu^+ \to e^+ \gamma\gamma$	(B.R. $\lesssim 1.6 \times 10^{-5}$)
$\mu^+ \to e^+ e^+ e^-$	(B.R. $\lesssim 6.2 \times 10^{-9}$)
$\mu^- + \mathrm{Cu} \to e^- + \mathrm{Cu}$	(B.R. $\lesssim 2.4 \times 10^{-7}$)
$K_L^o \to e^\pm \mu^\mp$	(B.R. $\lesssim 8.0 \times 10^{-6}$)
$K^+ \to \pi^+ e^- \mu^+$	(B.R. $\lesssim 3.0 \times 10^{-5}$) .

Again, none of these has ever been seen and the present upper limit on the relevant branching ratio (B.R.) is indicated beside each process.

2.5.2 NEUTRAL CURRENTS

Finally, the leptonic current hypothesis predicts that the leptonic current always carries one unit of charge. The operator form (2.264a) for L_α

plainly destroys one unit of electric charge or creates minus one unit (the neutrino fields $\psi_{(\nu)}$ are neutral fields). Similarly, the hermitian conjugate current $L^\dagger_\alpha$ creates one unit of charge, or destroys minus one unit. For this reason L_α, $L^\dagger_\alpha$ are called "charged" currents. If we assume that the Hamiltonian responsible for the semileptonic processes is linear in L_α and $L^\dagger_\alpha$, then the matrix elements of this Hamiltonian vanish whenever the total charge carried by the leptons in the initial and final states is the same. If we further assume that these processes may to good approximation be described by lowest order perturbation theory, then it follows that the above processes should not be seen in nature. In particular, this absence of "neutral currents" forbids the processes

$$K^o_L \rightarrow \mu^+ \mu^- \qquad (\text{B.R.} \lesssim 1.8 \times 10^{-5})$$

$$K^o_L \rightarrow e^+ e^- \qquad (\text{B.R.} \lesssim 1.0 \times 10^{-6})$$

$$K^+ \rightarrow \pi^+ e^+ e^- \qquad (\text{B.R.} \lesssim 4.0 \times 10^{-7})$$

$$K^+ \rightarrow \pi^+ \mu^+ \mu^- \qquad (\text{B.R.} \lesssim 2.4 \times 10^{-6})$$

$$K^+ \rightarrow \pi^+ \nu \bar{\nu}\ , \qquad (\text{B.R.} \lesssim 1.4 \times 10^{-6})\ ,$$

in all of which the total charge carried by the leptons is zero in both initial and final states. None of these has ever been seen, and the present upper limit on the relevant branching ratio is indicated beside each process. It is important to realize that none of these processes is absolutely forbidden. As we shall see, they are allowed in second order perturbation theory, and, except for the last process, they are also allowed as mixed electromagnetic and weak processes. The limits on the branching ratios imply that the decay amplitudes are reduced by a factor of at least 10^{-2} compared with those of the relevant allowed processes.

However, it is important to recognize that in all of the above processes hypercharge is not conserved; the initial state kaons have hypercharge $Y = \pm 1$, while the final states all have $Y = 0$. It is easy to think of neutral current processes which conserve hypercharge. If there are no neutral currents the following processes are also forbidden, since the leptons in the initial and final states have the same (zero) charge :

$$\nu_\mu \, p \to \nu_\mu \, p \, \pi^o$$

$$\nu_\mu \, p \to \nu_\mu \, n \, \pi^+$$

In fact both of these have recently been observed. Comparing with the cross sections for the "common" charged current analogue the branching ratios are[9]

$$\frac{\sigma(\nu_\mu \, p \to \nu_\mu \, p \, \pi^o)}{\sigma(\nu_\mu \, p \to \mu^- \, p \, \pi^+)} = 0.51 \pm 0.25 \qquad (2.284a)$$

$$\frac{\sigma(\nu_\mu \, p \to \nu_\mu \, n \, \pi^+)}{\sigma(\nu_\mu \, p \to \mu^- \, p \, \pi^+)} = 0.17 \pm 0.08 \qquad (2.284b)$$

Other processes involving neutral currents have also been observed (these will be discussed in Chapter 6).

Thus the existence of neutral currents now seems beyond doubt, although the reason for the suppression or absence of neutral current processes in which hypercharge is not conserved is still something of a mystery. The magnitude of the branching ratios (2.284) shows that the neutral current interaction has a strength which is entirely comparable with that of the charged current interaction. So there is no doubt that they must be put on a common footing. Our hypothesis is that the lepton fields enter the weak Hamiltonian in two independent combinations: the charged currents L_α, $L_\alpha^\dagger$ given in (2.264), and the neutral "currents" whose form is still unspecified; the results of 2.5.1 strongly suggest that both combinations conserve both electron number N_e and muon number N_μ. In the next two chapters we shall explore the consequences of our hypothesis in processes to which, almost exclusively, only L_α, $L_\alpha^\dagger$ can contribute: the charged leptonic current phenomena. These constitute the vast majority of known leptonic processes, which, of course, is why neutral currents were not discovered earlier. In Chapter 6 we shall discuss the theoretical attempts which have been made to unify what for the present we shall regard as independent phenomena.

Chapter 3

PURELY LEPTONIC PROCESSES

From a theoretical viewpoint the "cleanest" weak processes are those in which only the leptons participate. By definition any process involving hadrons is complicated by the unknown and incalculable effects of strong interactions. The leptons are defined as those particles which participate only in weak and electromagnetic interactions both of which, it is hoped, are calculable using perturbation theory. The simplest possibility for the Hamiltonian responsible for the purely leptonic processes is that it has a simple current-current form obtained by coupling the currents L_α, $L_\alpha^\dagger$ given in (2.264). Thus we assume that the Hamiltonian density responsible for these processes is given by

$$\mathcal{H}_W^L(x) = \frac{G}{\sqrt{2}} L_\alpha^\dagger(x) \; L^\alpha(x) \; , \tag{3.1}$$

where G is a constant and the factor $\sqrt{2}$ is included for historical reasons[1]. Of course, there are other possible ways of coupling L_α and $L_\alpha^\dagger$ by the use of derivatives, so the form (3.1) has no a priori merit other than its manifest simplicity. We shall now use this form to make physical predictions which may then be compared with the experimental data.

3.1 MUON DECAY

If we work in first order perturbation theory, the Hamiltonian $\mathcal{H}_W^L$ given in (3.1) implies the existence of a number of weak interactions among the leptons. The part of $\mathcal{H}_W^L$ which arises from the electronic part of L_α coupled to itself gives rise to the following scattering processes

$$e^- \nu_e \to e^- \nu_e \tag{3.2a}$$

$$e^- \bar{\nu}_e \rightarrow e^- \bar{\nu}_e \tag{3.2b}$$

$$e^+ \nu_e \rightarrow e^+ \nu_e \tag{3.2c}$$

$$e^+ \bar{\nu}_e \rightarrow e^+ \bar{\nu}_e \tag{3.2d}$$

$$e^+ e^- \leftrightarrow \nu_e \bar{\nu}_e \quad . \tag{3.2e}$$

For reasons which will become apparent, the experimental study of these processes in the laboratory is difficult, and none of them has yet been subjected to a thorough analysis. Similar considerations apply to the corresponding scattering processes which arise from the muonic piece of L_α coupled to itself, in which $e^\pm$, ν_e, $\bar{\nu}_e$ in the above processes are replaced by $\mu^\pm$, ν_μ, $\bar{\nu}_\mu$. The "non-diagonal" part of $\mathcal{H}_W^L$, which arises from coupling the electronic part of L_α to the muonic part, gives rise to similar scattering processes:

$$\mu^- \nu_e \leftrightarrow e^- \nu_\mu \tag{3.3a}$$

$$\mu^- \bar{\nu}_\mu \leftrightarrow e^- \bar{\nu}_e \tag{3.3b}$$

$$e^+ \nu_e \leftrightarrow \mu^+ \nu_\mu \tag{3.3c}$$

$$e^+ \bar{\nu}_\mu \leftrightarrow \mu^+ \bar{\nu}_e \tag{3.3d}$$

$$\nu_e \bar{\nu}_\mu \leftrightarrow e^- \mu^+ \tag{3.3e}$$

$$\bar{\nu}_e \nu_\mu \leftrightarrow e^+ \mu^- \quad . \tag{3.3f}$$

In addition, however, this non-diagonal part gives rise to two muon decay processes, which are allowed kinematically because the muon has a larger mass than the electron

$$\mu^- \rightarrow e^- \bar{\nu}_e \nu_\mu \tag{3.4a}$$

$$\mu^+ \rightarrow e^+ \nu_e \bar{\nu}_\mu \quad . \tag{3.4b}$$

Of all the processes predicted by $\mathcal{H}_W^L$ only muon decay has been studied in detail in the laboratory. Thus the process has a unique position theoretically and experimentally, and it has naturally been subjected to considerable experimental analysis in order to check the precise form of $\mathcal{H}_W^L$. Let us first see how experimental predictions are made with the form (3.1). To be precise we consider the decay of the μ^-, given in (3.4a). In general, the muon may be prepared in the laboratory in one of two spin states. The energy and momentum of the final state electron may be observed together with its spin. In practice, neither the electronic anti-neutrino $\bar{\nu}_e$ nor the muonic neutrino ν_μ are observed. Thus the object of the calculation about to be presented is to derive a formula giving the transition probability for the decay of a polarized muon as a function of the momentum and polarization of the final state electron.

The S-matrix may be expanded as a power series in the weak coupling constant G

$$S = 1 - i\int d^4x \, \mathcal{H}_W^L(x) + O(G^2) \quad . \tag{3.5}$$

The required S-matrix element is thus given by

$$< e^- \bar{\nu}_e \nu_\mu | S | \mu^- > = - i\int d^4x < e^- \bar{\nu}_e \nu_\mu | \mathcal{H}_W^L(x) | \mu^- > \quad , \tag{3.6}$$

if we neglect all terms of order G^2 and higher. We denote the momentum and spin vectors of the muon by k, t respectively, so that

$$| \mu^- > = a^\dagger_{(\mu)} (k, t) | 0 > \quad , \tag{3.7}$$

where $a^\dagger_{(\mu)}$ is the creation operator of the μ^-. Further, since there is no muon in the final state,

$$a_{(\mu)} (k, t) | e^- \bar{\nu}_e \nu_\mu > = 0 \tag{3.8a}$$

or

$$< e^- \bar{\nu}_e \nu_\mu | a^\dagger_{(\mu)} (k, t) = 0 \quad . \tag{3.8b}$$

Substituting (3.7) into (3.6) and using (3.8b) gives

$$< e^- \bar{\nu}_e \nu_\mu | \mathcal{H}_W^L(x) | \mu^- > = < e^- \bar{\nu}_e \nu_\mu | [\mathcal{H}_W^L(x), a^\dagger_{(\mu)} (k, t)] | 0 > \quad . \tag{3.9}$$

Using the form (3.1) for $\mathcal{H}_W^L$ we must evidently evaluate the commutator

$$\begin{aligned} [L^\dagger_\alpha(x) L^\alpha(x), a^\dagger_{(\mu)} (k, t)] &= [L^\dagger_\alpha(x), a^\dagger_{(\mu)} (k, t)] L^\alpha(x) \\ &+ L^\dagger_\alpha(x) [L^\alpha(x), a^\dagger_{(\mu)} (k, t)] \quad . \end{aligned} \tag{3.10}$$

To evaluate the right hand side of (3.10) we make use of the form (2.264) for $L_\alpha^\dagger$ and L_α and identities similar to that used in deriving (2.186). It is plain that

$$[L_\alpha(x), \; a^\dagger_{(\mu)}(k,t)] = 0 \quad , \tag{3.11}$$

since the only way a non-zero contribution can arise is from the anti-commutation of $a^\dagger_{(\mu)}(k,t)$ with an annihilation operator $a_{(\mu)}$ of the muon. We see that $a_{(\mu)}$ occurs only in $\psi_{(\mu)}(x)$ (not $\bar{\psi}_{(\mu)}(x)$) and that this operator does not appear in L_α. For this reason also,

$$[L^\dagger_\alpha(x), \; a^\dagger_{(\mu)}(k,t)] = [\bar{\psi}_{(\nu_\mu)}(x)\,\gamma_\alpha(1-\gamma_5)\,\psi_{(\mu)}(x), \; a^\dagger_{(\mu)}(k,t)]$$

$$= \bar{\psi}_{(\nu_\mu)}(x)\,\gamma_\alpha\,(1-\gamma_5)\,\{\psi_{(\mu)}(x), \; a^\dagger_{(\mu)}(k,t)\} \quad . \tag{3.12}$$

Using the decomposition (2.179a) and the anti-commutation relations (2.182)

$$\{\psi_{(\mu)}(x), \; a^\dagger_{(\mu)}(k,t)\} = \int d^3p\,(2p_o)^{-1} \sum_{\pm s} 2k_o\,\delta(\underline{p}-\underline{k})\,\delta_{st}\,u_{(\mu)}(p,s)\,e^{-ipx}$$

$$= u_{(\mu)}(k,t)\,e^{-ikx} \quad . \tag{3.13}$$

Substituting (3.13) into (3.12) and then back into (3.10) and (3.9) we conclude that

$$<e^-\bar{\nu}_e\nu_\mu|\mathcal{H}^L_W(x)|\mu^->$$

$$= \frac{G}{\sqrt{2}} <e^-\bar{\nu}_e\nu_\mu|\bar{\psi}_{(\nu_\mu)}(x)\gamma_\alpha(1-\gamma_5)u_{(\mu)}(k,t)\;L^\alpha(x)|0> e^{-ikx} \quad . \tag{3.14}$$

Proceeding similarly for the particles in the final state, and performing the integration in (3.6), we have finally

$$<e^-\bar{\nu}_e\nu_\mu|S|\mu^-> = -i(2\pi)^4\;\delta(k-p-q_1-q_2)\;M(\mu^-\to e^-\bar{\nu}_e\nu_\mu) \,, \tag{3.15a}$$

where the invariant matrix element M is given by

$$M(\mu^-\to e^-\bar{\nu}_e\nu_\mu) = \frac{G}{\sqrt{2}}[\bar{u}_{(e)}(p,s)\,\gamma^\alpha\,(1-\gamma_5)\,v_{(\nu_e)}(q_1,\tfrac{1}{2})]$$

$$[\bar{u}_{(\nu_\mu)}(q_2,-\tfrac{1}{2})\,\gamma_\alpha\,(1-\gamma_5)\,u_{(\mu)}(k,t)] \quad . \tag{3.15b}$$

p, s are the momentum and spin vectors of the final state electron, while

q_1, q_2 are the momenta of $\bar{\nu}_e$ and ν_μ respectively. To calculate the transition probability we need to evaluate the modulus squared of the S-matrix element and therefore of M. Now it follows from the results derived in Chapter 2 that

$$[\bar{u}_{(\nu_\mu)}(q_2, -\tfrac{1}{2})\gamma_\beta(1-\gamma_5)u_{(\mu)}(k,t)]^* = \bar{u}_{(\mu)}(k,t)\gamma_\beta(1-\gamma_5)u_{(\nu_\mu)}(q_2, -\tfrac{1}{2}) . \tag{3.16}$$

Thus, using (2.159) and (2.247), we find

$$\begin{aligned} M_{\alpha\beta} &\equiv [\bar{u}_{(\nu_\mu)}(q_2, -\tfrac{1}{2})\gamma_\alpha(1-\gamma_5)u_{(\mu)}(k,t)][\bar{u}_{(\nu_\mu)}(q_2, -\tfrac{1}{2})\gamma_\beta(1-\gamma_5)u_{(\mu)}(k,t)]^* \\ &= \mathrm{Tr}\,\{\gamma_\alpha(1-\gamma_5)(\not{k}+m_\mu)\tfrac{1}{2}(1+\gamma_5\not{t})\gamma_\beta(1-\gamma_5)\tfrac{1}{2}(1-\gamma_5)\not{q}_2\} \\ &= \mathrm{Tr}\,\{\gamma_\beta\not{q}_2\gamma_\alpha\not{k}(1+\gamma_5)\} - m_\mu\,\mathrm{Tr}\,\{\gamma_\beta\not{q}_2\gamma_\alpha\not{t}(1+\gamma_5)\} . \end{aligned} \tag{3.17}$$

In deriving the last line of (3.17) we have used the cyclic property of traces, the anti-commuting property of γ_5, and the fact that the trace of an odd number of γ-matrices is zero. Finally, using (2.168) we find

$$M_{\alpha\beta} = 4\{[\beta\, q_2\, \alpha\, (k-m_\mu t)] + i\,(\beta\, q_2\, \alpha\, (k-m_\mu t))\} , \tag{3.18a}$$

where for arbitrary vectors A, B we define

$$[\beta\, B\, \alpha\, A] \equiv B_\beta A_\alpha + B_\alpha A_\beta - g_{\alpha\beta}\, A.B \tag{3.19a}$$

$$(\beta\, B\, \alpha\, A) \equiv \epsilon_{\beta\gamma\alpha\delta}\, B^\gamma A^\delta . \tag{3.19b}$$

In precisely the same way we find

$$\begin{aligned} E^{\alpha\beta} &\equiv [\bar{u}_{(e)}(p,s)\gamma^\alpha(1-\gamma_5)v_{(\nu_e)}(q_1, \tfrac{1}{2})][\bar{u}_{(e)}(p,s)\gamma^\beta(1-\gamma_5)v_{(\nu_e)}(q_1, \tfrac{1}{2})]^* \\ &= 4\{[\alpha\, q_1\, \beta\, (p-m_e s)] + i\,(\alpha\, q_1\, \beta\, (p-m_e s))\} . \end{aligned} \tag{3.20}$$

Thus from (3.15b)

$$\begin{aligned} |M(\mu^- \to e^- \bar{\nu}_e \nu_\mu)|^2 &= \frac{G^2}{2} M_{\alpha\beta} E^{\alpha\beta} \\ &= 32\, G^2 [q_1 . (k-m_\mu t)][q_2 . (p-m_e s)] . \end{aligned} \tag{3.21}$$

The transition probability for the process is given by the modulus squared of the S-matrix element

$$|\langle e^- \bar{\nu}_e \nu_\mu | S | \mu^- \rangle|^2 = (2\pi)^8 \delta(k-p-q_1-q_2)\,\delta(0)\,|M(\mu^- \to e\bar{\nu}_e \nu_\mu)|^2 . \tag{3.22}$$

The appearance of the infinite quantity $\delta(0)$ on the right hand side, which results from squaring the delta-function, arises because we are considering the transition rate in the whole volume VT of space time. If we divide by $(2\pi)^4 \delta(0)$ we obtain an expression for the transition probability in a unit volume in unit time. This can be seen as follows

$$VT = \int d^4x = \lim_{k\to 0} \int d^4x\, e^{ikx} = (2\pi)^4 \delta(0) . \tag{3.23}$$

However, the normalization we have chosen for our wave functions is such that a unit volume contains $2k_o$ muons. Thus the transition rate for a <u>single</u> muon into a final state with momenta in the ranges $d^3\underline{p}$ around $\underline{p}$, d^3q_1 around $\underline{q}_1$, and d^3q_2 around $\underline{q}_2$ is

$$d\Gamma = (2\pi)^{-5} \frac{1}{(2k_o)} \delta(k-p-q_1-q_2)\,|M(\mu^- \to e^- \bar{\nu}_e \nu_\mu)|^2 \frac{d^3p\, d^3q_1\, d^3q_2}{(2p_o)(2q_{10})(2q_{20})} . \tag{3.24}$$

Since the neutrino ν_μ and the anti-neutrino $\bar{\nu}_e$ are not observed in the experiments with which we are concerned, we integrate over all values of q_1, q_2 consistent with the energy-momentum delta function in (3.24). The form of $|M|^2$ is given in (3.21). Thus to perform the required integration we need to know

$$I_{\alpha\beta} = \int d^3q_1\, d^3q_2\, (q_{10}\, q_{20})^{-1}\, q_{1\alpha}\, q_{2\beta}\, \delta(q - q_1 - q_2) , \tag{3.25a}$$

where

$$q \equiv k - p . \tag{3.25b}$$

Plainly, $I_{\alpha\beta}$ is covariant, since

$$q_{io}^{-1}\, d^3q_i = 2\delta(q_i^2)\, d^4q_i \qquad (i = 1,2) \tag{3.26}$$

is covariant. Thus, since $I_{\alpha\beta}$ depends only on q,

$$I_{\alpha\beta} = A(q^2)\, g_{\alpha\beta} + B(q^2)\, q_\alpha\, q_\beta . \tag{3.27}$$

Energy-momentum conservation, which is ensured by the delta function in (3.25), implies

$$q^2 = 2\, q_1 . q_2 , \tag{3.28a}$$

since

$$q_1^2 = q_2^2 = 0 . \tag{3.28b}$$

Thus

$$g^{\alpha\beta} I_{\alpha\beta} = 4A + q^2 B = \frac{q^2}{2} I , \tag{3.29a}$$

$$q^\alpha q^\beta I_{\alpha\beta} = q^2 A + (q^2)^2 B = (\frac{q^2}{2})^2 I , \tag{3.29b}$$

where

$$I = \int d^3 q_1 d^3 q_2 (q_{10} q_{20})^{-1} \delta (q - q_1 - q_2) .$$

Plainly, I is an invariant integral, so we may evaluate it in the centre-of-mass (c. m.) frame of the neutrinos in which $\underline{q} = 0$. Thus

$$\begin{aligned} I &= \int d^3 q_1 (q_{10})^{-2} \delta (q_0 - 2q_{10}) \\ &= \int 4\pi \, d q_{10} \, \delta (q_0 - 2q_{10}) \\ &= 2\pi . \end{aligned} \tag{3.30}$$

Substituting (3.30) into (3.29) we find

$$A(q^2) = \frac{1}{6} \pi q^2, \qquad B(q^2) = \tfrac{1}{3} \pi , \tag{3.31a}$$

so that

$$I_{\alpha\beta} = \frac{\pi}{6} (q^2 g_{\alpha\beta} + 2q_\alpha q_\beta) . \tag{3.31b}$$

Thus using (3.31b) we may integrate over the neutrinos' momenta in (3.24). We find the required transition rate is

$$d\Gamma = (2\pi)^{-4} \frac{G^2}{6} (k_o p_o)^{-1} \{q^2 (k-m_\mu t).(p-m_e s) + 2q.(k-m_\mu t) q.(p-m_e s)\} d^3 p . \tag{3.32}$$

Let us evaluate this quantity in the laboratory frame of reference in which the muon is at rest. In this frame we have

$$k = (m_\mu, \underline{0}) , \quad t = (0 , \underline{t}) \tag{3.33a}$$

$$p = (E, \underline{p}) , \quad s = \left(\frac{\underline{p}.\underline{n}}{m} , \underline{n} + \frac{(\underline{p}.\underline{n})\underline{p}}{m(E+m)} \right) , \tag{3.33b}$$

so that

$$d\Gamma = (2\pi)^{-4} \frac{G^2}{3} m_\mu \{ 3W - 2E - m_e^2 m_\mu^{-1} + (\underline{p}.\underline{t}) E^{-1} (W-2E+m_e^2 m_\mu^{-1})$$

$$- (\underline{p}.\underline{n}) E^{-1} (3W - 2E - m_e^2 m_\mu^{-1}) - m_e (\underline{n}.\underline{t}) E^{-1} (W-E)$$

$$- (\underline{p}.\underline{n}) (\underline{p}.\underline{t}) E^{-1} (E+m_e)^{-1} (W-2E-m_e) \} d^3p \ , \qquad (3.34a)$$

where

$$W \equiv \tfrac{1}{2} m_\mu^{-1} (m_\mu^2 + m_e^2) \qquad (3.34b)$$

is the maximum possible energy of the electron in this frame. This follows from energy-momentum conservation, as follows:

$$(k - p)^2 = (q_1 + q_2)^2 \geq 0 \ , \qquad (3.35)$$

because q_1, q_2 are light-like vectors. Thus in the laboratory frame we have

$$(k-p)^2 = m_\mu^2 - 2m_\mu E + m_e^2 \geq 0 \qquad (3.36a)$$

or

$$E \leq W \ . \qquad (3.36b)$$

We note that the transition probability (3.34) has the form (2.265) indicating that parity is not conserved in this interaction,as we anticipated. In principle the full structure of (3.34) can be explored experimentally. However, in practice, only one of the two spin polarization vectors is observed. In the case when the decaying muon is polarized, the experiment studies the electron energy spectrum and angular distribution about the direction defined by the muon's spin $\underline{t}$. Since in this case the electron's spin is unobserved, the relevant transition probability is obtained by summing the two distributions (3.34a) for $\pm \underline{n}$. This gives

$$d\Gamma = (2\pi)^{-3} \frac{2G^2}{3} m_\mu |\underline{p}| E\, dE \sin\theta\, d\theta \ \{ 3W - 2E - m_e^2 E^{-1}$$

$$+ |\underline{p}| E^{-1} \cos\theta \ (W - 2E + m_e^2 m_\mu^{-1}) \} \ , \qquad (3.37)$$

where θ is the angle between the momentum $\underline{p}$ of the electron and the spin polarization vector $\underline{t}$ of the decaying muon. Some of the experimental data on muon decay is accurate to less than the half of one per cent. The electromagnetic corrections to the process being considered, which arise because the charged particles participating interact electromagnetically with each other, are of a comparable magnitude. Thus before comparing the predicted distribution with experiment, it is essential to include the contribution from these electromagnetic corrections. The spectacular success of quantum electrodynamics leads us to suppose that the electromagnetic interaction of the electron and muon are known. Thus, in principle, the required corrections to the process we are considering can be computed. These calculations have been done[(2)] to order α (the fine structure constant), and they lead to a finite modification of the spectrum (3.37), at least for electron energies E not too close to the maximum W. (The corrections diverge logarithmically as E approaches W, where soft bremsstrahlung photons accompany the decay.) Including these corrections, then, the predicted spectrum is

$$d\Gamma = (2\pi)^{-3} \frac{2G^2}{3} m_\mu |\underline{p}| E\, dE \sin\theta\, d\theta \, \{3W - 2E - m_e^2 E^{-1} + \frac{\alpha}{2\pi} f(E)$$

$$+ |\underline{p}| E^{-1} \cos\theta \, (W - 2E + m_e^2 m_\mu^{-1} + \frac{\alpha}{2\pi} g(E))\} , \qquad (3.38)$$

where f, g are known functions. We shall see shortly that this prediction is <u>consistent</u> with the experimental data to better than half of a percent. This leads one to ask the converse question: how much can one actually deduce about the muon decay matrix element from the available data ? To answer this one must first consider a more general form for the matrix element than the single parameter form (3.15b) which we have postulated. What is usually done is to start from a matrix element involving 19 real constant parameters which characterize the ten possible complex Lorentz invariant, non-derivative couplings of the four spinors. This leads to a more general form for the transition probability we have calculated.

$$d\Gamma = (2\pi)^{-3} \frac{A}{6} m_\mu |\underline{p}| E\, dE \sin\theta\, d\theta \, \{3(W-E) + 2\rho(\frac{4}{3} E - W - \tfrac{1}{3} m_e^2 E^{-1})$$

$$+ 3 m_e E^{-1} \eta (W-E) + \frac{\alpha}{4\pi} f(E) - |\underline{p}| E^{-1} \cos\theta \, \xi [(W-E)$$

$$+ 2\delta (\frac{4}{3} E - W - \tfrac{1}{3} m_e^2 m_\mu^{-1}) + \frac{\alpha}{4\pi} g(E)]\} , \qquad (3.39)$$

where A, ρ, η, ξ, δ are different combinations of the 19 constants in the original matrix element. We see that the form (3.38) is obtained when the values

$$A = 8G^2, \rho = \delta = \tfrac{3}{4}, \xi = 1, \eta = 0 \qquad (3.40)$$

are substituted into (3.39). The first of these predictions cannot be tested without prior knowledge of the single constant G appearing in (3.15b). The remaining four constants may be determined from the analysis of the observed spectrum shape. The actual values obtained from such an analysis are as follows

$$\rho = 0.7517 \pm 0.0026 \qquad (3.41a)$$

$$\delta = 0.7551 \pm 0.0085 \qquad (3.41b)$$

$$\xi = 0.972 \pm 0.013 \qquad (3.41c)$$

$$\eta = -0.12 \pm 0.21 \ . \qquad (3.41d)$$

We see that these values are consistent with those predicted by the Hamiltonian (3.1).

The second type of experiment consists of measuring the helicity of the electron observed when an unpolarized muon decays. Since in this case the muon's spin is unobserved, the required transition probability is obtained by summing the two distributions (3.34a) for $\pm \underline{t}$. This gives

$$d\Gamma = (2\pi)^{-4} \frac{2G^2}{3} m_\mu \{3W-2E-m_e^2 E^{-1} - (\underline{p}.\underline{n}) E^{-1}(3W-2E-m_e^2 m_\mu^{-1})\} d^3p \ . \qquad (3.42)$$

The helicity h of the final state electron is defined as

$$h \equiv \frac{d\Gamma(P) - d\Gamma(A)}{d\Gamma(P) + d\Gamma(A)} , \qquad (3.43)$$

where $d\Gamma(P)$, $d\Gamma(A)$ are the transition probabilities for the emission of an electron with spin polarization $\underline{n}$ respectively parallel, anti-parallel to its momentum $\underline{p}$. Using (3.42) we find that h is given by

$$h = - |\underline{p}| E^{-1} (3W-2E-m_e^2 m_\mu^{-1}) (3W-2E-m_e^2 E^{-1})^{-1} \qquad (3.44a)$$

$$= -1 + O(m_e^2 E^{-2}) \ . \qquad (3.44b)$$

Thus at energies E for which $E^2 \gg m_e^2$, the electron is predicted to be in a totally left-handed polarization state. Since the maximum energy W is some 50 electron masses,plainly this condition is satisfied for almost all of the available energy range. The average helicity of the electron emitted in unpolarized muon decay is found experimentally to be

$$h = -1.00 \pm 0.13 \ . \tag{3.45}$$

This too is consistent with value (3.44) predicted from the Hamiltonian (3.1).

We may easily understand why (3.1) leads to the prediction that the emitted electron is essentially 100% left-handed polarized. The form (2.264) for L_α, $L_\alpha^\dagger$ which appear in (3.1) ensures that the electron field operator $\psi_{(e)}$ and its adjoint $\bar{\psi}_{(e)}$ occur only in the respective combinations $(1-\gamma_5)\psi_{(e)}$ and $\bar{\psi}_{(e)}(1+\gamma_5)$. We have already seen that the combinations $(1-\gamma_5)\psi_{(\nu)}$ and $\bar{\psi}_{(\nu)}(1-\gamma_5)$ ensure that only left-handed neutrinos and right-handed anti-neutrinos are observed in nature. Plainly, these considerations will also apply to the electron field provided we may neglect its mass m_e, as we can for most energies in muon decay. Thus the forms for L_α, $L_\alpha^\dagger$ ensure that at these energies only left-handed electrons and right-handed positrons will be observed in muon decay. We could, of course, calculate the transition probability for the decay of μ^+ just as we have done for the μ^--decay. However, this is unnecessary. The Hamiltonian (3.1) is CP-invariant and, as we shall see, this enables us to write down the decay probability for μ^+ directly from that for μ^-.

Consider first the behaviour of $L_\alpha(x)$ under the operation $\mathcal{C}\mathcal{P}$. Using (2.208), (2.210) and (2.261) we find

$$\begin{aligned}
&(\mathcal{C}\mathcal{P})\,\bar{\psi}_{(e)}(x)\,\gamma_\alpha\, a\,\psi_{(\nu_e)}(x)\,(\mathcal{C}\mathcal{P})^{-1} \\
&= -\epsilon_e\,\omega_e\,\epsilon^*_{\nu_e}\,\omega^*_{\nu_e}\,\psi^T_{(e)}(x^o,-\underline{x})\,C^{-1}\gamma^o\gamma_\alpha\gamma^o\,C\,[\bar{\psi}_{(\nu_e)}(x^o,-\underline{x})\,a']^T \\
&= \epsilon_e\,\omega_e\,\epsilon^*_{\nu_e}\,\omega^*_{\nu_e}\,\bar{\psi}_{(\nu_e)}(x^o,-\underline{x})\,[\gamma_\alpha - 2g_{o\alpha}\gamma_o]\,a\,\psi_{(e)}(x^o,-\underline{x})\ .
\end{aligned} \tag{3.46}$$

Thus, taking $\epsilon_e = \epsilon_{\nu_e}$ and $\omega_e = \omega_{\nu_e}$,

$$(\mathcal{C}\mathcal{P})\,L_\alpha(x^o,\underline{x})\,(\mathcal{C}\mathcal{P})^{-1} = -L^{\dagger\alpha}(x^o,-\underline{x}) \tag{3.47a}$$

and

$$(\mathcal{C}\mathcal{P})\,L^\dagger_\alpha(x^o,\underline{x})\,(\mathcal{C}\mathcal{P})^{-1} = -L^\alpha(x^o,-\underline{x})\ . \tag{3.47b}$$

Thus

$$(\mathcal{C}\,\mathcal{P})\,\mathcal{H}_W^L(x^o, \underline{x})\,(\mathcal{C}\,\mathcal{P})^{-1} = \mathcal{H}_W^L(x^o, -\underline{x}) \tag{3.48}$$

and our theory is therefore CP-invariant. Hence

$$< e^- \bar{\nu}_e \nu_\mu |(\mathcal{C}\,\mathcal{P})^{-1}\, S\,(\mathcal{C}\,\mathcal{P})|\,\mu^- > \; = \; < e^- \bar{\nu}_e \nu_\mu | S |\,\mu^- > \; . \tag{3.49}$$

Now,

$$\mathcal{C}\,\mathcal{P}|\mu^-(\underline{k}, \underline{t})> \; = \; |\mu^+(-\underline{k}, \underline{t})> \tag{3.50a}$$

$$\mathcal{C}\,\mathcal{P}|e^-(\underline{p}, \underline{n})> \; = \; |e^+(-\underline{p}, \underline{n})> \tag{3.50b}$$

$$\mathcal{C}\,\mathcal{P}|\bar{\nu}_e(q_1)> \; = \; |\nu_e(-\underline{q}_1)> \tag{3.50c}$$

$$\mathcal{C}\,\mathcal{P}|\nu_\mu(\underline{q}_2)> \; = \; |\bar{\nu}_\mu(-\underline{q}_2)> \; . \tag{3.50d}$$

Thus the matrix element, and hence the transition probability, for μ^+-decay is obtained from that for μ^--decay by changing the signs of the three-momenta, while leaving the corresponding spin vectors $\underline{t}$, $\underline{n}$ unaltered. The probability of observing an e^+ with momentum $\underline{p}$ and spin $\underline{n}$ after the decay of a μ^+ at rest with spin vector $\underline{t}$ is therefore

$$\begin{aligned} d\Gamma^{(+)} = (2\pi)^{-4}\frac{G^2}{3}\, m_\mu \{ & 3W-2E-m_e^2 E^{-1} - (\underline{p}.\underline{t})\, E^{-1}(W-2E+m_e^2 m_\mu^{-1}) \\ & + (\underline{p}.\underline{n})\, E^{-1}(3W-2E-m_e^2 m_\mu^{-1}) - m_e(\underline{n}.\underline{t})\, E^{-1}(W-E) \\ & - (\underline{p}.\underline{n})\,(\underline{p}.\underline{t})\, E^{-1}(E+m_e)^{-1}(W-2E-m_e)\}\, d^3p \; . \end{aligned} \tag{3.51}$$

Neglecting the mass m_e this takes the simple form

$$d\Gamma^{(+)} = (2\pi)^{-4}\frac{G^2}{3}\, m_\mu [1 + (\underline{p}.\underline{n})|\underline{p}|^{-1}]\,[3W-2E-(\underline{p}.\underline{t})|\underline{p}|^{-1}(W-2E)]\, d^3p \; . \tag{3.52}$$

In the same approximation the rate for μ^--decay is

$$d\Gamma^{(-)} = (2\pi)^{-4}\frac{G^2}{3}\, m_\mu [1 - (\underline{p}.\underline{n})|\underline{p}|^{-1}]\,[3W-2E+(\underline{p}.\underline{t})|\underline{p}|^{-1}(W-2E)]\, d^3p \; . \tag{3.53}$$

We see that (3. 52) implies that the helicity of the e^+ in μ^+-decay will have the value $h = +1$; that is to say the e^+ will be essentially 100% right-handed polarized, as anticipated. We may also obtain some physical understanding of the electron asymmetry predicted in (3. 53). At high electron energies ($E \sim W$) the $\bar{\nu}_e$, ν_μ pair will be emitted in the same direction, while the e^- is emitted in the opposite direction. Since $\bar{\nu}_e$, ν_μ have opposite helicities, their total angular momentum will be zero. Thus the e^- will have to carry the spin of the decaying μ^-. Since the e^- is essentially 100% left-handedly polarized it follows that the e^- will be emitted in a direction anti-parallel to the spin $\underline{t}$ of the μ^-. This is confirmed by the form (3.53), since when $E \sim W$

$$d\Gamma^{(-)} \propto 1 - (\underline{p}.\underline{t})|\underline{p}|^{-1} . \tag{3.54}$$

Thus $d\Gamma^{(-)}$ will be largest when $\underline{p}$ and $\underline{t}$ are anti-parallel as we predicted. On the other hand, at low energies $E \sim 0$ (but still high enough to neglect m_e) $\bar{\nu}_e$ and ν_μ will be emitted in <u>opposite</u> directions. Then similar angular momentum considerations show that the e^- will have to be emitted parallel to the $\bar{\nu}_e$. This direction is parallel to the spin of the ν_μ and also to that of the spin $\underline{t}$ of the μ^-. Again, this is confirmed by (3.53), since with $E \sim 0$

$$d\Gamma^{(-)} \propto 3 + (\underline{p}.\underline{t})|\underline{p}|^{-1} . \tag{3.55}$$

This is largest when $\underline{p}$ and $\underline{t}$ are parallel, as predicted.

Let us turn briefly to the converse question, already posed: how much about the muon decay matrix element can be deduced from the present experimental data ? To answer this question we must first parametrize the "general" matrix element and then see how the parameters are limited by the data. If we start from the 19 parameter form (which even so is not the most general) it turns out[3] that the data allows very substantial deviations from the form (3.15), which follows from our hypothesis (3.1). We may see how this comes about. Firstly, let us rewrite the matrix element (3.15b) in the so-called charge retention (CR) ordering.

$$M(\mu^- \to e^- \bar{\nu}_e \nu_\mu) = \frac{-G}{\sqrt{2}} [\bar{u}_{(e)}(p,s)\gamma^\alpha(1-\gamma_5)u_{(\mu)}(k,t)]$$

$$\times [\bar{u}_{(\nu_\mu)}(q_2, -\tfrac{1}{2})\, \gamma_\alpha (1-\gamma_5)\, v_{(\nu_e)}(q_1, \tfrac{1}{2})] . \tag{3.56}$$

This follows from the Fierz Identity for the γ-matrices

$$[\gamma^\alpha(1-\gamma_5)]_{ab}[\gamma_\alpha(1-\gamma_5)]_{cd} = -[\gamma^\alpha(1-\gamma_5)]_{ad}[\gamma_\alpha(1-\gamma_5)]_{cb} . \tag{3.57}$$

The proof of this and the other Fierz Identities[4] are given in the

Appendix. The form (3.56) is obtained from (3.15b) by interchanging the spinors for μ^- and $\bar{\nu}_e$, and we verify that $|M|^2$ given in (3.21) is indeed invariant if we interchange the vectors $(k - m_\mu t)$ and q_1. Instead of (3.56) let us consider the more general form

$$M = \frac{G}{\sqrt{2}} [\bar{u}_{(e)}(p,s) \gamma^\alpha (1-\epsilon \gamma_5) u_{(\mu)}(k,t)] [\bar{u}_{(\nu_\mu)}(q_2, -\tfrac{1}{2}) \gamma_\alpha (1-\gamma_5) v_{(\nu_e)}(q_1, \tfrac{1}{2})] , \tag{3.58}$$

where ϵ is an arbitrary constant. Note that we have retained the projection operator $(1-\gamma_5)$ in the second spinor product; this ensures that only left-handed ν_μ and right-handed $\bar{\nu}_e$ will participate. Starting from (3.58) we may calculate the transition probability for polarized muon decay as before. This gives an expression of the form (3.39) with

$$\rho = \delta = \tfrac{3}{4} \tag{3.59a}$$

$$\xi = (2 \operatorname{Re} \epsilon)(1 + |\epsilon|^2)^{-1} \tag{3.59b}$$

$$\eta = \tfrac{1}{2}(|\epsilon|^2 - 1)(|\epsilon|^2 + 1)^{-1} . \tag{3.59c}$$

Also, as before, we may calculate the helicity of the electron in the decay of an unpolarized muon. We find

$$h = -(2 \operatorname{Re} \epsilon)(1 + |\epsilon|^2)^{-1} . \tag{3.60}$$

Of course all of our previous results (3.40), (3.44) follow when we substitute $\epsilon = 1$ in (3.59), (3.60). Note that the values $\rho = \delta = \frac{3}{4}$ are obtained in our more general $(V - \epsilon A)$ theory as well as the hypothesised (V-A) theory. Thus although ρ and δ are the most accurately measured parameters of the spectrum the parameter ϵ is totally unrestricted. To restrict ϵ at all we must consider the remaining three measurements. In fact the large error in the experimental value of η in (3.41d) allows ϵ to take just about any value. If we make the further assumption that ϵ is <u>real</u>[5], then the data restrict ϵ to a value differing from unity by at most 20%. Thus while our hypothesis leads to predictions which are consistent with experiment, we see that the experimental data are some way from 'proving' the hypothesis; even a two or three parameter form of the matrix allows quite substantial uncertainties in the parameters.

The single unknown parameter G in our hypothesis (3.1) fixes the overall scale of the decay probability (3.38). Integrating this expression we obtain the total decay rate

$$\Gamma(\mu^- \to e^- \bar{\nu}_e \nu_\mu) = \frac{G^2 m_\mu^5}{192 \pi^3} [1-8y+8y^3-y^4-12y^2 \ln y + O(\alpha)] , \tag{3.61a}$$

where

$$y \equiv m_e^2/m_\mu^2 , \tag{3.61b}$$

and $O(\alpha)$ is a known number arising from the electromagnetic corrections. Notice that the decay rate Γ vanishes when $y = 1$; this reflects the fact that the muon cannot decay in the (academic) case that $m_\mu = m_e$, since there is no phase space available. The total decay rate Γ is known to an accuracy of 3×10^{-4} and, in order to determine G from (3.61), it is necessary to include at least the dominant contributions of the order α^2 electromagnetic corrections[6]. When this is done, the observed decay rate and the latest values of the fundamental constants α, m_μ, m_e yield the value

$$G = (1.4354 \pm 0.0003) \times 10^{-49} \text{ erg cm}^3 . \tag{3.62}$$

If we scale the value of G with the proton mass m_p we obtain a dimensionless measure of the coupling constant

$$G m_p^2 = 1.026 \times 10^{-5} . \tag{3.63}$$

As was noted in the Introduction, the smallness of this measure leads to the belief that it is an excellent approximation to calculate in first order perturbation theory.

3.2 ELECTRON-NEUTRINO SCATTERING

The leptonic Hamiltonian (3.1) also predicts the existence of the "diagonal" processes (3.2) as well as the corresponding muonic diagonal processes. It is therefore of considerable interest to see whether experiment confirms the existence of these processes and whether the data are consistent with the precise predictions which follow from (3.1). With the value (3.62) for the coupling constant G the cross-sections for the scattering processes (3.2) are fully determined in terms of the (known) lepton masses.

3.2.1 ELASTIC $\nu_e e$, $\bar{\nu}_e e$ SCATTERING

Consider firstly the process (3.2a)

$$e^-(p_1, s_1) + \nu_e(q_1) \to e^-(p_2, s_2) + \nu_e(q_2) \quad , \tag{3.64}$$

where p_i, q_i are the four momenta of the particles, s_i are the spin vectors of the electrons. Then, proceeding as before, the invariant matrix element for the process is

$$M(e^-\nu_e \to e^-\nu_e) = \frac{G}{\sqrt{2}} [\bar{u}_{(e)}(p_2, s_2)\gamma^\alpha(1-\gamma_5)u_{(\nu_e)}(q_1)]$$

$$\times [\bar{u}_{(\nu_e)}(q_2)\gamma_\alpha(1-\gamma_5)u_{(e)}(p_1, s_1)] \quad . \tag{3.65}$$

In practice neither of the spin vectors s_i is observed. Thus to calculate the relevant transition probability we sum $|M|^2$ over the values $\pm s_2$ and average over the values $\pm s_1$. It follows, then, from (3.21) that

$$\tfrac{1}{2} \sum_{s_1, s_2} |M(e^-\nu_e \to e^-\nu_e)|^2 = 64\, G^2 (q_1 \cdot p_1)(q_2 \cdot p_2) \quad . \tag{3.66}$$

Thus the transition rate for scattering from a single electron into a final state with momenta in the ranges d^3p_2 around $\underline{p}_2$ and d^3q_2 around $\underline{q}_2$ is

$$d\Gamma = (2\pi)^{-2} \frac{1}{8\, p_{10} p_{20} q_{20}} \delta(p_1 + q_1 - p_2 - q_2) \tfrac{1}{2} \sum_{s_i} |M|^2 d^3p_2\, d^3q_2 \,. \tag{3.67}$$

The cross section is obtained by dividing $d\Gamma$ by the incident neutrino flux. Our normalization is such that there are $2q_{10}$ neutrinos in unit volume, and in the laboratory frame their velocity relative to the target electron is clearly 1. Thus the incident neutrino flux is $2q_{10}$, and the laboratory cross section $d\sigma_L$ is given by

$$d\sigma_L = (2\pi)^{-2} (p_{10} p_{20} q_{10} q_{20})^{-1}\, 4G^2 (q_1 \cdot p_1)(q_2 \cdot p_2)\, \delta(p_1 + q_1 - p_2 - q_2) d^3p_2\, d^3q_2 \,. \tag{3.68}$$

Now in the laboratory frame

$$p_1 = (m_e, \underline{0}), \; q_1 = E_\nu(1, \underline{n}), \; p_2 \equiv (E, \underline{p}) \quad . \tag{3.69}$$

Also, energy-momentum conservation gives

$$(p_1 + q_1)^2 = (p_2 + q_2)^2 \quad ,$$

so that

$$p_1 \cdot q_1 = p_2 \cdot q_2 = m_e E_\nu \ , \tag{3.70}$$

since $p_1^2 = p_2^2 = m_e^2$. The final state neutrino is not observed, so, as before, we integrate over its momentum $\underline{q}_2$. Using (3.69), (3.70) we find then

$$d\sigma_L = \frac{G^2}{\pi^2} \frac{m_e E_\nu}{E|\underline{p} - E_\nu \underline{n}|} \delta(m_e + E_\nu - E - |\underline{p} - E_\nu \underline{n}|)\, d^3p \ . \tag{3.71}$$

We write

$$|\underline{p} - E_\nu \underline{n}| = [p^2 + E_\nu^2 - 2p E_\nu x]^{\frac{1}{2}} \ , \tag{3.72}$$

where x is the cosine of the angle between $\underline{p}$ and the direction $\underline{n}$ of the incident neutrino. Thus

$$d^3p = 2\pi p E dE dx \ . \tag{3.73}$$

The argument of the δ-function is a function $f(x)$

$$f(x) \equiv m_e + E_\nu - E - [p^2 + E_\nu^2 - 2p E_\nu x]^{\frac{1}{2}} \ . \tag{3.74}$$

To carry out the integration over x we change variables to

$$y = f(x) \ . \tag{3.75a}$$

Then

$$\delta[f(x)]\, dx = \delta(y) \frac{dx}{dy} dy = \delta(y)[f'(x)]^{-1} dy \ . \tag{3.75b}$$

Using (3.74) we have

$$f'(x) = p E_\nu [p^2 + E_\nu^2 - 2p E_\nu x]^{-\frac{1}{2}} = p E_\nu (m_e + E_\nu - E)^{-1} \ .$$

Thus performing the x-integration in (3.71) gives

$$d\sigma_L = \frac{2G^2}{\pi} \frac{m_e E_\nu}{E(m_e + E_\nu - E)} [p E_\nu (m_e + E_\nu - E)^{-1}]^{-1} p E dE = \frac{2G^2}{\pi} m_e dE \ , \tag{3.76}$$

<u>provided</u> $y = f(x) = 0$ vanishes for some value of x in the range $-1 \leq x \leq 1$. This places restrictions upon the kinematically accessible values of E. For example, when $x = -1$, $f(x) = 0$ gives

$$m_e + E_\nu - E = E_\nu + p \ ,$$

so that

$$E = m_e \ .$$

For $x = +1$ we require

$$m_e + E_\nu - E = |p - E_\nu| \ ,$$

so that either

$$E = m_e$$

or

$$E = \tfrac{1}{2}\left[m_e + 2E_\nu + \frac{m_e^2}{m_e + 2E_\nu} \right] .$$

Thus, defining $\omega = E_\nu m_e^{-1}$,

$$\frac{d\sigma_L}{dE}(\nu_e e \to \nu_e e) = \frac{2G^2}{\pi} m_e \qquad \left(1 \le \frac{E}{m_e} \le \frac{1 + 2\omega + 2\omega^2}{1 + 2\omega} \right) . \tag{3.77}$$

The total cross section for the process is obtained by integrating over the allowed range of E.

$$\sigma_L(\nu_e e \to \nu_e e) = \sigma_o \frac{2\omega^2}{2\omega + 1} \ , \tag{3.78a}$$

where

$$\sigma_o \equiv \frac{2G^2}{\pi} m_e^2 = 8.8 \times 10^{-45} \text{ cm}^2 \ . \tag{3.78b}$$

In the same way we find

$$\frac{d\sigma_L}{dE}(\bar{\nu}_e e \to \bar{\nu}_e e) = \frac{2G^2}{\pi} m_e (E_\nu + m - E)^2 E_\nu^{-2} \ , \tag{3.79}$$

where E_ν is now the energy of the incident anti-neutrino. Kinematically, the two processes are identical, so the energy E of the scattered electron is restricted by the same constraint as obtained previously. For this process the total cross section is

$$\sigma_L(\bar{\nu}_e e \to \bar{\nu}_e e) = \tfrac{1}{3} \sigma_o \, \omega [1 - (1 + 2\omega)^{-3}] \ . \tag{3.80}$$

At high neutrino energies $(\omega \gg 1)$ we see that both cross sections increase linearly with E_ν

$$\sigma_L(\nu_e e \to \nu_e e) \simeq 3\sigma_L(\bar{\nu}_e e \to \bar{\nu}_e e) \simeq \sigma_o \omega \ . \tag{3.81}$$

It is easy to see how this comes about, without going through the precise details of the calculation as we have done. Plainly the cross sections must be proportional to G^2, which has the dimensions $[M^{-4}]$, as is apparent from (3.63). The total cross section σ_L has the dimensions $[L^2] = [M^{-2}]$ in natural units. Thus on dimensional grounds

$$\sigma_L = G^2 s \ , \tag{3.82}$$

where s has dimensions $[M^2]$. The quantity s must be a Lorentz invariant, since σ_L is, and at high energies we will expect σ_L to be independent of the masses of the participating particles. Thus the only possibility is that s is a constant multiple of the (Lorentz invariant) square of the centre-of-mass energy

$$s = nW^2 = n(p_1 + q_1)^2 \tag{3.83}$$

(n is a constant). In the laboratory frame

$$W^2 = m_e^2 + 2m_e E_\nu \simeq 2m_e E_\nu \ , \quad \text{if } \omega \gg 1 \ . \tag{3.84}$$

Thus we have

$$\sigma_L = 2n\, G^2 m_e^2\, \omega \ , \tag{3.85}$$

as found already in (3.81).

For incident neutrino energies of the order of 5 MeV ($\omega \sim 10$) we see that the predicted cross section is about $10^{-45}\ cm^2$. Thus the scattering cross section for neutrinos emitted from atomic piles is hardly measurable. On the other hand, for 1 GeV neutrinos the cross section is about $10^{-41}\ cm^2$. Neutrinos of such energy may be produced in an accelerator, but the flux is so small that very large amounts of machine time are needed. For these reasons the detection of these scattering processes is predicted to be extremely difficult. As we shall see in the next chapter, neutrinos produced in accelerators are almost entirely muonic, whilst those produced in piles are electronic. Thus it is important that experiments of both varieties are pursued.

The required accuracy is beginning to be attained in the pile experiments[7]. Anti-neutrinos ($\bar{\nu}_e$) from the Savannah River reactor scatter the electrons in a plastic scintillator. The experiment looks for fairly high energy recoil electrons with $3.6 < E < 5$ MeV, and measures the difference in the counting rates with the reactor on or off. The observed difference is consistent with the background effects, the chief of which is $\bar{\nu}_e p \to e^+ n$. The results thus provide an upper limit on the cross section; the partial cross section for producing recoil electrons in the above energy range is found to be less than 6×10^{-47} cm^2. In terms of the predicted cross sections this gives at the 90% confidence limit

$$\sigma_L(\bar{\nu}_e e \to \bar{\nu}_e e)\big|_{exp} \leq 3\sigma_L(\bar{\nu}_e e \to \bar{\nu}_e e)\big|_{theory} \quad . \tag{3.86}$$

Thus although the theory we are using is consistent with the present data it has not been severely tested upon these diagonal processes. In fact, there is no direct evidence that these processes even exist.

There is a certain amount of indirect evidence for their existence which has been inferred from astrophysical data. Neutrino emission is important astrophysically, because of the enormous mean free path of the neutrino. This means that in some circumstances neutrino emission carries away far more energy than the more likely photon emission, since the latter are more easily reabsorbed. There are a number of mechanisms for the emission of $\nu_e \bar{\nu}_e$ pairs, all of which utilize the coupling of the electronic part of the leptonic current to itself, sometimes in conjunction with the electromagnetic interaction. The effect of these processes is to make heavy stars grow hot more rapidly than they would otherwise have done, and to make light stars cool more rapidly[(8)].

Heavy stars, say those of ten solar masses, after hydrogen burning, enter a helium burning stage (blue supergiant) and subsequently a carbon burning stage (red supergiant), getting hotter at each stage. The self-coupling mechanism has the effect of making the red supergiant phase shorter; thus at any particular time the ratio of the blue supergiants to red supergiants will be larger than it would otherwise have been. It is claimed[(9)] that the observed statistics require the existence of the self-coupling interaction even when the maximum known observational and theoretical uncertainties are included. Thus this test provides a lower bound on the strength of the self-coupling. In all, nine astrophysical tests have been studied, and these provide both upper and lower limits on the strength $G_{e\nu}$ of the electronic current self-coupling constant. (The interaction is assumed to be of the form (3.1) with the coupling constant $G_{e\nu}$ as a parameter.) On the basis of these astrophysical tests Stothers[(9)] concluded that

$$|G_{e\nu}/G| = 10^{0\pm1} \ . \tag{3.87}$$

Thus the astrophysical data are consistent with our hypothesis that

$$G_{e\nu} = G \ , \tag{3.88}$$

and in addition provide the only positive evidence that the self-coupling interaction actually exists. However, the result (3.87) should be viewed with some caution[(8)]. Only the surface of a star is observable, and to relate the surface state to the state of the core demands the use of hydrodynamic approximations. The validity of these approximations is not well understood so that estimating the maximum theoretical uncertainty in a given calculation is very difficult.

3.2.2 ELASTIC $\nu_\mu e$, $\bar{\nu}_\mu e$ SCATTERING

In addition to the predicted processes (3.2), (3.3) and (3.4), our hypothesis also makes the negative predictions

$$\nu_\mu e^\pm \not\rightarrow \nu_\mu e^\pm \tag{3.89a}$$

$$\bar{\nu}_\mu e^\pm \not\rightarrow \bar{\nu}_\mu e^\pm \ . \tag{3.89b}$$

There are similar negative predictions for the scattering of ν_e, $\bar{\nu}_e$ on $\mu^\pm$.

These negative predictions from (3.1) follow because the muonic part of L_α, for example, destroys ν_μ but creates μ^- rather than e^-. Both processes conserve electron-number and muon-number, but if we try to write a Hamiltonian which allows them it must have the structure $(\bar{e}\, e)(\bar{\nu}_\mu \nu_\mu)$ when expressed in terms of "currents" which conserve N_e, N_μ. Each of these is a neutral leptonic current, so the search for the processes (3.89) provides a rigorous test of our hypothesis that the leptons enter the weak Hamiltonian only in a charged current combination. Moreover, the test is free of any dynamical complications from the hadrons.

Both of these processes have been looked for using accelerator produced ν_μ and $\bar{\nu}_\mu$'s in the heavy liquid Gargamelle bubble chamber[10] at CERN. Altogether 375,000 ν and 360,000 $\bar{\nu}$ pictures have been analysed, and just two examples of the process $\bar{\nu}_\mu e^- \to \bar{\nu}_\mu e^-$ have been identified. The background is minute, so that experiment provides further firm evidence that the leptons participate in the weak interactions in a neutral, as well as a charged, current combination. At the 90% confidence level (C. L.) the observed cross sections satisfy

$$\sigma(\nu_\mu e^- \to \nu_\mu e^-) < (0.15)\sigma_o \omega \tag{3.90a}$$

$$(0.05)\tfrac{1}{3}\sigma_o \omega < \sigma(\bar{\nu}_\mu e^- \to \bar{\nu}_\mu e^-) < (0.51)\tfrac{1}{3}\sigma_o \omega \;. \tag{3.90b}$$

The existence of the process $\bar{\nu}_\mu e^- \to \bar{\nu}_\mu e^-$ indicates that the neutral current has terms with the structure $(\bar{e}\, e)$ and $(\bar{\nu}_\mu \nu_\mu)$. If, as seems likely, it also has a term with the structure $(\bar{\nu}_e \nu_e)$, then this and the $(\bar{e}\, e)$ term may couple to give a contribution to the $\nu_e e^\pm$, $\bar{\nu}_e e^\pm$ scattering processes over and above the contribution from the charged currents discussed in (3.2.1). It is therefore most desirable that all of these extremely difficult neutrino scattering experiments should be pursued with continued vigour so that the precise form of the neutral leptonic current can be tied down. Obviously it will be some time before this can be done.

Finally, we note that it is unlikely that the neutral current contributes to muon decay, discussed in (3.1); to do so the current would have to include a $(\bar{e}\,\mu)$ term which does not conserve N_e or N_μ. The experimental data given in (2.5.1) effectively rule out this possibility which again explains why neutral currents were not discovered earlier.

3.3 INTERMEDIATE VECTOR BOSON HYPOTHESIS

The similarity between the leptonic current $L_\alpha(x)$ and the electromagnetic current $j_\mu(x)$ given in (2.222b) leads one to wonder whether the leptonic current might be coupled to a vector[11] field $W_\alpha(x)$ in a way analogous to the coupling of $j_\mu(x)$ the electromagnetic field $A_\mu(x)$ in

(2.220). More precisely, this hypothesis states that the fundamental coupling of the lepton fields is given by the following (semi-weak) Hamiltonian:

$$\mathcal{H}^L_{SW}(x) = f\, L^{\dagger}_{\alpha}(x)\, W^{\alpha}(x) + f\, L^{\alpha}(x)\, W_{\alpha}(x)^{\dagger} \,. \tag{3.91}$$

The field $W_{\alpha}(x)$ is a vector field and therefore, like the photon, it describes a spin-one particle. Further, since $L_{\alpha}(x)$ destroys one unit of charge, the field $W^{\dagger}_{\alpha}(x)$ creates one unit, so that charge is conserved. Thus $W_{\alpha}(x)$ is the field operator of a particle W^+ with charge plus one unit. If there is such a particle, the coupling (3.91) indicates that it decays via the mode

$$W^+ \rightarrow e^+ \bar{\nu}_e \,.$$

No spin-one particle with such a decay mode has ever been observed. Thus, if it exists, we conclude that its mass must be too high for it to have been created in the machines at present available.

If (3.91) is the basic Hamiltonian, we may expand the S-matrix in a power series in f.

$$S = 1 - i\int dx\, \mathcal{H}^L_{SW}(x) + \frac{(i)^2}{2!}\int dx\, dy\, T\{\mathcal{H}^L_{SW}(x)\, \mathcal{H}^L_{SW}(y)\} + 0(f^3)\,, \tag{3.92}$$

where $T\{\ \}$ is the time ordered product of the two operators. Plainly, the term proportional to f in (3.92) will give rise to processes involving the W boson in the initial or final state. The first term which can contribute to the purely leptonic processes we have been considering is that proportional to f^2. The W-particle created by one operator may be destroyed by the other thereby giving rise to processes involving only the leptons.

3.3.1 MUON DECAY VIA W

Consider, for example, muon decay. Its matrix element is given by

$$< e^- \bar{\nu}_e \nu_\mu | S | \mu^- > = -\tfrac{1}{2} f^2 \int dx\, dy < e^- \bar{\nu}_e \nu_\mu | T\{ L^\dagger_\alpha(x)\, W^\alpha(x)\, L_\beta(y)\, W^{\dagger\beta}(y)$$

$$+ L_\alpha(x)\, W^{\dagger\alpha}(x)\, L^\dagger_\beta(y)\, W^\beta(y) \} | \mu^- >$$

$$= - f^2 \int dx\, dy < e^- \bar{\nu}_e \nu_\mu | T\{ L^\dagger_\alpha(x)\, W^\alpha(x)\, L_\beta(y)\, W^{\dagger\beta}(y) \} | \mu^- >. \tag{3.93}$$

Proceeding as before, we find

$$< e^- \bar{\nu}_e \nu_\mu | S | \mu^- > = - f^2 \int dx\, dy\, e^{i(q_2 - k)x}\, e^{i(q_1 + p)y}$$

$$\times\ [\bar{u}_{(\nu_\mu)}(q_2)\gamma_\alpha(1-\gamma_5) u_{(\mu)}(k)][\bar{u}_{(e)}(p)\gamma_\beta(1-\gamma_5)\, v_{(\nu_e)}(q_1)]$$

$$\times < 0 | T\{ W^\alpha(x)\, W^{\dagger\beta}(y) \} | 0 > \quad . \tag{3.94}$$

We may evaluate the vacuum expectation value in (3.94) by expanding the field $W_\alpha(x)$ in terms of creation and annihilation operators analogous to that given in (2.179) for the electron field operator. Thus

$$W_\alpha(x) = (2\pi)^{-3} \int d^3K\, (2K_o)^{-1} \sum_{\lambda=1}^{3} \epsilon_\alpha(K,\lambda) [\, a(K,\lambda)\, e^{-iKx} + b^\dagger(K,\lambda) e^{iKx}] \tag{3.95a}$$

$$W^\dagger_\beta(y) = (2\pi)^{-3} \int d^3L\, (2L_o)^{-1} \sum_{\mu=1}^{3} \epsilon_\beta(L,\mu) [\, b(L,\mu)\, e^{-iLy} + a^\dagger(L,\mu) e^{iLy}], \tag{3.95b}$$

where

$$K_o = + (\underline{K}^2 + M_W^2)^{\frac{1}{2}} \tag{3.95c}$$

and

$$K.\,\epsilon(K,\lambda) = 0, \qquad \epsilon(K,\lambda).\,\epsilon(K,\mu) = -\delta_{\lambda\mu} \qquad \lambda,\mu = 1,2,3 \quad . \tag{3.95d}$$

We see that the field W_α satisfies

$$(\partial^\mu \partial_\mu + M_W^2)\, W_\alpha(x) = 0 \quad .$$

Thus, W_α, $W^\dagger_\alpha$ describe particles of mass M_W. The three vectors $\epsilon(K,\lambda)$ satisfying (3.95d) are the vectors specifying the spin polarization of the $W^\pm$ created or destroyed by the operators $a^\dagger$, $b^\dagger$ or a,b. The fact that there are three such vectors corresponds to the fact that a particle with total spin one may be observed in any one of three eigenstates of the third component of angular momentum. As before, we assume canonical commutation relations for the creation and annihilation operators. Specifically, we take

$$[a(K,\lambda), a^{\dagger}(L,\mu)] = [b(L,\mu), b^{\dagger}(K,\lambda)] = (2\pi)^3\, 2K_o\, \delta(\underline{K}-\underline{L})\, \delta_{\lambda\mu} , \tag{3.96}$$

and all other commutators are zero. Note that in this case we take commutation relations rather than anti-commutation relations, since spin-one particles have Bose statistics. With the aid of the above decomposition, and using (3.96), we find

$$\begin{aligned} \langle 0 | W_\alpha(x)\, W_\beta^{\dagger}(y) | 0 \rangle &= (2\pi)^{-6} \int d^3K\, d^3L\, (4K_o L_o)^{-1}\, \epsilon_\alpha(K,\lambda)\, \epsilon_\beta(L,\mu) \\ &\qquad \times e^{-iKx}\, e^{iLy} \langle 0 | a(K,\lambda)\, a^{\dagger}(L,\mu) | 0 \rangle \\ &= (2\pi)^{-3} \int d^3K\, (2K_o)^{-1} \sum_{\lambda=1}^{3} \epsilon_\alpha(K,\lambda)\, \epsilon_\beta(K,\lambda)\, e^{iK(x-y)} . \end{aligned} \tag{3.97}$$

The quantity $\sum_\lambda \epsilon_\alpha(K,\lambda)\, \epsilon_\beta(K,\lambda)$ is an invariant, in the sense that it has the same value for all choices of $\epsilon(K,\lambda)$ for a given K. This is easily verified by considering a different set of vectors $\epsilon'(K,\lambda)$ satisfying the conditions (3.95d). Thus we may write

$$\epsilon'(K,\rho) = \sum_{\lambda=1}^{3} a_{\rho\lambda}\, \epsilon(K,\lambda) , \tag{3.98a}$$

where the matrix $A = (a_{\rho\lambda})$ satisfies

$$AA^T = A^T A = I_3 . \tag{3.98b}$$

It therefore follows that

$$\sum_\rho \epsilon'_\alpha(K,\rho)\, \epsilon'_\beta(K,\rho) = \sum_{\rho,\lambda,\mu} a_{\rho\lambda}\, a_{\rho\mu}\, \epsilon_\alpha(K,\lambda)\, \epsilon_\beta(K,\mu) = \sum_\lambda \epsilon_\alpha(K,\lambda)\, \epsilon_\beta(K,\lambda). \tag{3.99}$$

Thus we may write

$$\sum_{\lambda=1}^{3} \epsilon_\alpha(K,\lambda)\, \epsilon_\beta(K,\lambda) = B\, K_\alpha K_\beta + C\, g_{\alpha\beta} , \tag{3.100}$$

where B, C are constants. Using (3.95d) we deduce

$$B K^2 + C = 0 \tag{3.101a}$$

$$B K^2 + 4C = - 3 \ , \tag{3.101b}$$

so that

$$\sum_{\lambda=1}^{3} \epsilon_\alpha (K,\lambda) \ \epsilon_\beta (K,\lambda) = - g_{\alpha\beta} + M_w^{-2} K_\alpha K_\beta \ . \tag{3.101c}$$

Thus for $x_o > y_o$

$$< 0 | T \{ W_\alpha(x) W_\beta^\dagger(y) \} | 0 > \equiv < 0 | W_\alpha(x) W_\beta^\dagger(y) | 0 >$$

$$= (2\pi)^{-3} \int d^3 K \ (2K_o)^{-1} [- g_{\alpha\beta} + K_\alpha K_\beta / M_w^2] \ e^{-iK(x-y)} \ , \tag{3.102a}$$

and in the same way for $x_o < y_o$

$$< 0 | T \{ W_\alpha(x) W_\beta^\dagger(y) \} | 0 > \equiv < 0 | W_\beta^\dagger(y) W_\alpha(x) | 0 >$$

$$= (2\pi)^{-3} \int d^3 K \ (2K_o)^{-1} [- g_{\alpha\beta} + K_\alpha K_\beta / M_w^2] \ e^{iK(x-y)} \ . \tag{3.102b}$$

We may combine these two values and write the W-propagator in the form

$$< 0 | T \{ W_\alpha(x) W_\beta^\dagger(y) \} | 0 > = - i \ (2\pi)^{-4} \int d^4 K \ e^{-iK(x-y)}$$

$$\times \ [g_{\alpha\beta} - K_\alpha K_\beta / M_w^2] \ (K^2 - M_w^2 + i\epsilon)^{-1} \ . \tag{3.103}$$

This may be seen by observing that the integrand in (3.103) has poles at $K_o = \pm [(\underline{K}^2 + M_w^2) - i\epsilon]^{\frac{1}{2}}$. When $x_o > y_o$ we may complete the contour in the lower half of the K_o-plane; this picks out the pole at $K_o = + [(\underline{K}^2 + M_w^2) - i\epsilon]$ and the calculus of residues gives us the form (3.102a). Actually, as written, the form on the right of (3.103) is not quite correct. There should be an additional non-covariant term $(- i M_w^{-2} g_{\alpha o} g_{\beta o} \delta(x-y))$ arising from a more careful definition of the θ-functions in the T-product. Plainly this term only contributes when $x_o = y_o$, which is why we did not encounter it in our informal analysis. However, the contribution from this term to any matrix element is always cancelled by a similar non-covariant contribution from the interaction Hamiltonian $\mathcal{H}_I$ when it is constructed from the Lagrangian using the canonical formalism(12). Thus a consistent procedure, and the one we follow, is to drop such terms in both the propagator (3.103) and the

Hamiltonian (3.91).

Returning now to (3.94) and substituting (3.103) we may write the S-matrix element in the form (3.15a) where now we have

$$M(\mu^- \to e\bar{\nu}_e \nu_\mu) = f^2 (K^2 - M_W^2)^{-1} [\bar{u}_{(e)}(p)\,\gamma_\beta(1-\gamma_5)\,v_{(\nu_e)}(q_1)]$$

$$\times\ [g^{\alpha\beta} - K^\alpha K^\beta/M_W^2]\,[\bar{u}_{(\nu_\mu)}(q_2)\,\gamma_\alpha(1-\gamma_5)\,u_{(\mu)}(k)] \tag{3.104a}$$

with

$$K \equiv q_2 - k = -q_1 - p\ . \tag{3.104b}$$

We see that for large values of M_W the form (3.104a) reduces to that obtained previously in (3.15b) with

$$\frac{G}{\sqrt{2}} = f^2\, M_W^{-2}\ . \tag{3.105}$$

Thus in this case, and in general to this order, the current-current Hamiltonian (3.1) is just a limiting case of the intermediate vector boson hypothesis. The difference between the matrix elements (3.104a) and (3.15b) may be expected to lead to differences in the observable quantities in muon decay. In the approximation that we neglect m_e it is easy to see that (3.104a) leads to the prediction that the helicity h of the decay electron is -1. This is because the $K_\alpha K_\beta/M_W^2$ term in the W-propagator gives rise to a term proportional to m_e, by use of the Dirac equation, and our previous argument applies to the remainder of the matrix element. We are therefore led to consider the electron energy spectrum and angular distribution in the decay of a polarized muon. This may be calculated in a way quite analogous to that described previously in deriving (3.37). In the approximation that we neglect m_e and the electromagnetic corrections, and retain only the leading correction term of order m_μ^2/M_W^2, we obtain in this case[13]

$$d\Gamma = (2\pi)^{-3}\,\frac{4f^4}{3M_W^4}\, m_\mu\, E^2\, dE\, \sin\theta\, d\theta \left\{ 3W - 2E + \frac{m_\mu^2}{M_W^2}\,\frac{E}{W}\left(2W-E\right) + \cos\theta \left[W - 2E - \frac{m_\mu^2}{M_W^2}\,\frac{E^2}{W} \right] \right\}, \tag{3.106}$$

using the same notation as in (3.37). As expected, in the limit $M_W \to \infty$ this reduces to (3.37), if we make the above approximations and substitute (3.105). Thus the effect of the W-boson is to alter the shape of the electron's energy spectrum from the "general" form (3.39). Nevertheless we may find effective values for the parameters by using a least squares fit to (3.106) using the form (3.39) with the above approximations. This gives

$$A/(8G^2) = 1 + \frac{3}{5}(m_\mu/M_W)^2 \qquad (3.107a)$$

$$\rho - \tfrac{3}{4} = \frac{13}{40}(m_\mu/M_W)^2 \qquad (3.107b)$$

$$\xi - 1 = \frac{3}{5}(m_\mu/M_W)^2 \qquad (3.107c)$$

$$\delta - \tfrac{3}{4} = \frac{7}{40}(m_\mu/M_W)^2 \ . \qquad (3.107d)$$

We see that the W-boson produces deviations from the values (3.40) predicted for these parameters by our hypothesis (3.1), and the measured deviations enable us to deduce information on the mass of the W-boson. Of course (3.107) cannot be entirely correct, since we have neglected electromagnetic effects which in any case are affected by the presence of the charged W-boson[(14)]. When these effects are included, it is found that[(15)]

$$M_W = 1.3^{+\infty}_{-0.4} \text{ GeV} \ . \qquad (3.108)$$

Thus the one standard deviation bound upon M_W is

$$M_W > 0.9 \text{ GeV} \ . \qquad (3.109)$$

(Actually, this result is obtained from a least squares fit to the high momentum end of the energy spectrum for unpolarized muon decay, which is more sensitive to W-boson effects, as can be seen from (3.106).) In order to avoid the unobserved fast decay $K \to W\gamma$, the mass M_W must be greater than the kaon's mass. This is compatible with (3.109).

3.3.2 W-PRODUCTION AND DECAY

The most common method used to search for the existence of the W-boson is to try and produce it using high energy neutrinos. The process used is

$$\nu_\mu + (Z,A) \to \mu^- W^+ + (Z,A) \quad , \qquad (3.110)$$

in which the muon or W scatters in the Coulomb field of the nucleus (Z, A). The cross section for the production can be computed fairly reliably[16], the only unknowns being the mass and electromagnetic moments of the W. If M_W is about 2 GeV, the expected cross section is of the order of $10^{-36}\ \mathrm{cm}^2$ with incident neutrinos of energy 10 GeV. The calculation is not particularly sensitive to the electromagnetic moments of the W. Cross sections of this size can be measured fairly easily, but the problem lies in detecting the W in the first place. It will certainly decay via the reaction

$$W^+ \to \ell^+ \nu_\ell \quad , \qquad (3.111)$$

where ℓ is e or μ. The invariant matrix element for this process is

$$M(W_{\ell 2}) = f\, \epsilon_\alpha^{(\lambda)}\, \bar{u}_{(\nu_\ell)}(q)\, \gamma^\alpha (1-\gamma_5)\, v_{(\ell)}(p) \quad , \qquad (3.112)$$

where $\epsilon_\alpha^{(\lambda)}$ is the polarization vector of the W^+, and p, q are the momenta of ℓ^+, ν_ℓ respectively. To find the transition probability we need $|M|^2$, and if the spin s of ℓ^+ is not measured, we may sum over the two values $\pm$s. Then, as in muon decay, we find, using the notation (3.19),

$$\sum_s |M|^2 = f^2\, \epsilon_\alpha^{(\lambda)}\, \epsilon_\beta^{(\lambda)}\, 8\, [\alpha\, q\, \beta\, p] \quad . \qquad (3.113)$$

The antisymmetric part cannot contribute in this case. Further, if we do not observe the polarization ϵ of the decaying W we must average over the three polarization states. Thus using (3.101c) we have

$$\tfrac{1}{3} \sum_{\lambda, s} |M|^2 = \frac{8f^2}{3} [\, p.q + 2\, M_W^{-2}\, K.p\ K.q\,] \quad , \qquad (3.114)$$

where K is the momentum of the W. Energy-momentum conservation requires

$$K = p + q \quad . \qquad (3.115a)$$

Thus

$$K.q = p.q = \tfrac{1}{2}(M_W^2 - m_\ell^2) \quad , \qquad (3.115b)$$

and

$$K.p = m_\ell^2 + p.q = \tfrac{1}{2}(M_W^2 + m_\ell^2) \,. \tag{3.115c}$$

Hence

$$\tfrac{1}{3}\sum_{\lambda,s} |M|^2 = \frac{8f^2}{3}\left(M_W^2 - m_\ell^2\right)\left(1 + \frac{m_\ell^2}{2M_W^2}\right) . \tag{3.116}$$

The differential transition probability for the decay of a single W is then

$$d\Gamma = (2\pi)^{-2} \frac{1}{8K_o p_o q_o} \tfrac{1}{3}\sum_{\lambda,s} |M|^2 \,\delta(K-p-q)\, d^3p\, d^3q \quad . \tag{3.117}$$

As before, this may be integrated to give the total decay rate, and in the rest frame of the W this is

$$\Gamma(W_{\ell 2}) = \frac{f^2 M_W}{6\pi}\left(1 - \frac{m_\ell^2}{M_W^2}\right)^2 \left(1 + \frac{m_\ell^2}{2M_W^2}\right) . \tag{3.118}$$

Now it follows from (3.109) that $m_\ell^2 << M_W^2$. Thus for either decay mode the decay rate satisfies

$$\Gamma(W_{\ell 2}) \simeq \frac{f^2 M_W}{6\pi} = \frac{G\, M_W^3}{6\sqrt{2}\,\pi} \simeq 10^{18}\left(\frac{M_W}{M_p}\right)^3 s^{-1} \quad , \tag{3.119}$$

using (3.105) and the value (3.63) for G. Thus a W having a mass of 2 GeV will decay in at least 10^{-19} s, and this is too short a time for it to leave a visible track in a bubble chamber. The detection of the W^+ is therefore performed by looking for its decay products, of which the μ^+, via the mode (3.111), is the easiest to observe.

The experiment thus consists of looking for $\mu^+ \mu^-$ pairs produced by the reactions (3.110) and (3.111) sequentially. Of course, pairs are also likely to be produced via the direct reaction

$$\nu_\mu + (Z,A) \rightarrow \mu^+ \mu^- \nu_\mu + (Z,A) \tag{3.120}$$

if no W-boson exists. However, if the W is created, its decay products will have no "memory" of the beam direction, and $\mu^+ \mu^-$ pairs separated by wide angles are much more likely to be produced than in the direct reaction (3.120). The experiments so far performed reach the conclusion[(17)] that M_W is greater than 1.8 GeV.

Chapter 4

SEMILEPTONIC PROCESSES

The semileptonic processes are defined as those weak interactions which involve both hadrons and leptons. If we now assume that the leptons enter the weak Hamiltonian in the form of the leptonic current (2.264), then the simplest possibility for a local Hamiltonian for the semileptonic processes is that it has a current-current form analogous to (3.1) :

$$\mathcal{H}_W^{SL} = \frac{G}{\sqrt{2}} (\mathcal{J}_\lambda^+ L^\lambda + \mathcal{J}_\lambda^- L^{\lambda\dagger}) . \qquad (4.1)$$

$\mathcal{J}_\lambda^+$ is a mixture of vector and axial vector currents constructed entirely from the hadronic fields, and $\mathcal{J}_\lambda^-$ is its hermitian conjugate. There is no loss of generality in writing the same coupling constant $G/\sqrt{2}$ as in (3.1), since we have not yet specified the scale for $\mathcal{J}_\lambda^\pm$. The assumptions made in (4.1) are first that the leptons and hadrons interact at a point (locality) and second that the leptonic current L_λ enters the Hamiltonian only linearly; any derivative of L_λ, for example, could be recast in the form (4.1) by dropping a four-divergence, which has no effect on the equations of motion. In addition (4.1) ignores the existence of the neutral currents discussed in (2.5.2) and (3.2.2). Of course, these cannot contribute to any of the processes predicted by (4.1), at least in lowest order perturbation theory; by definition the leptons carry away zero charge in the neutral current phenomena, whereas the charged currents L_λ, $L_\lambda^\dagger$ imply that in the interactions allowed by (4.1) the leptons must carry away (plus or minus) one unit of charge. In this chapter, we shall be concerned solely with charged current semileptonic phenomena, and our hypothesis is that these are all described by (4.1).

The precise form of $\mathcal{J}_\lambda^\pm$ in terms of the hadrons' fields is not known, and in the absence of a theory of strong interactions it is not clear how useful that information would be even if were available. The theory of semileptonic processes thus rests upon the specification of more general properties of $\mathcal{J}_\lambda^\pm$, including its description in terms of quantities which are conserved by the strong interactions. The most powerful hypothesis so far advanced specifies the properties of $\mathcal{J}_\lambda^\pm$ in terms of the group SU(3), which we have already noted is an approximate symmetry of the strong interactions[1]. Before stating the hypothesis we therefore review the group theory we shall require.

4.1 SU(3) AND THE CABIBBO HYPOTHESIS

SU(3) is the group consisting of all unitary unimodular transformations A of the complex 3-dimensional vector space C_3

$$z'^i = A_{ij} z^j \quad (i, j = 1,2,3) \ , \tag{4.2a}$$

where

$$AA^\dagger = A^\dagger A = I_3 \tag{4.2b}$$

$$\det A = 1 \ . \tag{4.2c}$$

As for the Lorentz group, let us consider an infinitesimal transformation

$$A = \delta_{ij} + i\epsilon\,\alpha_{ij} \ , \tag{4.3}$$

where ϵ is real and infinitesimal. Clearly (4.2b) requires α to be hermitian

$$\alpha^\dagger = \alpha \ , \tag{4.4a}$$

and from (4.2c) we find that it is traceless

$$\mathrm{tr}\ \alpha = 0 \ . \tag{4.4b}$$

Plainly there are 9 linearly independent hermitian 3 x 3 matrices of which just 8 may be chosen to be traceless. Proceeding as before we define 9 traceless 3 x 3 matrices

$$(F^k_\ell)_{ij} = \delta_{\ell i}\,\delta_{kj} - \tfrac{1}{3}\,\delta_{k\ell}\,\delta_{ij} \ . \tag{4.5}$$

So

$$(F^\ell_k)^\dagger = F^k_\ell \ . \tag{4.6}$$

Moreover the 9 matrices are not linearly independent, since

$$F^1_1 + F^2_2 + F^3_3 = 0 \ . \tag{4.7}$$

The commutation relations satisfied by the matrices F^k_ℓ are

$$[F^k_\ell, F^m_n] = \delta^k_n F^m_\ell - \delta^m_\ell F^k_n , \qquad (4.8)$$

as may be verified easily using (4. 5). It follows from (4. 6) that we may define a basis of eight linearly independent hermitian matrices λ^i(i=1,..8) from the F^k_ℓ as follows

$$\lambda^1 = F^2_1 + F^1_2 \qquad \lambda^2 = -i(F^2_1 - F^1_2)$$

$$\lambda^4 = F^1_3 + F^3_1 \qquad \lambda^5 = i(F^1_3 - F^3_1)$$

$$\lambda^6 = F^3_2 + F^2_3 \qquad \lambda^7 = -i(F^3_2 - F^2_3)$$

$$\lambda^3 = F^1_1 - F^2_2 \qquad \lambda^8 = -\sqrt{3}\, F^3_3 = +\sqrt{3}\,(F^1_1 + F^2_2) . \qquad (4.9)$$

Then the commutation relations of the λ^i are

$$\left[\frac{\lambda^i}{2}, \frac{\lambda^j}{2}\right] = if^{ijk} \frac{\lambda^k}{2} , \qquad (4.10)$$

where f^{ijk} are real and totally antisymmetric structure constants. They may be evaluated with the aid of (4. 8), (4. 9) and (4.10). The non-zero values are as follows :

ijk	f^{ijk}
123	1
147	$\frac{1}{2}$
156	$-\frac{1}{2}$
246	$\frac{1}{2}$
257	$\frac{1}{2}$
345	$\frac{1}{2}$
367	$-\frac{1}{2}$
458	$\sqrt{3}/2$
678	$\sqrt{3}/2$

(4.11)

In addition the λ^i also satisfy the anti-commutation relations

$$\{\lambda^i, \lambda^j\} = \frac{4}{3}\delta^{ij} I_3 + 2d^{ijk}\lambda^k , \qquad (4.12)$$

where the non-zero values of d^{ijk} are found to be as follows

ijk	d^{ijk}
118	$1/\sqrt{3}$
146	$\frac{1}{2}$
157	$\frac{1}{2}$
228	$1/\sqrt{3}$
247	$-\frac{1}{2}$
256	$\frac{1}{2}$
338	$1/\sqrt{3}$
344	$\frac{1}{2}$
355	$\frac{1}{2}$
366	$-\frac{1}{2}$
377	$-\frac{1}{2}$
448	$-1/(2\sqrt{3})$
558	$-1/(2\sqrt{3})$
668	$-1/(2\sqrt{3})$
778	$-1/(2\sqrt{3})$
888	$-1/\sqrt{3}$. (4.13)

It is occasionally useful to introduce $\lambda^0 = \sqrt{\frac{2}{3}}\, I_3$. The anti-commutation relations (4.12) remain true for $i, j, k = 0 \ldots 8$ with

$$d^{0jk} = \sqrt{\frac{2}{3}}\, \delta^{ik} \quad . \tag{4.14}$$

Equation (4.10) is the Lie Algebra of the group SU(3). The representations are found by finding <u>all</u> matrices F^i $(i = 1 \ldots 8)$ satisfying the commutation relations

$$[F^i, F^j] = i f^{ijk} F^k \quad , \tag{4.15}$$

and in any particular n-dimensional representation the general infinitesimal group transformation is given by the matrix

$$A = I_n + i\epsilon^i F^i \quad . \tag{4.16}$$

The problem of constructing all irreducible representations of the group is outside the scope of this book. The interested reader is referred to the many excellent treatments of this topic in the literature[(2)]. We merely note that the matrices

$$F^i F^i \ , \ d^{ijk} F^i F^j F^k \tag{4.17}$$

commute with all of the generators F^i. (The proof of this statement is left as an exercise for the reader.) Thus, by Schur's lemma, these matrices are multiples of the unit matrix in any irreducible representation, and these eigenvalues may be used to classify the irreducible representations. To specify a basis within an irreducible representation we must select a complete set of commuting matrices, and the basis is then the set of all eigenvectors of these matrices. We choose the complete set as follows; the matrices F^1, F^2, F^3 define an SU(2) sub-algebra of SU(3), since $f^{ijk} = \epsilon^{ijk}$ for $(i,j,k = 1,2,3)$. Thus, just as for the angular momentum algebra, we may choose $(F^1)^2 + (F^2)^2 + (F^3)^2$ and F^3 as two of the commuting matrices. With these choices the only remaining matrix which commutes with these is F^8. Then the basis of states within an irreducible representation are chosen to be $|Y, I, I^3>$, where

$$F^8 \,|Y, I, I^3> \ = \frac{\sqrt{3}}{2} Y|Y, I, I^3> \ , \tag{4.18a}$$

$$\underline{F}^2 \,|Y, I, I^3> \ = \ I(I+1)|Y, I, I_3> \ , \tag{4.18b}$$

with $\underline{F}^2 \equiv (F^1)^2 + (F^2)^2 + (F^3)^2$ and $I = 0, \frac{1}{2}, 1, \frac{3}{2}, \ldots,$

and

$$F^3 |Y, I, I^3> = I^3 |Y, I, I^3> \ , \tag{4.18c}$$

where $\qquad I^3 = I, I\text{-}1, \ldots \ \text{-}I \quad .$

The restrictions on the eigenvalues I, I^3 follow since the corresponding matrices belong to an SU(2) algebra. As we shall see, we may identify I, I^3 with the isospin quantum numbers of the state, and the eigenvalue Y with its hypercharge.

4.1.1 OCTET REPRESENTATION OF SU(3)

Let us illustrate these remarks by considering the 8-dimensional representation or octet representation, $\underline{8}$, of the group. It is easy to see from what we have already done that such a representation exists. Consider the Jacobi identity

$$\left[[\lambda^i, \lambda^j], \ \lambda^a \right] + \left[[\lambda^j, \lambda^a], \lambda^i \right] + \left[[\lambda^a, \lambda^i], \ \lambda^j \right] \equiv 0 \, . \tag{4.19}$$

The commutation relations of the λ^i, given in (4.10), then yield

$$f^{ijb} f^{bac} + f^{jab} f^{bic} + f^{aib} f^{bjc} = 0 \quad . \tag{4.20}$$

Thus if we define the eight 8 x 8 hermitian (traceless) matrices

$$(F^i)_{ab} = if^{aib} \quad , \tag{4.21}$$

then (4.20) may be rewritten as

$$(F^i)_{ab} (F^j)_{bc} - (F^j)_{ab} (F^i)_{bc} = if^{ijb} (F^b)_{ac} \quad , \tag{4.22}$$

which is just the Lie Algebra (4.15). (Plainly the existence of this representation is not dependent upon the actual values (4.11) of the structure constants - any Lie Group will possess an anlogous representation known as the "regular" representation.) It is easy to verify that the Casimir operators (4.17) are multiples of the unit matrix in this representation, and it follows that the $\underline{8}$ representation is irreducible.

Let us construct the basis states of the octet representation as defined in (4.18). With the definition (4.21) it is easy to see that the matrices $(F^8)^2$, $\underline{F}^2$, $(F^3)^2$ are diagonal

$$(F^8)^2 = dg\,(0,0,0,\tfrac{3}{4},\tfrac{3}{4},\tfrac{3}{4},\tfrac{3}{4},0) \tag{4.23a}$$

$$\underline{F}^2 = dg\,(2,2,2,\tfrac{3}{4},\tfrac{3}{4},\tfrac{3}{4},\tfrac{3}{4},0) \tag{4.23b}$$

$$(F^3)^2 = dg\,(1,1,0,\tfrac{1}{4},\tfrac{1}{4},\tfrac{1}{4},\tfrac{1}{4},0) \quad . \tag{4.23c}$$

Thus the only permissible values of Y, I, I^3 are

$$Y = \pm 1, 0 \tag{4.24a}$$

$$I = 1, \tfrac{1}{2}, 0 \tag{4.24b}$$

$$I^3 = \pm 1, 0, \pm \tfrac{1}{2} \quad . \tag{4.24c}$$

Evidently there are three states with $|Y, I\rangle = |0, 1\rangle$ and one state with $|Y, I\rangle = |0, 0\rangle$. Of the four remaining states there are two with $|Y, I\rangle = |1, \tfrac{1}{2}\rangle$ and two with $|Y, I\rangle = |-1, \tfrac{1}{2}\rangle$. Let us denote by $\varphi^{(a)}$ the column vector with components

$$(\varphi^{(a)})_b = \delta_{ab} \quad . \tag{4.25}$$

The the required normalized basis states of the octet representation are given by

$$- | 8, 0, 1, 1> = \frac{1}{\sqrt{2}} (\varphi^{(1)} + i\varphi^{(2)}) = | \pi^{+} > \quad (4.26a)$$

$$| \underline{8}, 0, 1, -1> = \frac{1}{\sqrt{2}} (\varphi^{(1)} - i\varphi^{(2)}) = | \pi^{-} > \quad (4.26b)$$

$$| \underline{8}, 0, 0, 0> = \varphi^{(3)} = | \pi^{o} > \quad (4.26c)$$

$$| \underline{8}, 1, \tfrac{1}{2}, \tfrac{1}{2}> = \frac{1}{\sqrt{2}} (\varphi^{(4)} + i\varphi^{(5)}) = | K^{+} > \quad (4.26d)$$

$$| \underline{8}, 1, \tfrac{1}{2}, -\tfrac{1}{2}> = \frac{1}{\sqrt{2}} (\varphi^{(6)} + i\varphi^{(7)}) = | K^{o} > \quad (4.26e)$$

$$| \underline{8}, -1, \tfrac{1}{2}, \tfrac{1}{2}> = \frac{1}{\sqrt{2}} (\varphi^{(6)} - i\varphi^{(7)}) = | \bar{K}^{o} > \quad (4.26f)$$

$$- | \underline{8}, -1, \tfrac{1}{2}, -\tfrac{1}{2}> = \frac{1}{\sqrt{2}} (\varphi^{(4)} - i\varphi^{(5)}) = | K^{-} > \quad (4.26g)$$

$$| \underline{8}, 0, 0, 0> = \varphi^{(8)} = | \eta > \quad . \quad (4.26h)$$

The right-hand side of the equations (4.26) give the pseudoscalar meson states having the quantum numbers specified on the left-hand side, provided we identify I, I^3 with isospin and Y with hypercharge. The minus sign in front of the two states (4.26a), (4.26g) is necessitated by the Condon-Shortley phase convention[(3)]

$$(F^1 \pm iF^2) | I, I^3> = + [(I \mp I^3)(I \pm I^3 + 1)]^{\frac{1}{2}} | I, I_3 \pm 1> \quad .$$

For example, this convention requires

$$(F^1 \pm i F^2) | 1, 0> = \sqrt{2} | 1, \pm 1> \quad . \quad (4.28)$$

With the definition (4.21)

$$(F^1 \pm i F^2) \varphi^{(3)} = \mp \varphi^{(1)} - i\varphi^{(2)} \quad . \quad (4.29)$$

Thus with the definition (4.26c) the signs in (4.26a,b) are fixed as shown. The above noted identity of quantum numbers between the $\underline{8}$ representation of SU(3) and the pseudoscalar meson states suggests that these mesons do indeed constitute an irreducible representation of the group. Consider a general 8 component wave function $\varphi_a(x)$ which we may suppose defines one of the above meson states. Just as for the Lorentz group, the SU(3) invariance of the theory requires that the wave function

$$\varphi'_a(x) = \varphi_a(x) + i\epsilon^i (F^i)_{ab}\varphi_b(x) \tag{4.30}$$

satisfies the same equations of motion as does $\varphi_a(x)$. If so, then (using matrix notation)

$$(\partial_\mu \partial^\mu I + m^2)\, \varphi(x) = 0 = (\partial_\mu \partial^\mu I + m^2)\, \varphi'(x) \ , \tag{4.31a}$$

which implies

$$[m^2, F^i] = 0 \ ; \tag{4.31b}$$

this follows because (4.31a) is true for arbitrary ϵ^i and the matrices F^i are independent of x. But if the F^i constitute an irreducible representation of SU(3), Schur's lemma shows that the only matrices satisfying (4.31b) are multiples of the unit matrix: $(m^2)_{ab} = m^2 \delta_{ab}$. Thus, if the pseudoscalar mesons belonged to an 8 representation of SU(3) they would have to have the same mass. Experimentally, the masses differ substantially :

$$m_{\pi^\circ} \simeq 135 \text{ MeV}; \quad m_\eta \simeq 549 \text{ MeV} .$$

Thus SU(3) cannot be an exact symmetry of nature. Nevertheless, the agreement of the quantum numbers suggests that, in some sense, the symmetry is approximate. We shall return to this point later.

4.1.2 SU(3) INVARIANCE IN QUANTUM FIELD THEORY

Let us now, as for the Lorentz group, formulate SU(3) symmetry within the framework of quantum field theory. The general infinitesimal group transformation (4.3) is effected by a unitary operator $U(\epsilon)$ which acts on the state vectors. As before we may write

$$U(\epsilon) = I + i\epsilon^i \mathbb{F}^i \ , \tag{4.32}$$

where $\mathbb{F}^i$ are 8 hermitian operators. Suppose we have a set of field operators $\varphi_a(x)$. To show the SU(3) invariance of the field theory we must demonstrate the existence of the hermitian operators $\mathbb{F}^i$ with the property

$$[\mathbb{F}^i, \varphi_a(x)] = -(F^i)_{ab}\varphi_b(x) \ , \tag{4.33}$$

which leaves the Lagrangian invariant. F^i is one of the matrix representations of the group, which we have already discussed. (4.33) follows from

(4.32) in precisely the way that we proved (2.202). In the case that $\varphi_a(x)$ are an <u>octet</u> of operators the F^i are given by (4.21), so

$$[\mathbb{F}^i, \varphi_j(x)] = i f^{ijk} \varphi_k(x) \ . \tag{4.34}$$

Thus the field operators transform just as the generators $\mathbb{F}^i$ which, it is easy to see, satisfy

$$[\mathbb{F}^i, \mathbb{F}^j] = i f^{ijk} \mathbb{F}^k \ . \tag{4.35}$$

Now suppose we have an SU(3) invariant field theory involving a set of scalar fields $\varphi_a(x)$. That is to say $\mathcal{L}(\varphi_a, \partial_\mu \varphi_a)$ is invariant under the substitution

$$\varphi_a \to \varphi_a + \delta \varphi_a \ , \tag{4.36a}$$

where

$$\delta\varphi_a = i \epsilon^i (F^i)_{ab} \varphi_b \ . \tag{4.36b}$$

Thus

$$\delta\mathcal{L} = \frac{\partial\mathcal{L}}{\partial\varphi_a} \delta\varphi_a + \frac{\partial\mathcal{L}}{\partial(\partial^\mu \varphi_a)} \delta(\partial^\mu \varphi_a) = 0 \ . \tag{4.37}$$

The Euler-Lagrange equations of motion for the system are

$$\partial^\mu \pi_{\mu a} = \frac{\partial\mathcal{L}}{\partial\varphi_a} \ , \tag{4.38a}$$

where

$$\pi_{\mu a} \equiv \frac{\partial\mathcal{L}}{\partial(\partial^\mu \varphi_a)} \ . \tag{4.38b}$$

Then substituting into (4.37) gives

$$\delta\mathcal{L} = \partial^\mu (\pi_{\mu a} \delta\varphi^a) = 0 \ , \tag{4.39a}$$

since

$$\delta(\partial^\mu \varphi_a) = \partial^\mu (\delta\varphi_a) \ . \tag{4.39b}$$

Since (4.39) is true for arbitrary ϵ^i, plainly the currents

$$V_\mu^i \equiv (-i) \pi_{\mu a} (F^i)_{ab} \varphi_b \tag{4.40}$$

are conserved. That is

$$\partial^{\mu} V^{i}_{\mu} = 0 \ . \tag{4.41}$$

Thus it follows that the operator

$$\mathbb{F}^{i}(t) \equiv \int d^{3}x \, V^{i}_{o}(t, \underline{x}) \tag{4.42}$$

is a constant of the motion. This is because

$$\frac{d}{dt} \mathbb{F}^{i}(t) = \int d^{3}x \, \frac{\partial}{\partial t} V^{i}_{o}(\underline{x}, t) = 0 \ , \tag{4.43}$$

neglecting the integral of the spatial divergence. Further, we may verify that the operator $\mathbb{F}^{i}$ does indeed possess the required property (4.33). Since $\mathbb{F}^{i}$ is independent of t, we may write

$$\mathbb{F}^{i} = (-1) \int d^{3}y \, \pi_{od}(x^{o}, \underline{y}) \, (F^{i})_{db} \varphi_{b}(x^{o}, \underline{y})$$

The operator π_{od} is just the canonical conjugate of φ_{d} (see (2.176)). It therefore obeys canonical commutation relations (since the φ_{a} are boson fields)

$$[\varphi_{a}(x^{o}, \underline{x}), \pi_{od}(x^{o}, \underline{y})] = i \delta_{ad} \, \delta(\underline{x} - \underline{y}) \tag{4.44a}$$

$$[\varphi_{a}(x^{o}, \underline{x}), \varphi_{d}(x^{o}, \underline{y})] = 0 \ . \tag{4.44b}$$

Thus

$$[\mathbb{F}^{i}, \varphi_{a}(x)] = -(F^{i})_{ab} \varphi_{b}(x) \qquad \text{(Q.E.D.)} \tag{4.45}$$

It is thus apparent that any theory which is SU(3) invariant in the classical sense ($\delta \mathcal{L} = 0$) also provides an SU(3) invariant quantum field theory, since the required operator $\mathbb{F}^{i}$ can always be constructed. The same is true for any "internal" symmetry.

Now SU(3) is not an exact symmetry of nature, as we have already observed. Thus $\delta \mathcal{L} \neq 0$, and going back to (4.39a) we have

$$\delta \mathcal{L} = i \epsilon^{i} \partial^{\mu} [\pi_{\mu a} (F^{i})_{ab} \varphi_{b}] \neq 0 \ . \tag{4.46}$$

Using the definition (4.40) this gives[4]

$$\partial^{\mu} V^{i}_{\mu} = - \frac{\partial \mathcal{L}}{\partial \epsilon^{i}} \neq 0 \ . \tag{4.47}$$

So the current V^{i}_{μ} is no longer conserved, and in general

$$\frac{d}{dt} \mathbb{F}^{i}(t) \neq 0 \ . \tag{4.48}$$

Nevertheless the commutation relation (4.45) remains true at equal times[1,5], since the derivation used only the canonical commutation relations whose validity is unaffected by whether or not SU(3) is an exact symmetry. Thus when SU(3) is not exact we have

$$[\,\mathbb{F}^i(x^o),\, \varphi_a(x)\,] = -(F^i)_{ab}\,\varphi_b(x)\,. \tag{4.49}$$

4.1.3 THE CABIBBO HYPOTHESIS

We are now in a position to state the Cabibbo Hypothesis[6]. This is that the hadronic current $\mathcal{J}^{\pm}_{\lambda}(x)$, introduced in (4.1), may be written in the form

$$\mathcal{J}^{\pm}_{\lambda} = J^{\pm}_{\lambda}\cos\theta_c + S^{\pm}_{\lambda}\sin\theta_c\,, \tag{4.50a}$$

where

$$J^{\pm}_{\lambda} = V^1_{\lambda} \pm i\,V^2_{\lambda} + A^1_{\lambda} \pm i\,A^2_{\lambda}\,, \tag{4.50b}$$

$$S^{\pm}_{\lambda} = V^4_{\lambda} \pm i\,V^5_{\lambda} + A^4_{\lambda} \pm i\,A^5_{\lambda}\,, \tag{4.50c}$$

and θ_c is the Cabibbo angle. V^i_{λ}, A^i_{λ} $(i = 1\ldots 8)$ are respectively vector and axial vector currents transforming as octet representations of SU(3), in the equal time sense (4.49). Further, the SU(3) generators (which we have been calling $\mathbb{F}^i$) are the space integrals of the quantities V^i_o, as in (4.42). Thus, if we define

$$v^i(t) \equiv \int d^3x\, V^i_o(t,\underline{x}), \quad (i = 1,\ldots 8)\,, \tag{4.51}$$

this statement may be written as

$$\mathbb{F}^i(t) = v^i(t)\,, \tag{4.52a}$$

and

$$[\,v^i(t),\, v^j(t)\,] = i f^{ijk}\, v^k(t)\,. \tag{4.52b}$$

The statements that V^i_{μ}; A^i_{μ} transform as octets gives

$$[\,v^i(x^o),\, V^j_{\lambda}(x)\,] = i f^{ijk}\, V^k_{\lambda}(x) \tag{4.53a}$$

$$[\,v^i(x^o),\, A^j_{\lambda}(x)\,] = i f^{ijk}\, A^k_{\lambda}(x)\,, \tag{4.53b}$$

since the matrices $(F^i)_{ab}$ in (4.49) are given by (4.21) in the $\underline{8}$ representation. It is important to realize how powerful the preceding statements are, in principle. The belief is that they are absolutely exact with no corrections to be made owing to symmetry breaking, electromagnetic effects or anything else. However, as we shall see, it is very difficult to test the hypothesis without the use of further assumptions which are known to be only approximate.

In the same notation as (4.50) the electromagnetic current of the hadrons is written

$$j_\lambda = V^3_\lambda + \frac{1}{\sqrt{3}} V^8_\lambda \ . \tag{4.54}$$

Thus the electric charge operator

$$Q = I^3 + \tfrac{1}{2} Y \ , \tag{4.55a}$$

where, as already noted, the third component of isospin I^3, the hyperchange and the total isospin $\underline{I}^2$ are identified as

$$I^3 = v^3 \tag{4.55b}$$

$$Y = \frac{2}{\sqrt{3}} v^8 \tag{4.55c}$$

$$\underline{I}^2 = (v^1)^2 + (v^2)^2 + (v^3)^2 \ . \tag{4.55d}$$

It then follows from the commutation rules given that, for example

$$[\, Q, \mathcal{J}^\pm_\lambda \,] = \pm \mathcal{J}^\pm_\lambda \ , \tag{4.56a}$$

$$[\, Y, J^\pm_\lambda \,] = 0 \ . \tag{4.56b}$$

(4.56a) means that $\mathcal{J}^+_\lambda$ has the net effect of creating one unit of charge, as it must since L_λ creates minus one unit net. In the same way it follows as in (4.26a,b), (4.26d,g) that

$$J^\pm_\lambda \text{ has } (Y, I, I^3) = (0, 1, \pm 1) \tag{4.57a}$$

$$S^\pm_\lambda \text{ has } (Y, I, I^3) = (\pm 1, \tfrac{1}{2}, \pm\tfrac{1}{2}) \ . \tag{4.57b}$$

Thus it is a consequence of the Cabibbo Hypothesis that two types of semileptonic processes are predicted in which the quantum numbers of the hadrons in the initial and final states differ by the amounts given in (4.57). Of course this statement is only meaningful inasmuch as isospin and hypercharge <u>are</u> good quantum numbers for the hadrons. It is believed that all strong interactions conserve both quantities, that conservation of hypercharge is violated only by the weak interactions and that isospin symmetry is violated by weak and electromagnetic interactions.

Another consequence of the Cabibbo Hypothesis is that it predicts that the isovector part of j_λ, namely V^3_λ, and the vector part of $J^\pm_\lambda$, namely $V^1_\lambda \pm i V^2_\lambda$ form an isospin triplet. That is to say

$$[v^i(t), v^j(t)] = i\epsilon^{ijk} v^k(t) \quad (i,j,k = 1,2,3) \tag{4.58a}$$

$$[v^i(x^o), V^j_\lambda(x)] = i\epsilon^{ijk} V^k_\lambda(x) \quad (i,j,k = 1,2,3) \tag{4.58b}$$

and, to the extent that isospin is conserved,

$$\partial^\lambda (V^1_\lambda \pm i V^2_\lambda) = 0 \quad . \tag{4.58c}$$

Equations (4.58) constitute the Conserved Vector Current Hypothesis[7] (CVC). We shall use the above implications of the Cabibbo Hypothesis to make concrete physical predictions in the succeeding sections. But the Cabibbo Hypothesis is a much stronger statement than just these consequences, which would themselves follow simply from the requirement that the vector and axial vector parts of $\mathcal{J}^\pm_\lambda$ belong to an octet. The requirement that $\mathcal{J}^\pm_\lambda$ has the most general form for a charged current constructed from the octets V^i_λ, A^i_λ gives

$$\begin{aligned} \mathcal{J}^\pm_\lambda = {} & R_V \{ (V^1_\lambda \pm i V^2_\lambda) \cos\theta_V + (V^4_\lambda \pm i V^5_\lambda) \sin\theta_V \} \\ & + R_A \{ (A^1_\lambda \pm i A^2_\lambda) \cos\theta_A + (A^4_\lambda \pm i A^5_\lambda) \sin\theta_A \} , \end{aligned} \tag{4.59}$$

since only the $(1 \pm i\,2)$ and $(4 \pm i\,5)$ combinations of the octet are charged, as is apparent from (4.26). The form (4.59) is sufficient to ensure the selection rules implied by (4.57). Thus the Cabibbo Hypothesis makes a statement on the overall strength of the vector and axial vector currents by requiring

$$R_V = R_A = 1 \quad . \tag{4.60a}$$

It also makes a statement about the relative directions in SU(3) space of the vector and axial vector currents by requiring

$$\theta_V - \theta_A = 0 \quad . \tag{4.60b}$$

Actually the statement (4. 60a) about the overall strength of the axial vector current is not yet meaningful, since we have not yet specified the scale of the A_λ^i. The scale of the V_λ^i is fixed by (4. 51) and (4. 53a), but evidently (4. 53b) is true for arbitrarily scaled A_λ^i.

To define the scale of A_λ^i we define quantities a^i analogous to v^i :

$$a^i(t) \equiv \int d^3x\, A_o^i(t, \underline{x}) \quad (i, = 1, \ldots 8) \quad . \qquad (4.61)$$

Thus (4. 53b) implies that

$$[v^i(t), \; a^j(t)] = i f^{ijk} a^k(t) \quad . \qquad (4.62)$$

The scale of the a^i, or equivalently the A_λ^i, is fixed by assuming one further equal time commutation relation

$$[a^i(t), a^j(t)] = i f^{ijk} v^k(t) \quad . \qquad (4.63)$$

The commutation relations previously written down follow in any canonical quantum field theory in which we make the identification (4. 52a), which, as we have seen, is an obvious generalization of CVC to the SU(3) group. No such statement is true of the commutation relation (4. 63). It is easy to construct models in which the previous commutation relations are true but in which the right-hand side of (4. 63) is zero. The 'gradient coupling model' of Gell-Mann and Levy(4) has just this property. Thus (4. 63) constitutes a major additional dynamical assumption. To justify it we must not only find models in which it is true but also explain why these models are more plausible than, for example, the gradient coupling model. In the end, of course, the use of (4. 63) is justified by the fact that it yields physical predictions which are consistent with experiment.

4.1.4 QUARKS AND THE SU(3) x SU(3) CURRENT ALGEBRA

The simplest way of justifying it is to use the quark model(1). The undoubted relevance of SU(3) symmetry to particle physics is most naturally explained by assuming the existence of a fundamental triplet of fermion fields constituting a 3-dimensional representation of the group. This is the 'fundamental' representation we derived in (4. 9) and (4. 10). With the previous identification of isospin and hypercharge it is easy to see that the three basis states of the $\underline{3}$ representation have

$$(Y, I, I^3) = (\tfrac{1}{3}, \tfrac{1}{2}, \tfrac{1}{2}),\ (\tfrac{1}{3}, \tfrac{1}{2}, -\tfrac{1}{2}),\ (-\tfrac{2}{3}, 0, 0), \tag{4.64a}$$

so that

$$Q = \tfrac{2}{3}\ ,\ -\tfrac{1}{3}\ ,\ -\tfrac{1}{3}\ . \tag{4.64b}$$

These fractionally-charged particles are called "quarks" - to date they have not been found experimentally. Nevertheless the observed meson states may be understood by supposing that they are bound states of quark/ anti-quark pairs, in which picture the baryons are bound states of three quarks. We denote the triplet of quark field operators by the column vector $q(x)$. Then the simplest form for the octets of vector and axial vector currents is

$$V^i_\lambda(x) = \bar{q}(x)\, \gamma_\lambda\, \tfrac{1}{2}\, \lambda^i\, q(x)\ , \tag{4.65a}$$

$$- A^i_\lambda(x) = \bar{q}(x)\, \gamma_\lambda\, \gamma_5\, \tfrac{1}{2}\, \lambda^i\, q(x)\ , \tag{4.65b}$$

where the matrices λ^i are given in (4.9). If the quark fields satisfy free field canonical anti-commutation relations analogous to (2.177),

$$\{ q(x), \bar{q}(y) \}_{x^o = y^o} = \gamma^o\, I_3\, \delta(\underline{x} - \underline{y}) \tag{4.66a}$$

$$\{ q(x), q(y) \}_{x^o = y^o} = 0 = \{ \bar{q}(x), \bar{q}(y) \}_{x^o = y^o}\ , \tag{4.66b}$$

we may evaluate a general equal time commutator. Using the identity

$$[AB, CD] \equiv A\{B,C\}D - AC\{B,D\} + \{A,C\}BD - C\{A,D\}B\ , \tag{4.67}$$

and (4.66) it is easy to see that

$$[\bar{q}(x)\,\Gamma\,\Lambda q(x),\ \bar{q}(y)\,\gamma\,\lambda\, q(y)]_{x^o = y^o} = \tfrac{1}{2}\,\delta(\underline{x}-\underline{y})\,\bar{q}(x)\,\gamma_o \Big(\{\gamma_o \Gamma, \gamma_o \gamma\}[\Lambda, \lambda] + [\gamma_o \Gamma, \gamma_o \gamma]\{\Lambda, \lambda\} \Big)\, q(x)\ , \tag{4.68}$$

where Γ, γ are arbitrary 4 x 4 matrices in spinor space and Λ, λ are arbitrary 3 x 3 matrices in SU(3) space. By specializing these arbitrary matrices to those in (4.65) it is straightforward to obtain the relations (4.52b), (4.62) and (4.63) by integration over $\underline{x}$ and $\underline{y}$. This approach yields much more than just the commutators we are attempting to justify; it yields the equal time commutators of all components (space and time) of all of the currents. In particular, at equal times, $x^o = y^o$, we have

$$[V_o^i(x), V_\lambda^j(y)] = [A_o^i(x), A_\lambda^j(y)] = if^{ijk} V_\lambda^k(x)\, \delta(\underline{x}-\underline{y}) \tag{4.69a}$$

$$[V_o^i(x), A_\lambda^j(y)] = [A_o^i(x), V_\lambda^j(y)] = if^{ijk} A_\lambda^k(x)\, \delta(\underline{x}-\underline{y}) \quad . \tag{4.69b}$$

Actually the commutator $[V_o^i, V_o^j]$ has a more general validity, since it follows from (4.40) with the canonical commutation relations (4.44). However, some of the space-time commutators $[V_o^i, V_r^j]$ cannot be correct. The right-hand sides must be augmented by the addition of a "Schwinger term"[(8)] proportional to $\partial_r\, \delta(\underline{x}-\underline{y})$. Such terms disappear if we perform one spatial integration of (4.69) to yield the "charge-current" commutators

$$[v^i(x^o), V_\lambda^j(x)] = [a^i(x^o), A_\lambda^j(x)] = if^{ijk} V_\lambda^k(x) \tag{4.70a}$$

$$[a^i(x^o), V_\lambda^j(x)] = [v^i(x^o), A_\lambda^j(x)] = if^{ijk} A_\lambda^k(x) \quad . \tag{4.70b}$$

Defining

$$\chi_\pm^i(t) \equiv \tfrac{1}{2}\,(v^i(t) \pm a^i(t)) \quad , \tag{4.71}$$

we may rewrite the "charge" algebra (4.52b), (4.62), (4.63) in the equivalent form

$$[\chi_\pm^i(t), \chi_\pm^j(t)] = if^{ijk} \chi_\pm^k(t) \tag{4.72a}$$

$$[\chi_+^i(t), \chi_-^j(t)] = 0 \quad . \tag{4.72b}$$

Thus χ_+^i and χ_-^i are two commuting sets of generators of SU(3) algebras, and the above commutation relations are therefore called those of the SU(3) x SU(3) algebra.

4.1.5 UNIVERSALITY

We have now completely specified the scale of the currents V_λ^i, A_λ^i. Let us now attempt to justify the Cabibbo Hypothesis (4.50), as opposed to the more general hypothesis contained in (4.59). This is evidently equivalent to justifying the appearance of the coupling constant $\frac{G}{\sqrt{2}}$ in (4.1), which also appears in (3.1), since (4.50) and (4.59) differ in the strengths with which the various currents (whose scale now is defined) are coupled to the leptonic current in $\mathcal{H}_W^{SL}$. The argument we shall use is a modified form of "universality" (see Gell-Mann[9]). Universality connects the form of the hadronic current $\mathcal{J}_\lambda^{\pm}$ with that of the leptonic current L_λ in a way which we shall now specify.

We define a weak leptonic "charge" W^ℓ

$$W_+^\ell (x^o) = \int d^3x\, L_o^\dagger (x) \quad , \tag{4.73}$$

where L_λ and $L_\lambda^\dagger$ are defined in (2.264). We define "leptonic spin" spinors as follows:

$$E(x) \equiv \begin{pmatrix} \psi_{(\nu_e)}(x) \\ \psi_{(e)}(x) \end{pmatrix} \qquad M(x) \equiv \begin{pmatrix} \psi_{(\nu_\mu)}(x) \\ \psi_{(\mu)}(x) \end{pmatrix} . \tag{4.74}$$

We also define for $i = 1,2,3$

$$\xi_\pm^i(x^o) \equiv \int d^3x \,\{ \overline{E}(x)\, \gamma_o (1\pm \gamma_5) \tfrac{1}{2} \tau^i\, E(x) + \overline{M}(x) \gamma_o (1\pm\gamma_5) \tfrac{1}{2}\tau^i M(x) \} , \tag{4.75}$$

where the τ^i $(i = 1,2,3)$ are 2 x 2 Pauli matrices.

Then it is easy to see that

$$\tfrac{1}{2} W_+^\ell (t) = \xi_-^1(t) + i\, \xi_-^2(t) \tag{4.76}$$

and that the $\xi_\pm^i$ satisfy an SU(2) x SU(2) algebra

$$[\, \xi_\pm^i(t),\ \xi_\pm^j(t)\,] = i\epsilon^{ijk}\, \xi_\pm^k(t) \quad (i,j,k = 1,2,3) \tag{4.77a}$$

$$[\, \xi_+^i(t),\ \xi_-^j(t)\,] = 0 \quad . \tag{4.77b}$$

Of course, there is a current corresponding to the neutral generator ξ_-^3, namely

$$L_\lambda^o = \overline{\psi}_{(\nu_e)} \gamma_\lambda (1-\gamma_5)\, \psi_{(\nu_e)} - \overline{\psi}_{(e)}\, \gamma_\lambda\, (1-\gamma_5)\, \psi_{(e)} + \overline{\psi}_{(\nu_\mu)}\, \gamma_\lambda (1-\gamma_5)\, \psi_{(\nu_\mu)} - \overline{\psi}_{(\mu)} \gamma_\lambda (1-\gamma_5)\, \psi_{(\mu)} . \tag{(4.78)}$$

It may be that this neutral current is relevant to the recently discovered phenomena discussed in (2. 5. 2) and (3. 2. 2). We shall encounter it again in Chapter 6.

By "universality" we shall mean that if we define the weak hadronic charge $W^h(t)$ in a way analogous to $W^\ell(t)$,

$$W^h_{\pm}(t) \equiv \int d^3x\ \mathcal{J}^{\pm}_o(t, \underline{x}) \quad , \tag{4.79}$$

then we may write W^h_+ also in the form

$$\tfrac{1}{2}\ W^h_{\pm}(t) = \zeta^1_- \pm i\zeta^2_- \quad , \tag{4.80}$$

where $\zeta^i_{\pm}$ $(i = 1,2,3)$ also satisfy an SU(2) x SU(2) algebra.

Consider first the identity

$$e^{-2i\theta v^7(t)}\left[v^1(t) \pm iv^2(t)\right] e^{2i\theta v^7(t)} = \cos\theta\left(v^1(t) \pm iv^2(t)\right) + \sin\theta\left(v^4(t) \pm iv^5(t)\right) . \tag{4.81}$$

This may be proved by rewriting the left-hand side as a power series in θ:

$$\text{LHS} = \sum_{n=o}^{\infty} \frac{(2i\theta)^n}{n!} \underbrace{[\,[\ldots[}_{n\text{ - times}}\ v^1(t) \pm i v^2(t),\ v^7(t)\,],\ v^7(t)\,]\ldots,\ v^7(t)\,] \ . \tag{4.82}$$

Now using (4. 52b) it follows that

$$[v^1 \pm i v^2,\ v^7] = -\tfrac{1}{2} i\ (v^4 \pm i v^5) \tag{4.83a}$$

$$[v^4 \pm i v^5,\ v^7] = \tfrac{1}{2} i\ (v^1 \pm i v^2) \ . \tag{4.83b}$$

Thus the even powers of θ in (4. 82) are proportional to $v^1 \pm i v^2$, and odd powers to $v^4 \pm i v^5$. The identity (4. 81) follows a summing of these terms separately. A similar identity follows for the axial charges $a^1 \pm i a^2$ using (4. 62). If we now suppose that the current $\mathcal{J}^{\pm}_{\lambda}$ has the general form (4. 59), we may use the identity (4. 81) and its axial analogue to write $W^h_{\pm}$ in the form (4. 80), where

$$2\,\zeta^i_{\pm}(t) = R_V\ e^{-2i\theta_V v^7(t)}\ v^i(t)\ e^{2i\theta_V v^7(t)}$$

$$\mp R_A\ e^{-2i\theta_A v^7(t)}\ a^i(t)\ e^{2i\theta_A v^7(t)} \qquad (i = 1,2,3)\ . \tag{4.84}$$

Thus, if we now impose the requirement that $\zeta^i_{\pm}(t)$ satisfy the SU(2) x SU(2) algebra, we deduce, equating scalar and pseudoscalar parts ,

$$\theta_V = \theta_A \tag{4.85a}$$

$$R^2_V + R^2_A = 2\ R_V \tag{4.85b}$$

$$R_V R_A = R_A \ . \tag{4.85c}$$

Solving (4.85b, c) we find that the only solutions with non-zero values for both R_V and R_A are

$$R_V = 1 \ , \quad R_A = \pm 1 \ . \tag{4.86}$$

Thus the Cabibbo Hypothesis, which corresponds to the solution $R_V = R_A = 1$ and $\theta_V = \theta_A = \theta_c$, satisfies "universality" as we have defined it, whereas the more general hypothesis (4.59) will not in general do so. The ambiguity in the sign of R_A is unavoidable and arises from the invariance of the SU(3) x SU(3) algebra under the transformation $A^i_\lambda \to - A^i_\lambda$. We shall return to this point shortly.

The neutral current $\mathcal{J}^o_\lambda$ corresponding to the neutral generator ζ^3_- is

$$\mathcal{J}^o_\lambda = (1 - \tfrac{1}{2}\sin^2\theta_c)(V^3_\lambda + A^3_\lambda) - \tfrac{1}{2}\sin 2\theta_c (V^6_\lambda + A^6_\lambda) + \frac{\sqrt{3}}{2}\sin^2\theta_c (V^8_\lambda + A^8_\lambda) \ . \tag{4.87}$$

As we have seen in (2.5.2), there is experimental evidence that a neutral current is coupled to the leptons. But the current proportional to $V^6_\lambda + A^6_\lambda$, which would induce processes in which the hadronic hypercharge changes by one unit, certainly appears to be absent.

If we accept the above argument, the 'universal' hadronic current $\mathcal{J}^\pm_\lambda$ still has one undetermined parameter θ_c. As we shall see, this is determined from the experimental data to have a value of about 0.2, which, by construction, explains the observed inhibition of the semi-leptonic decays arising from $S^\pm_\lambda$ compared with those arising from $J^\pm_\lambda$. So far, there has been no generally accepted argument as to why θ_c has the value it has. The value of θ_c is at present generally believed to be an intrinsic parameter of the weak interaction, in just the way that G is. (It seems to be just a tantalizing accident that θ_c has a value remarkably close to m_π/m_K: the ratio of the masses of the pseudoscalar mesons having the internal quantum numbers (4.57a,b) which characterize $J^\pm_\lambda$, $S^\pm_\lambda$.)

In the next two sections of this Chapter we shall see how the Cabibbo Hypothesis is used to make concrete physical predictions and to what extent these predictions are consistent with the known experimental data.

4.2 HYPERCHARGE CONSERVING SEMILEPTONIC PROCESSES

This section is devoted to a study of the semileptonic processes which arise from the part $J_\lambda^{\pm}$ of $\mathcal{J}_\lambda^{\pm}$. The hadronic selection rules characterizing these processes are given in (4.57a). Thus, in all of these processes the hypercharge of the hadrons is conserved, while their charge and isospin change by one unit.

4.2.1 $\pi_{\ell 2}$

The simplest processes allowed by the selection rules (4.57a) are the leptonic decays of the charged pion

$$\pi^- \to \ell^- \bar{\nu}_\ell \tag{4.88a}$$

$$\pi^+ \to \ell^+ \nu_\ell \quad , \tag{4.88b}$$

where $\ell = e$ or μ. Both the electronic and muonic modes have been observed. As before we work in first order perturbation theory. The S-matrix element for the process (4.88a) is then

$$< \ell^- \bar{\nu}_\ell | S | \pi^- > = - i \int d^4x < \ell^- \bar{\nu}_\ell | \mathcal{H}_W^{SL}(x) | \pi^- > \quad . \tag{4.89}$$

Substituting the form (4.1), and proceeding with the lepton states as before we find

$$< \ell^- \bar{\nu}_\ell | \mathcal{H}_W^{SL}(x) | \pi^- > = \frac{G}{\sqrt{2}} \bar{u}_{(\ell)}(p,s) \gamma^\lambda (1-\gamma_5) v_{(\nu_\ell)}(q, \tfrac{1}{2}) e^{i(p+q).x}$$

$$\times < 0 | \mathcal{J}_\lambda^+(x) | \pi^- > \quad , \tag{4.90}$$

where p, s are the momentum and spin of ℓ^-, q is the momentum of $\bar{\nu}_\ell$. To proceed further we must evaluate the matrix element $< 0 | \mathcal{J}_\lambda^+(x) | \pi^- >$. Translation invariance enables us to determine the dependence upon x. Quite generally

$$[P_\mu , \mathcal{J}_\lambda^+(x)] = - i \partial_\mu \mathcal{J}_\lambda^+(x) \quad , \tag{4.91}$$

where P_λ is the energy-momentum operator; since (2.202) is true for any field operator, it is also true of any local function of field operators.

It follows that

$$< 0|[P_\mu, \mathscr{J}^+_\lambda(x)]|\pi^-> = -k_\mu < 0|\mathscr{J}^+_\lambda(x)|\pi^-> = -i\,\partial_\mu < 0|\mathscr{J}^+_\lambda(x)|\pi^->, \qquad (4.92)$$

where k is the momentum of the π^-. Solving (4.92) we have

$$< 0|\mathscr{J}^+_\lambda(x)|\pi^-> = e^{-ikx} < 0|\mathscr{J}^+_\lambda(0)|\pi^-> \quad . \qquad (4.93)$$

Next we use Lorentz Invariance. Since $\mathscr{J}^+_\lambda$ transforms as a vector under restricted Lorentz transformations,

$$U(\Lambda)^{-1}\,\mathscr{J}^{+\lambda}\,U(\Lambda) = \Lambda^\lambda_{\ \mu}\,\mathscr{J}^{+\mu} \quad , \qquad (4.94)$$

where $U(\Lambda)$ is the unitary operator representing the Lorentz transformation. Taking the matrix element of (4.94) between $<0|$ and $|\pi^->$ gives

$$< 0|\mathscr{J}^{+\lambda}\,U(\Lambda)|\pi^-(k)> = \Lambda^\lambda_{\ \mu} < 0|\mathscr{J}^{+\mu}|\pi^-(k)> \quad . \qquad (4.95)$$

The state $U(\Lambda)|\pi^-(k)>$ is plainly a single pion state of momentum (Λk), and the state normalization is unchanged by the Lorentz transformation :

$$U(\Lambda)|\pi^-(k)> = |\pi^-(k')> \quad , \qquad (4.96a)$$

where

$$k'^\alpha = \Lambda^\alpha_{\ \beta}\,k^\beta \quad . \qquad (4.96b)$$

To verify this we note that the pion number density operator $dN(\pi, k)$ is Lorentz invariant. Thus

$$U(\Lambda)^{-1}\,dN(\pi,k)\,U(\Lambda) = dN(\pi,k') \quad , \qquad (4.97a)$$

where

$$dN(\pi,k) = a^\dagger(\underline{k})\,a(\underline{k})\,d^3k\,(2\pi)^{-3}\,(2k_o)^{-1} \quad . \qquad (4.97b)$$

$a(\underline{k})$, $a^\dagger(\underline{k})$ are the annihilation, creation operators of the pion, As in the case of the electron (see (2.185)), $dN(\pi,k)$ is the operator representing the number of pions having momenta in the range d^3k around $\underline{k}$. Now the quantity $d^3k(2k_o)^{-1}$ is Lorentz invariant, since

$$\int \frac{d^3k}{2k_o} = \int d^4k\,\delta(k^2 - m^2) \qquad (4.98)$$

and both d^4k and $\delta(k^2 - m^2)$ are manifestly Lorentz invariant. Then from (4.97) we deduce

$$U(\Lambda)^{-1}\, a(\underline{k})^{\dagger}\, a(\underline{k})\, U(\Lambda) = a(\underline{k}')^{\dagger}\, a(k') \ , \tag{4.99}$$

from which (4.96a) follows.

Returning to the problem in hand, substituting (4.96) into (4.95) we find

$$<0|\mathcal{J}^{+\lambda}|\pi^-(k')> = \Lambda^{\lambda}_{\mu} <0|\mathcal{J}^{+\mu}|\pi^-(k)> . \tag{4.100}$$

Thus the quantity $<0|\mathcal{J}^{+\lambda}|\pi^-(k)>$ transforms as a vector under restricted Lorentz transformations, just as k does in (4.96b). Since this quantity depends only on k it must be a scalar multiple of k^{λ}. Then it follows that

$$<0|\mathcal{J}^{+}_{\lambda}|\pi^-(k)> = C\, k_{\lambda} \ . \tag{4.101}$$

Now plainly only the part J^+_λ of $\mathcal{J}^+_\lambda$ (see (4.50)) can contribute, since the pion and the vacuum have zero hypercharge and the current S^+_λ carries unit hypercharge. Further, under parity reversal

$$\mathscr{P}^{-1}\,(V^1_o + i\,V^2_o)\,\mathscr{P} = V^1_o + i\,V^2_o \ , \tag{4.102}$$

since the currents V^i_λ are vector rather than axial vector quantities. Hence

$$<0|\mathscr{P}^{-1}(V^1_o + iV^2_o)\,\mathscr{P}|\pi^-(\underline{k})> = -<0|V^1_o + i\,V^2_o|\pi^-(-\underline{k})>$$

$$= <0|V^1_o + iV^2_o|\pi^-(\underline{k})> \ , \tag{4.103}$$

since the pion is a pseudoscalar particle

$$\mathscr{P}|\pi^-(\underline{k})> = -|\pi^-(-\underline{k})> \ . \tag{4.104}$$

Then using (4.101) it follows that the vector part $V^1_\lambda + i\,V^2_\lambda$ of J^+_λ makes no contribution to the matrix element (4.101). Hence, using (4.93), we have finally

$$<0|\mathcal{J}^+_\lambda(x)|\pi^-(k)> = \cos\theta_c <0|A^1_\lambda(x) + i\,A^2_\lambda(x)|\pi^->$$

$$\equiv i f_\pi \cos\theta_c\, k_\lambda\, e^{-ikx} \ , \tag{4.105}$$

where f_π is an undetermined constant. Thus substituting back into (4.90) and (4.89)

$$<\ell^-\,\bar{\nu}_\ell|S|\pi^-> = -\,i(2\pi)^4\ M(\pi^-_{\ell 2})\ \delta\,(k-p-q) \ , \tag{4.106a}$$

where

$$M(\pi^-_{\ell 2}) = \frac{G}{\sqrt{2}} \, i f_\pi \cos\theta_c \, k_\lambda \, \bar{u}_{(\ell)}(p,s) \, \gamma^\lambda (1-\gamma_5) \, v_{(\nu_\ell)}(q,\tfrac{1}{2}) \, , \tag{4.106b}$$

is the invariant matrix element for the decay. The decay probability may be calculated just as in muon decay, with the result:

$$|M|^2 = \tfrac{1}{2} G^2 f_\pi^2 \cos^2\theta_c \, k_\lambda k_\mu E^{\lambda\mu} \, , \tag{4.107}$$

where $E^{\lambda\mu}$ is given in (3.20). Since the decay products in this process consist of just two particles their energies and momenta are completely determined. Since

$$k = p + q \, , \tag{4.108a}$$

$$0 = q^2 = (k-p)^2 = m_\pi^2 + m_\ell^2 - 2k.p \tag{4.108b}$$

$$m_\ell^2 = p^2 = (k-q)^2 = m_\pi^2 - 2k.q \quad . \tag{4.108c}$$

Thus in the rest frame of the pion $(k = (m_\pi, \underline{0}))$

$$p_o = (m_\pi^2 + m_\ell^2)\,(2m_\pi)^{-1} \tag{4.109a}$$

$$q_o = |\underline{q}| = |\underline{p}| = (m_\pi^2 - m_\ell^2)\,(2m_\pi)^{-1} \quad . \tag{4.109b}$$

We find then

$$d\Gamma = \frac{G^2}{64\pi^2} f_\pi^2 \cos^2\theta_c \frac{m_\ell^2}{m_\pi^3} (m_\pi^2 - m_\ell^2)^2 \, (1 + |\underline{p}|^{-1} \, \underline{p}.\underline{n}) \, d\Omega \, , \tag{4.110}$$

where $\underline{n}$ is a unit vector specifying the direction of the lepton's spin in its rest frame. $d\Omega$ is the element of solid angle around the lepton momentum $\underline{p}$. It follows that the lepton is 100% right-handed polarized, since

$$h \equiv \frac{d\Gamma(P) - d\Gamma(A)}{d\Gamma(P) + d\Gamma(A)} = 1 \, , \tag{4.111}$$

where, as before, $d\Gamma(P,A)$ are the differential transition rates with $\underline{p}$ and $\underline{n}$ parallel, anti-parallel. This is quite obvious physically; the pion has zero spin and thus the total angular momentum of the decay products must also be zero. In the rest frame of the pion ℓ^- and $\bar{\nu}_\ell$ have equal and oppositely directed momenta. Since the $\bar{\nu}_\ell$ is 100% right-handed polarised, it follows that ℓ^- must also have 100% right-handed polarization to ensure that the final state does indeed have zero angular momentum. But recall now the argument, given in the context of muon decay, that the form of the leptonic current L_α ensures that only left-handed ℓ^- (and right-handed ℓ^+) may be coupled. This argument, it will be recalled, is valid in so far as we may neglect the mass m_ℓ of the lepton. Thus angular momentum conservation requires ℓ^- to have $h = 1$, while the form of L_α requires ℓ^- to have $h = -1$, if $m_\ell = 0$. Plainly the only way these statements can be reconciled is if the transition rate $d\Gamma$ vanishes when $m_\ell = 0$. We note that the form (4.110) has just this property. Indeed the matrix element M also vanishes in this limit, since substituting (4.108a) for k in (4.106b) and using the Dirac equation for $\bar{u}_{(\ell)}$ and $v_{(\nu_\ell)}$ gives

$$M(\pi^-_{\ell 2}) = \frac{G}{\sqrt{2}}\, i f_\pi \cos\theta_c\, m_\ell\, \bar{u}_{(\ell)}(p,s)\,(1-\gamma_5)\, v_{(\nu_\ell)}(q,\tfrac{1}{2}) \; . \tag{4.112}$$

Thus M is proportional to m_ℓ and $d\Gamma$ to m_ℓ^2, as we found in (4.110). As before, we may find the total decay rate for this mode by adding the two forms for $\pm \underline{n}$, and doing the angular integration we find

$$\Gamma(\pi^-_{\ell 2}) = \frac{G^2}{8\pi}\, f_\pi^2\, \cos^2\theta_c\, \frac{m_\ell^2}{m_\pi^3}\, (m_\pi^2 - m_\ell^2)^2 \; . \tag{4.113}$$

This formula is valid for both of the leptonic modes, electronic and muonic, and the constant f_π is determined by the pion alone, as is apparent from its definition in (4.105). Thus the branching ratio of the two leptonic modes is independent of f_π

$$R_o \equiv \Gamma(\pi^-_{e2}) / \Gamma(\pi^-_{\mu 2}) = x\,(1-x)^2\, y^{-1}\, (1-y)^{-2} \; , \tag{4.114a}$$

where

$$x \equiv (m_e/m_\pi)^2 \; , \quad y \equiv (m_\mu/m_\pi)^2 \; . \tag{4.114b}$$

(The suffix '0' is to remind us that we have not yet included electro-magnetic corrections to the processes which arise because the charged particle participating in the processes also interact with each other electromagnetically.)

Thus using the experimentally known values of x and y, our theory predicts the value of the ratio R_o. Using the values[10]

$$m_e = 0.5110 \text{ MeV} \tag{4.115a}$$

$$m_\mu = 105.66 \text{ MeV} \tag{4.115b}$$

$$m_\pi = 139.57 \text{ MeV} \tag{4.115c}$$

gives

$$R_o = 1.278 \times 10^{-4} . \tag{4.116}$$

The smallness of this ratio arises from the proportionality of M to m_ℓ and the smallness of the ratio x. It is the more striking since on phase space considerations alone we should a priori have supposed that the electronic mode would proceed most rapidly. Experimentally[11] the ratio observed is

$$R_{exp} = (1.247 \pm 0.028) \times 10^{-4} . \tag{4.117}$$

Thus there is no doubt that in nature the muonic mode is overwhelmingly preponderant. We anticipate that the small difference between the prediction (4.116) and the experimental result (4.117) can be attributed to the electromagnetic corrections mentioned above. The calculation of these effects is non-trivial since the pion, unlike the lepton, is known not to have a point-like charge distribution; the strong interactions of the pion mean that it is surrounded by a "cloud" of virtual particles, which themselves interact with the electromagnetic field. We shall return subsequently to the problem of calculating electromagnetic corrections. For the present we simply quote the results[12] of Berman and Kinoshita. With the approximation that the pion is point-like they calculate the branching ratio R, including electromagnetic effects

$$R = R_o \left\{ 1 - \frac{\alpha}{2\pi} \left[\gamma\left(\frac{m_e}{m_\pi}\right) - \gamma\left(\frac{m_\mu}{m_\pi}\right) \right] \right\} , \tag{4.118a}$$

where

$$\gamma(x) = 4 \ln(1-x^2) - \frac{4(1+x^2)}{1-x^2}\left[L(1-x^2) - \ln x \ln(1-x^2)\right]$$

$$- (1-x^2)^{-2}(11x^4 - 20x^2 + 6)\ln x + \tfrac{3}{2}(1-x^2)^{-1} . \quad (4.118b)$$

Then substituting the values (4.115) yields the (corrected) prediction

$$R = 1.228 \times 10^{-4} . \quad (4.119)$$

Thus theory and experiment are consistent, and the excellent agreement lends strong support to the hypothesis contained in (2.264) that the electronic and muonic parts of L_λ have equal strength (μ - e universality).

As before, the decay rate for the decays of the π^+

$$\pi^+ \to \ell^+ \nu_\ell \quad (4.120)$$

may be deduced using CP-invariance. If $\mathcal{H}_W^{SL}$ is CP-invariant, then evidently we require that the Cabibbo current $\mathcal{J}_\lambda^\pm$ satisfies

$$(\mathcal{C}\mathcal{P})\, \mathcal{J}_\alpha^\pm (x^o, \underline{x})\, (\mathcal{C}\mathcal{P})^{-1} = - \mathcal{J}^{\mp\alpha}(x^o, -\underline{x}) , \quad (4.121)$$

since L_α, $L_\alpha^\dagger$ transform in just this way (see (3.47)). It is easy to see that, if $\mathcal{J}_\alpha^\pm$ are constructed from quark fields as in (4.65), then (4.121) is satisfied. The effect of $\mathcal{C}\mathcal{P}$ upon the lepton states is given in (3.50), and

$$\mathcal{C}\mathcal{P}\, |\pi^- (\underline{k})> = - |\pi^+(-\underline{k})> . \quad (4.122)$$

The minus sign arises since the pion is a pseudoscalar. Aside from this, the prescription for calculating the matrix element for (4.120) is precisely that used before - change the sign of all three momenta $\underline{k}$, $\underline{p}$, $\underline{q}$, and leave $\underline{s}$ alone. Thus the transition rate (4.113) is also that for the decay (4.120). So the CP-invariance of our theory predicts

$$\Gamma(\pi^-_{\ell 2}) = \Gamma(\pi^+_{\ell 2}) . \quad (4.123)$$

That is to say the same constant f_π appears in both decay rates. If we take the experimental value for $\Gamma(\pi_{\mu 2})$, we may deduce the value of the unknown quantity $f_\pi \cos\theta_c$. Dimensional considerations show that it has dimension [M] and we find

$$\cos\theta_c \, |f_\pi| = (0.9198 \pm 0.0003) \; m_{\pi^+} \; . \tag{4.124}$$

4.2.2. π_{e3}

The only other weak decay mode of the pion is its beta decay

$$\pi^- \rightarrow \pi^o e^- \bar{\nu}_\ell \; . \tag{4.125}$$

The corresponding muonic decay mode is not allowed kinematically. The derivation of the S-matrix element proceeds quite similarly to the derivation just given. The hadronic part of the matrix element is

$$< \pi^o | \mathcal{J}_\lambda^+ | \pi^- > \; = \; \cos\theta_c \; < \pi^o | \, J_\lambda^+ \, | \pi^- > \; . \tag{4.126}$$

In this case, however, only the vector part of J_λ^+ can contribute. Thus

$$< \pi^o | J_\lambda^+ | \pi^- > \; = \; < \pi^o | V_\lambda^1 + i V_\lambda^2 | \pi^- > \; . \tag{4.127}$$

The general form of this matrix element is determined as before using Lorentz covariance. In this case, since there are two (spinless) pions involved, the form of the matrix element is a linear combination of their momenta. Thus denoting the momenta of π^- and π^o by K and k respectively, we have

$$< \pi^o(k) | V_\lambda^1 + i V_\lambda^2 | \pi^-(K) > \; = \; g_+(q^2) \, (K+k)_\lambda + g_-(q^2) \, q_\lambda \; , \tag{4.128a}$$

where

$$q = K - k \; . \tag{4.128b}$$

$g_\pm(q^2)$ are arbitrary scalar functions of the Lorentz invariant quantity q^2. We may learn something of these form factors by using CVC, which is contained in the Cabibbo Hypothesis. The charge-current commutators (4.70) imply

$$[\, j_\lambda, \, v^1(0) + i v^2(0) \,] \; = \; V_\lambda^1 + i V_\lambda^2 \; , \tag{4.129}$$

where j_λ is the electromagnetic current given in (4.54). In the approximation that isospin symmetry is exact, $v^i(t)$ are independent of t for i = 1,2,3, and $v^1 + i v^2$ is the isospin 'raising' operator. Hence, using (4.26) and (4.27),

$$(V^1 + iV^2)\,|\,\pi^-(K)\rangle = \sqrt{2}\,|\,\pi^o(K)\rangle \tag{4.130a}$$

$$\langle\pi^o(k)\,|\,(V^1 + i\,V^2) = \sqrt{2}\,\langle\pi^-(k)\,| \;. \tag{4.130b}$$

It follows then from (4.129) that

$$\langle\pi^o(k)|\,V^1_\lambda + i\,V^2_\lambda\,|\,\pi^-(K)\rangle = \sqrt{2}\,\langle\pi^o(k)|j_\lambda|\,\pi^o(K)\rangle$$

$$-\sqrt{2}\,\langle\pi^-(k)\,|\,j_\lambda|\,\pi^-(K)\rangle + O(\alpha)\;, \tag{4.131}$$

where the $O(\alpha)$ term represents terms arising from the fact that isospin is not an exact symmetry; it is violated by electromagnetic interactions which presumably are of order α, the fine structure constant. The quantities on the right of (4.131) are determined by the electromagnetic properties of the pion. The standard notation is

$$\langle\pi^-(k)|j_\lambda|\,\pi^-(K)\rangle = -F_\pi(q^2)\,(k + K)_\lambda \;, \tag{4.132}$$

where $F_\pi(q^2)$ is the electromagnetic form factor of the pion whose Fourier transform determines the spatial charge distribution of the pion. (There is no term proportional to q_λ on the right of (4.132), since the electromagnetic current j_λ is conserved.) The corresponding matrix element for the π^o is identically zero, as may be seen directly from C-invariance.

$$\mathcal{C}\, j_\lambda\, \mathcal{C}^{-1} = -\,j_\lambda \tag{4.133a}$$

and

$$\mathcal{C}\,|\,\pi^o\rangle = |\,\pi^o\rangle \;, \tag{4.133b}$$

so

$$\langle\pi^o(k)\,|\,j_\lambda|\,\pi^o(K)\rangle = 0 \;. \tag{4.133c}$$

Substituting (4.132) and (4.133) into (4.131) and comparing with (4.128) we find

$$g_+(q^2) = \sqrt{2}\,F_\pi(q^2) + O(\alpha) \tag{4.134a}$$

$$g_-(q^2) = O(\alpha) \;. \tag{4.134b}$$

Thus, provided we are able to calculate the electromagnetic corrections $O(\alpha)$, the matrix element for π_{e3} is completely determined by the (independent) knowledge of $F_\pi(q^2)$. Experimentally[13] it is found that the q^2-dependence of F_π is determined by the ρ-meson pole, at least for $q^2 > 0$,

$$F_\pi(q^2) \sim (q^2 - m_\rho^2)^{-1} . \tag{4.135}$$

Now in π_{e3}

$$q^2 \leq \Delta^2 , \tag{4.136a}$$

where

$$\Delta \equiv m_{\pi^+} - m_{\pi^0} = 4.6 \text{ MeV} . \tag{4.136b}$$

Thus since

$$m_\rho = 770 \text{ MeV} \tag{4.137a}$$

and

$$\Delta^2 m_\rho^{-2} = 3.6 \times 10^{-5} , \tag{4.137b}$$

we may to excellent approximation ignore the q^2-dependence of the form factors in π_{e3}. Hence

$$g_+(q^2) \simeq g_+(0) = \sqrt{2} + O(\alpha) , \tag{4.138}$$

and the matrix element (4.128a) is entirely known, apart from the electromagnetic effects $O(\alpha)$. The decay rate may thus be calculated as before. We find (ignoring $O(\alpha)$)

$$\Gamma_0(\pi_{e3}) = \frac{G^2 \cos^2\theta_c}{30\pi^3} \Delta^5 \{1 - (3\Delta)(2m_\pi)^{-1} - 5m_e^2 \Delta^{-2} \cdots \} . \tag{4.139}$$

Before comparing this with the experimental data we must evaluate the $O(\alpha)$ terms. These are calculated, as before, under the assumption that the pion has a point-like electromagnetic interaction. They turn out to depend upon an ultraviolet cut-off Λ, although the dependence is only logarithmic and therefore not sensitive to the value chosen for the cut-off. With the value $\Lambda = m_n$ Chang[14] finds that the electromagnetic corrections amount to about 1%. The relatively large experimental errors in the decay rate do not make this an ideal testing ground for the theory, so that nothing very much is lost by taking this particular value of Λ. By comparing the theoretical prediction with the experimental result[15]

$$\Gamma(\pi_{e3})\Big|_{exp} = 0.38^{+0.03}_{-0.04} , \tag{4.140}$$

we may deduce a value for $\cos\theta_c$, since G is given in (3.62). We find

$$\cos\theta_c = 0.97 \pm 0.05 . \tag{4.141}$$

The large experimental errors make this process less than ideal for determining θ_c, though we shall see that it is at least consistent with the value derived from nuclear beta decay, which we shall deal with shortly.

4.2.3 NEUTRON DECAY

The beta decay of the neutron,

$$n \to p e^- \bar{\nu}_e \tag{4.142}$$

can occur either for a free neutron or for one bound in a nucleus. We consider first the decay of the free neutron. The S-matrix element is completely determined once we know the hadronic part

$$< p | \mathcal{J}^+_\lambda | n > = \cos\theta_c < p | J^+_\lambda | n > . \tag{4.143}$$

Unlike the previous processes considered, <u>both</u> the vector and the axial-vector parts of J^+_λ can contribute to the matrix element. In this case Lorentz covariance considerations show that the above matrix element may be expressed in terms of six complex form factors

$$\begin{aligned} <p| J^+_\lambda | n> = \Big[\bar{p} \Big\{ & g^V_1(q^2)\,\gamma_\lambda + g^V_2(q^2)\, i\sigma_{\lambda\mu} q^\mu + g^V_3(q^2)\, q_\lambda \\ & + g^A_1(q^2)\,\gamma_\lambda\gamma_5 + g^A_2(q^2)\, q_\lambda\gamma_5 + g^A_3(q^2)\, i\sigma_{\lambda\mu} q^\mu \Big\} n \Big], \end{aligned} \tag{4.144a}$$

where

$$q = n - p , \quad \sigma_{\lambda\mu} = \frac{i}{2}[\gamma_\lambda, \gamma_\mu] . \tag{4.144b}$$

Here, and elsewhere, we simplify our notation by using the same symbol to denote a particle, its momentum, and where appropriate, its spinor. It is easy to see that any other covariants constructed using the γ-matrices and the momenta n, p may be reduced to linear combinations of those given above. The form factors g^V_i, g^A_i $(i = 1,2,3)$ are complex scalar functions arising respectively from the vector and axial vector parts of J^+_λ. As before, we may use CVC to relate the vector form factors to the electromagnetic form factors of the proton and neutron. These are defined as follows

$$<p'|j_\lambda|p> \equiv [\bar{p}'\{F_1^p(q^2)\gamma_\lambda + F_2^p(q^2)\, i\sigma_{\lambda\mu} q^\mu\} p] \ , \tag{4.145a}$$

where

$$q = p' - p \ . \tag{4.145b}$$

This general form for the matrix element of the electromagnetic current j_λ follows from the fact that j_λ is a vector current and that it is conserved ($\partial^\lambda j_\lambda = 0$). Thus the analogue of g_3^V in (4.144) is not present. By considering the non-relativistic limit of (4.145) we obtain the physical interpretation of $F_1^p(q^2)$ as the charge form factor of the proton whose Fourier transform describes the spatial distribution of the proton's charge. Likewise, $F_2^p(q^2)$ is the magnetic form factor describing the distribution of the proton's anomalous magnetic moment. A precisely similar form defines the charge and magnetic moment form factors of the neutron, $F_1^n(q^2)$, $F_2^n(q^2)$. Using CVC as before we find

$$g_1^V(q^2) = F_1^p(q^2) - F_1^n(q^2) + O(\alpha) \tag{4.146a}$$

$$-g_2^V(q^2) = F_2^p(q^2) - F_2^n(q^2) + O(\alpha) \tag{4.146b}$$

$$g_3^V(q^2) = O(\alpha) \ . \tag{4.146c}$$

It is possible to measure the q^2-dependence of the form factors by doing high-energy neutrino scattering experiments

$$\nu_\ell n \to \ell^- p \ . \tag{4.147}$$

Then, with independent knowledge of the electromagnetic form factors, we may check directly the validity of the CVC predictions (4.146). However in neutron decay, and nuclear beta decay in general, the recoil momentum q is small compared with the nucleon's mass. Thus as a first approximation we may neglect the q-dependence in (4.144). The matrix element (4.144) is therefore determined by the two parameters $g_1^V(0)$, $g_1^A(0)$. We shall see now how the theory we have described predicts the magnitude of both of these parameters.

Firstly, from the CVC predictions (4.146a) we have

$$g_1^V(0) = F_1^p(0) - F_1^n(0) + O(\alpha) = 1 + O(\alpha) \;, \tag{4.148}$$

since $F_1^{p,n}(0)$ are just the charges of the nucleons.

PCAC and the Goldberger Treiman Formula

The evaluation of $g_1^A(0)$ is much more involved. Consider the axial part of (4.144):

$$\langle p| A_\lambda^1 + i A_\lambda^2 |n\rangle = [\bar{p} \; \{ g_1^A(q^2)\gamma_\lambda \gamma_5 + g_2^A(q^2) q_\lambda \gamma_5 + g_3^A(q^2) i\sigma_{\lambda\mu} q^\mu \gamma_5 \} n] \;. \tag{4.149}$$

Translation invariance implies that

$$[\mathbb{P}^\lambda, A_\lambda^1 + i A_\lambda^2] = -i\partial^\lambda (A_\lambda^1 + i A_\lambda^2) \quad , \tag{4.150}$$

so that, using the Dirac equation, we have

$$\langle p|\partial^\lambda (A_\lambda^1 + i A_\lambda^2)|n\rangle = [(m_n + m_p) g_1^A(q^2) - q^2 g_2^A(q^2)](\bar{p} i\gamma_5 n) \;. \tag{4.151}$$

The divergence of the axial vector current, which appears on the left-hand side of (4.151), has the same quantum numbers as the pion's field operator. Thus it will certainly couple to a pion which in turn has a non-zero interaction with the nucleon states. It follows that the matrix element on the left of (4.151) is non-zero, so the axial vector current is certainly <u>not</u> conserved. We make the approximation that the dominant contribution to the required matrix element is that arising from the <u>single</u> pion state, as opposed to the three-pion state and any other states which may couple to the divergence. This approximation is known as the Partially Conserved Axial Vector Current hypothesis (PCAC) and was proposed by Gell-Mann and Levy [(4)]. It may be expressed mathematically by saying that the divergence is proportional to the pion's field operator $\varphi_{(\pi^-)}$, which destroys a single π^- state

$$\partial^\lambda (A_\lambda^1 + i A_\lambda^2) = c\,\varphi_{(\pi^-)} \;. \tag{4.152}$$

The constant of proportionality, c, is determined by the strength of the coupling of the pion to the axial vector current. Using the notation used previously in (4.105):

$$<0|\, A^1_\lambda + i\, A^2_\lambda \,|\, \pi^- > = i f_\pi\, \pi_\lambda \quad , \tag{4.153}$$

so that with (4.150), we have

$$<0|\, \partial^\lambda (A^1_\lambda + i\, A^2_\lambda)\,|\, \pi^- > = f_\pi\, m^2_\pi \quad . \tag{4.154}$$

On the other hand, by definition ,

$$<0| \varphi_{(\pi^-)} |\pi^- > \; = \; 1 \; . \tag{4.155}$$

Comparing (4.154) and (4.155) with our hypothesis (4.152) we deduce the value of the constant of proportionality

$$c \; = \; f_\pi \;\; m^2_\pi \quad . \tag{4.156}$$

To the extent that isospin is an exact symmetry we may write

$$D^i(x) \equiv \partial^\lambda A^i_\lambda (x) = \frac{f_\pi}{\sqrt{2}}\, m^2_\pi\, \pi^i(x) \qquad (i = 1, 2, 3) \tag{4.157a}$$

where

$$\varphi_{(\pi^-)} = \frac{1}{\sqrt{2}}\, (\pi^1 + i\, \pi^2) \quad . \tag{4.157b}$$

Returning to the problem in hand, we wish to use PCAC to evaluate the left-hand side of (4.151), so we need to know the matrix element of $\varphi_{(\pi^-)}$ between the nucleon states. Translation invariance gives

$$< N' |\, \pi^i \,|\, N > \; = \; (m^2_\pi - q^2)^{-1} < N' |\, j^i_\pi \,|\, N > \quad , \tag{4.158a}$$

where

$$j^i_\pi (x) = (\,\Box_x + m^2_\pi\,)\, \pi^i(x) \qquad (i = 1, 2, 3) \tag{4.158b}$$

is the source of the pion's field. The matrix element of j^i_π defines the pion-nucleon coupling constant

$$<N' |\, j^i_\pi \,|\, N > \; = \; g_{\pi NN}\, K(q^2)\, \bar{N}' \tau^i\, i\gamma_5\, N \quad , \tag{4.159}$$

where τ^i are the 2 x 2 Pauli matrices and $K(m^2_\pi) = 1$. Hence

$$<p| \varphi_{(\pi^-)} |n> = \left(m^2_\pi - q^2\right)^{-1} g_{\pi NN}\, K(q^2) \sqrt{2}\; (\bar{p}\; i\, \gamma_5\; n) \tag{4.160}$$

Thus, using (4.152) and (4.156) together with (4.151) and (4.160), we deduce the formulae of Goldberger and Treiman(16)

$$\tfrac{1}{2}[(m_n+m_p)g_1^A(q^2) - q^2 g_2^A(q^2)] = \frac{f_\pi}{\sqrt{2}} \frac{m_\pi^2}{m_\pi^2 - q^2} g_{\pi NN} K(q^2) \qquad (4.161a)$$

$$m_N g_1^A(0) = \frac{f_\pi}{\sqrt{2}} g_{\pi NN} K(0) . \qquad (4.161b)$$

The latter is just the $q^2 = 0$ value of the first equation. To test these consequences of PCAC we need further information on the behaviour of the function $K(q^2)$. We assume the function is slowly varying, as is plausible if its variation arises from three pion and higher mass states with the quantum numbers of the pion. Thus with the approximation

$$K(0) \simeq K(m_\pi^2) \equiv 1 , \qquad (4.162)$$

(4.161b) gives

$$m_N g_1^A(0) = \frac{f_\pi}{\sqrt{2}} g_{\pi NN} . \qquad (4.163)$$

We shall see later that this relation, in which all quantities appearing are measurable and known, is satisfied to an accuracy of about 8%. This is about the accuracy to be expected in making the approximation (4.162) (since $m_\pi^2 (3\, m_\pi)^{-2} \sim 10\%$), and so we conclude that PCAC makes predictions which are consistent with experiment.

Adler-Weisberger Sum Rule

However, (4.163) only predicts the ratio of the two weak coupling parameters f_π and $g_1^A(0)$ in terms of the pion-nucleon coupling constant. It does not determine the overall scale of these quantities individually in terms of strong interaction quantities. Indeed it is obvious that this must be the case, since PCAC does not fix the scale of the axial vector current. Its scale is fixed by the commutation relation (4.63), and this fact was utilized by Adler and Weisberger[(17)] to determine $g_1^A(0)$ purely in terms of strong interaction quantities. We prefer to follow the derivation of Fubini et al[(18)]. The technique has been applied to many other parts of the subject, so we shall follow the derivation in detail. The first part consists in deriving a "low energy theorem". Consider first the quantity $T_{\mu\nu}^{ij}$.

$$T^{ij}_{\mu\nu} \equiv \int dx\, e^{iqx} < N | \, T\{A^i_\mu(x), A^j_\nu\} | N> \quad (i,j = 1,2,3) \;, \tag{4.164}$$

where $|N>$ is a single nucleon state of momentum N. $T\{\,\}$ denotes the time ordered product of the operators, so

$$T\{A^i_\mu(x), A^j_\nu\} \equiv \theta(x_o)\, A^i_\mu(x)\, A^j_\nu + \theta(-x_o)\, A^j_\nu\, A^i_\mu(x) \;. \tag{4.165}$$

Using Gauss' theorem to drop the total divergence we deduce

$$-\, iq^\mu\, T^{ij}_{\mu\nu} = -\int dx\, (\partial^\mu e^{iqx}) < N | \, T\{A^i_\mu(x)\, A^j_\nu\} | N> \tag{4.166a}$$

$$= \int dx\, e^{iqx} <N| \{ T\{D^i(x)\, A^j_\nu\} + \delta(x_o) [\, A^i_o(x), A^j_\nu\,] \} |N> \;. \tag{4.166b}$$

The equal time commutator arises from the differentiation ∂_o and the identity

$$\partial_o\, \theta(x_o) = \delta(x_o) \;. \tag{4.167}$$

We assume the equal time commutation relation (4. 69a)

$$\delta(x_o) [\, A^i_o(x), A^j_\nu\,] = i\,\epsilon^{ijk}\, V^k_\nu\, \delta(x) \;, \tag{4.168}$$

and, using translation invariance to rewrite the first term, we find

$$q^\mu\, T^{ij}_{\mu\nu}\, q^\nu = i\, q^\nu \int dx\, e^{-iqx} < N | \, T\{ D^i A_\nu(x) \} | N> - \epsilon^{ijk} q^\nu <N|V^k_\nu|N> \;. \tag{4.169}$$

Now we use Gauss' theorem for the first term and obtain

$$q^\mu\, T^{ij}_{\mu\nu}\, q^\nu = D^{ij} - < N | C^{ij} | N> - \epsilon^{ijk} q^\nu < N | V^k_\nu | N> \;, \tag{4.170a}$$

where

$$D^{ij} \equiv \int dx\, e^{iqx} < N | T\{ D^i(x)\, D^j \} | N> \;, \tag{4.170b}$$

and we assume an equal time commutation relation of the form

$$\delta(x_o)\, [\, D^i(0), A^j_o(x)\,] = C^{ij}\, \delta(x) \;. \tag{4.170c}$$

The basic assumption contained in both (4.168) and (4.170c) is the absence of Schwinger terms[8] on the right-hand sides. Feynman[19] has conjectured that any of these non-covariant derivatives of δ-functions will cancel, when we use Gauss' theorem, with the "seagull" terms necessary to make $T^{ij}_{\mu\nu}$ covariant. We therefore follow the, by now, traditional course of ignoring such complications. The interested reader is referred to Adler and Dashen[20], and references therein, for further discussion of these problems. It is easy to see, by reversing the order of application of Gauss' theorem, that

$$C^{ij} = C^{ji} . \tag{4.171}$$

Using translation invariance we may establish the "crossing" behaviour of D^{ij}

$$D^{ij}(q) = D^{ji}(-q) . \tag{4.172}$$

We may write D^{ij} in terms of the Lorentz invariants q^2 and ν, where

$$\nu = N.q/m_N . \tag{4.173}$$

In terms of these variables (4.172) requires that

$$D^{ij}(\nu, q^2) = D^{ji}(-\nu, q^2) . \tag{4.174}$$

Thus, without loss of generality, we may write

$$D^{ij} = i\,\overline{N}\left[d_1(\nu^2, q^2)\left\{\frac{\tau^i}{2}, \frac{\tau^j}{2}\right\} + \nu d_2(\nu^2, q^2)\left[\frac{\tau^i}{2}, \frac{\tau^j}{2}\right]\right] N . \tag{4.175}$$

τ^i $(i = 1,2,3)$ are the 2 x 2 Pauli matrices which operate in isospin space upon the nucleon isospinors N, $\overline{N}$. Suppose now we consider the limit $q \to 0$. Expanding D^{ij} in powers of ν and q^2 gives

$$D^{ij} = i\,\overline{N}\left[d_1(0,0)\left\{\frac{\tau^i}{2}, \frac{\tau^j}{2}\right\} + \nu\, d_2(0,0)\left[\frac{\tau^i}{2}, \frac{\tau^j}{2}\right]\right] N + O(\nu^2, q^2) . \tag{4.176}$$

We now make the same expansion for all terms appearing in (4.170a). We write

$$T^{ij}_{\mu\nu} = U^{ij}_{\mu\nu} + E^{ij}_{\mu\nu} , \tag{4.177}$$

where $U^{ij}_{\mu\nu}$ is the contribution from the single nucleon intermediate state and $E^{ij}_{\mu\nu}$ is the contribution from all other (excited) states. This latter contribution is finite as $q \to 0$, so that

$$q^{\mu} E^{ij}_{\mu\nu} q^{\nu} = O(\nu^2, q^2) . \tag{4.178}$$

Thus, to the approximation we are working, the entire contribution to $T^{ij}_{\mu\nu}$ arises from $U^{ij}_{\mu\nu}$, which is completely determined by the matrix elements of A^i_μ and A^j_ν between single nucleon states. Evidently $U^{ij}_{\mu\nu}$ is just the sum of the Feynman Diagrams shown below.

Fig. 4.1 Single nucleon pole terms contributing to $U^{ij}_{\mu\nu}$.

Thus,

$$U^{ij}_{\mu\nu} = \bar{N}\left[\Gamma^{A_i}_{\mu}(q)\frac{i}{\not{N}+\not{q}-m_N}\Gamma^{A_j}_{\nu}(-q) + \Gamma^{A_j}_{\nu}(-q)\frac{i}{\not{N}-\not{q}-m_N}\Gamma^{A_i}_{\mu}(q)\right]N, \tag{4.179a}$$

where

$$\Gamma^{A_i}_{\mu}(q) \equiv [\, g^A_1(q^2)\gamma_\mu\gamma_5 + g^A_2(q^2)q_\mu\gamma_5 + g^A_3(q^2)\, i\sigma_{\mu\rho}q^\rho\gamma_5\,]\,\frac{\tau^i}{2} \tag{4.179b}$$

specify the matrix elements of A^i_μ between single nucleon states. The above form for the axial current matrix element is just that in (4.149) generalized for all three components of the isovector A^i_μ. We are only concerned with identifying the leading contributions to $q^\mu U^{ij}_{\mu\nu} q^\nu$ and for these purposes we need only the contributions of the Γ_μ when $q = 0$. Thus, since

$$(\not{N} \pm \not{q} - m_N)^{-1} = (\not{N} \pm \not{q} + m_N)\,(q^2 \pm 2\, m_N \nu)^{-1}, \tag{4.180a}$$

$$q^\mu U^{ij}_{\mu\nu} q^\nu = i\,|g^A_1(0)|^2\, \bar{N}\not{q}\gamma_5\,(\not{N} + m_N)\not{q}\gamma_5 \times \{\tfrac{1}{4}\tau^i\tau^j(q^2 + 2\, m_N\nu)^{-1} + \tfrac{1}{4}\tau^j\tau^i\,(q^2 - 2\, m_N\nu)^{-1}\}\,N\,. \tag{4.180b}$$

In the limit that $q^2 \to 0$ followed by $\nu \to 0$,

$$q^\mu U^{ij}_{\mu\nu} q^\nu = i\,|g^A_1(0)|^2\, \bar{N}\nu\left[\frac{\tau^i}{2}, \frac{\tau^j}{2}\right] N + O(\nu^2, q^2)\,. \tag{4.181}$$

Also, in the same way

$$\langle N|\, V^k_\nu\, |N\rangle = \bar{N}\gamma_\nu\, \tfrac{1}{2}\tau^k\, N\,. \tag{4.182}$$

Thus

$$- \epsilon^{ijk} q^{\nu} < N | V_{\nu}^{k} | N > = i \bar{N} \nu \left[\frac{\tau^i}{2}, \frac{\tau^j}{2} \right] N \quad . \tag{4.183}$$

Finally, since C^{ij} is symmetric in i and j, as seen in (4.171), it makes no contribution to the antisymmetric part. So, from (4.170a), equating coefficients for the antisymmetric part, we deduce

$$| g_1^A (0) |^2 = d_2 (0,0) + 1 \quad . \tag{4.184}$$

This is the "low energy theorem" for the quantity D^{ij}. In principle this can be tested experimentally, since D^{ij} is determined by the matrix elements of D^i and these are measurable quantities. We shall return shortly to the way in which D^{ij} is measured directly, see 4.2.11 . For the present we observe that if we now use PCAC, as formulated in (4.157a), we are able to relate D^{ij} to the forward pion-nucleon scattering amplitude. To see this we substitute for D^i and D^j in (4.170b) using (4.157a). This gives

$$\begin{aligned} D^{ij} &= \tfrac{1}{2} f_\pi^2 m_\pi^4 \int dx \, e^{iqx} < N | T \{ \pi^i(x) \pi^j \} | N > \\ &= \tfrac{1}{2} f_\pi^2 m_\pi^4 (m_\pi^2 - q^2)^{-2} \int dx \, e^{iqx} K_x K_y < N | T \{ \pi^i(x) \pi^j(y) \} | N > |_{y=0} \, , \end{aligned} \tag{4.185a}$$

where

$$K_x \equiv \quad {}_x + m_\pi^2 \quad . \tag{4.185b}$$

Using the LSZ reduction formalism[(21)] it is easy to see that the integral appearing in (4.185a) is essentially the forward scattering amplitude for pion-nucleon scattering. Consider for example the process

$$\pi^+ (q) \, p \rightarrow \pi^+ (q) \, p \quad . \tag{4.186}$$

The invariant scattering amplitude is given by

$$M(\pi^+ p \rightarrow \pi^+ p) = i \int dx \, e^{iqx} K_x K_y < p | T \{ \varphi^*_{(\pi^-)}(x) \varphi_{(\pi^-)}(y) \} | p > |_{y=0} \, , \tag{4.187}$$

where $\varphi_{(\pi^-)}$ is given by (4.157b). Thus it follows, using (4.185a) and (4.175), that

$$\tfrac{1}{2} f_\pi^2 m_\pi^4 (m_\pi^2 - q^2)^{-2} M(\pi^+ p \rightarrow \pi^+ p) = m_N [d_1 (\nu^2, q^2) + \nu \, d_2 (\nu^2, q^2)] \, . \tag{4.188}$$

We may treat $\pi^- p$ scattering in the same way and deduce

$$\tfrac{1}{2} f_\pi^2 m_\pi^4 (m_\pi^2 - q^2)^{-2} [M(\pi^+ p \to \pi^+ p) - M(\pi^- p \to \pi^- p)] = 2m_N \nu d_2(\nu^2, q^2) . \tag{4.189}$$

This is the connection between D^{ij} and the forward pion-nucleon scattering amplitudes. The low energy theorem (4.184) thus predicts the value of the left-hand side of (4.189) in the limit when ν^2 and q^2 are zero. Since q is the momentum of the pion, we see that the low energy theorem is a "soft pion" theorem; that is to say, one concerning "pions" of zero mass. Such results are typically obtained by the combined use of current algebra and PCAC. For the most part, they are tested by making the further assumption that the extrapolation from $q^2 = 0$ to $q^2 = m_\pi^2$ does not have much effect. With this assumption the low energy theorems then make statements about processes involving physical pions. However, for the case we are considering Adler[17] has attempted to include off-mass shell effects.

The low energy theorem is first converted to a sum rule by assuming that $d_2(\nu^2, q^2)$ satisfies an unsubtracted dispersion relation in the variable ν^2 for fixed q^2. That is to say

$$d_2(\nu^2, q^2) = \frac{1}{\pi} \int_0^\infty d\nu'^2 (\nu'^2 - \nu^2 - i\epsilon)^{-1} \, \mathrm{Im}\, d_2(\nu'^2, q^2) . \tag{4.190}$$

The content of this assumption is first that $d_2(\nu^2, q^2)$ is an analytic function in the complex ν^2-plane except for possible singularities on the positive real axis, and secondly that $\mathrm{Im}\, d_2(\nu^2, q^2)$ falls off sufficiently fast for large real ν^2 for the integral in (4.190) to converge. Since d_2 is related by (4.189) to a difference of two pion-nucleon invariant scattering amplitudes, the above statement leads to a similar one for these scattering amplitudes which are defined in (4.187). The required analyticity properties of these amplitudes has been proved for physical pions by Goldberger[22]. We shall return shortly to the crucial question of whether the integral actually converges.

Let us, for simplicity, work in the rest frame of the nucleon. Then using translation invariance we may rewrite (4.187) as

$$M(\pi^+ p \to \pi^+ p) = i \int dx\, e^{iqx} K_x K_y \langle p | T\{ \varphi^*_{(\pi^-)}(\tfrac{x}{2}) \varphi_{(\pi^-)}(-\tfrac{x}{2}) \} | p \rangle \ . \quad (4.191)$$

Then

$$M^*(\pi^+ p \to \pi^+ p) = -i \int dx\, e^{-iqx} K_x K_y \langle p | \tilde{T}\{ \varphi^*_{(\pi^-)}(-\tfrac{x}{2}) \varphi_{(\pi^-)}(\tfrac{x}{2}) \} | p \rangle \ , \quad (4.192)$$

where $\tilde{T}\{\ \}$ is the anti-time ordered product. Replacing $x \to -x$

$$M^*(\pi^+ p \to \pi^+ p) = -i \int dx\, e^{iqx} K_x K_y \langle p | \tilde{T}\{ \varphi^*_{(\pi^-)}(\tfrac{x}{2}) \varphi_{(\pi^-)}(-\tfrac{x}{2}) \} | p \rangle \ . \quad (4.193)$$

Thus

$$\mathrm{Im}\, M(\pi^+ p \to \pi^+ p) = \tfrac{1}{2} \int dx\, e^{iqx} \langle p | \{ J^*_{(\pi^-)}(\tfrac{x}{2}), J_{(\pi^-)}(-\tfrac{x}{2}) \} | p \rangle \ , \quad (4.194a)$$

where

$$K_x \varphi_{(\pi^-)}(x) \equiv J_{(\pi^-)}(x) = \frac{1}{\sqrt{2}} (j^1_\pi(x) + i j^2_\pi(x)) \quad (4.194b)$$

and the j^i_π are defined in (4.158b). Inserting a complete set of intermediate states between the operators in (4.194) gives

$$\mathrm{Im}\, M(\pi^+ p \to \pi^+ p) = \tfrac{1}{2} \sum_{|r\rangle} \int dx \Big[e^{i(q+p-r)x} |\langle r | J_{(\pi^-)} | p \rangle|^2 + e^{i(q+r-p)x} |\langle r | J^*_{(\pi^-)} | p \rangle|^2 \Big] \ , \quad (4.195)$$

where we have used translation invariance to separate out the explicit dependence upon x. Performing the x-integration then gives

$$\mathrm{Im}\, M(\pi^+ p \to \pi^+ p) = \tfrac{1}{2} (2\pi)^4 \sum_{|r\rangle} \Big[\delta(q+p-r) |\langle r | J_{(\pi^-)} | p \rangle|^2 + \delta(q+r-p) |\langle r | J^*_{(\pi^-)} | p \rangle|^2 \Big] \ . \quad (4.196)$$

Consider first the contribution to Im M arising from the single nucleon intermediate state (the Born term). By inspection we see that only the second term of (4.196) receives such a contribution, in this case from the single neutron state n. For this term

$$\sum_{|r\rangle} = \sum_{\pm s} \int d^3n\, (2\pi)^{-3} (2n_0)^{-1} = \sum_{\pm s} \int d^4n\, \delta(n^2 - m_n^2)\, (2\pi)^{-3} \ . \quad (4.197)$$

Performing the integration over n, its contribution to Im M is

$$\text{Im } M^{(B)}(\pi^+ p) = \pi \sum_{\pm s} \delta(q^2 - 2m_p \nu) \, |\langle p-q, s \,|\, J^*_{(\pi^-)} \,|\, p\rangle|^2 , \tag{4.198}$$

ignoring the neutron-proton mass difference which is of electromagnetic origin. Now from (4.159)

$$\langle p-q, s \,|\, J^*_{(\pi^-)} \,|\, p\rangle = g_{\pi NN} K(q^2) \bar{u}(p-q, s) \, i\gamma_5 \sqrt{2} \, u(p) \quad . \tag{4.199}$$

Thus

$$\sum_{\pm s} |\langle p-q, s \,|\, J^*_{(\pi^-)} \,|\, p\rangle|^2 = g^2_{\pi NN} K^2(q^2)(-4m_n \nu) \quad , \tag{4.200a}$$

where we have used the identity

$$\bar{u}(p) \, i\gamma_5 \sqrt{2} \, (\not{p} - \not{q} + m_n) \, i\gamma_5 \sqrt{2} \, u(p) = -\, 4m_n \nu \quad . \tag{4.200b}$$

In the same way we may calculate the Born term for $\pi^- p$ scattering and find eventually that

$$\text{Im } M^{(B)}(\pi^- p) - \text{Im } M^{(B)}(\pi^+ p)$$

$$= 4m_n \pi g^2_{\pi NN} K^2(q^2) \, \nu \left\{ \delta(q^2 + 2m_n \nu) + \delta(q^2 - 2m_n \nu) \right\} . \tag{4.201}$$

Then from (4.189) we find that the Born contribution to $\text{Im } d_2$ is given by

$$\text{Im } d_2^{(B)}(\nu^2, q^2) = -\tfrac{1}{2} f_\pi^2 m_\pi^4 (m_\pi^2 - q^2)^{-2} \, 2\pi g^2_{\pi NN} K^2(q^2) \frac{q^2}{2m_n^2} \, \delta\left(\frac{(q^2)^2}{4m_n^2} - \nu^2 \right) . \tag{4.202}$$

The remaining contributions to Im M arise from $\pi N'$ and higher mass states (the continuum contribution). For these states

$$r^2 \geq (m_N + m_\pi)^2 . \tag{4.203}$$

Now the <u>first</u> term of (4.196) plainly receives contributions from states with

$$r^2 = (p + q)^2 = m_p^2 + q^2 + 2m_p \nu \quad . \tag{4.204}$$

Thus these continuum contributions are non zero only when

$$\nu \geq \nu_t(q^2) \equiv m_\pi + (m_\pi^2 - q^2)(2m_N)^{-1} . \tag{4.205}$$

Further,

$$<r|\, J_{(\pi^-)} |p> = i <r| \pi^+(q) p> \quad . \tag{4.206}$$

By definition, the total cross section for $\pi^+ p$ scattering is given by

$$4m_p \, \underline{|q|} \, \sigma(\pi^+ p) = \sum_{|r>} \, |<r| \pi^+(q) p>|^2 \, (2\pi)^4 \, \delta(p+q-r) \, . \tag{4.207}$$

Thus the continuum contribution is related to the total $\pi^+ p$ cross section. This result is just the famous Optical Theorem which follows from the unitarity of the S-matrix. Treating the $\pi^- p$ term in the same way we may deduce that <u>in the region $\nu \geq \nu_t$</u> the continuum contribution to Im M is given by

$$\mathrm{Im}\, M^{(C)}(\pi^+ p) - \mathrm{Im}\, M^{(C)}(\pi^- p) = 2m_n \, (\nu^2 - q^2)^{\frac{1}{2}} [\sigma(\pi^+ p) - \sigma(\pi^- p)] \, , \tag{4.208}$$

since the second term of (4.196) contributes only in the region $\nu \leq -\nu_t$. However the symmetry properties of this combination, which we have already derived, enable us to deduce Im d_2 directly

$$\mathrm{Im}\, d_2^{(C)}(\nu^2, q^2) = \tfrac{1}{2} f_\pi^2 \, m_\pi^4 \, (m_\pi^2 - q^2)^{-2} \left(1 - \frac{q^2}{\nu^2}\right)^{\frac{1}{2}} [\sigma(\pi^+ p) - \sigma(\pi^- p)] \, . \tag{4.209}$$

Thus, finally, substituting (4.202) and (4.209) into the dispersion relation (4.190) gives

$$d_2(\nu^2, q^2) = \tfrac{1}{2} f_\pi^2 \, m_\pi^4 \, (m_\pi^2 - q^2)^{-2} \Bigg\{ g_{\pi NN}^2 \, K(q^2)^2 \, \frac{-\nu_B \, m_n^{-1}}{\nu_B^2 - \nu^2} +$$

$$+ \frac{1}{\pi} \int_{\nu_t^2}^{\infty} d\nu'^2 \, (\nu'^2 - \nu^2 - i\epsilon)^{-1} \left(1 - \frac{q^2}{\nu^2}\right)^{\frac{1}{2}} [\sigma(\pi^+ p) - \sigma(\pi^- p)] \Bigg\} , \tag{4.210a}$$

where

$$\nu_B \equiv q^2/(2m_n) \quad . \tag{4.210b}$$

We now take the limit $q^2 \to 0$ followed by $\nu \to 0$, so that

$$d_2(0,0) = f_\pi^2 \, \frac{1}{\pi} \int_{\nu_t(0)}^{\infty} \frac{d\nu'}{\nu'} \, [\sigma^0(\pi^+ p) - \sigma^0(\pi^- p)] \quad . \tag{4.211}$$

$\sigma^0(\pi^\pm p)$ are the total pion nucleon cross section for <u>zero-mass</u> pions ($q^2 = 0$). Finally, we use the low energy theorem (4.184) and the

Goldberger-Treiman formula (4.161b) which gives

$$|g_1^A(0)|^{-2} = 1 + \frac{2}{\pi}\left(\frac{m_n}{g_{\pi NN}K(0)}\right)^2 \int_{\nu_t(0)}^{\infty} \frac{d\nu'}{\nu'}\left[\sigma^o(\pi^- p,\nu') - \sigma^o(\pi^+ p,\nu')\right]. \quad (4.212)$$

This is the famous Adler-Weisberger Sum Rule(17), which expresses the axial vector coupling constant entirely in terms of strong interaction quantities. The convergence of the sum rule is assured if the difference between the cross sections approaches zero as $\nu \to \infty$. The difference between the cross section for <u>physical</u> pions ($q^2 = m_\pi^2$) can be shown quite generally to have just this property(23). It therefore seems quite plausible to suppose that the same is true for the zero-mass pions, and that this integral consequently converges. This assumption, which is just the assumption of an unsubtracted dispersion relation for d_2 given in (4.190), is absolutely crucial. Since we are only predicting one number, even one subtraction would be fatal.

Even if the sum rule does converge, as we shall now assume, we still need to estimate the integral on the right-hand side. To do this we need to know the cross sections' difference for zero mass pions at <u>all</u> energies. The success of the Goldberger-Treiman formula in the form (4.163) indicates that we might reasonably assume that the off-mass-shell extrapolation does not alter things very much, that is

$$K(0)^{-2}\sigma^o(\pi^\pm p,\nu) \simeq \sigma(\pi^\pm p,\nu) \ , \quad (4.213)$$

where $\sigma(\pi^\pm p)$ are the physical cross sections. Of course even these are not known at all energies, but we may parametrize the high energy data with a form consistent with the assumed convergence properties. With these assumptions Weisberger(17) obtains

$$|g_1^A(0)| = 1.15 \pm 0.03 \ . \quad (4.214)$$

Adler(17) has attempted to include off-mass-shell corrections to this leading term which are of course model-dependent. He finds

$$|g_1^A(0)| = 1.24 \pm 0.03 \ . \quad (4.215)$$

As we shall see, this last value is in spectacular agreement with the latest experimental value, and, in view of the approximation inherent in using

PCAC, both must be regarded as being in excellent agreement with the data. Notice however that, although the sum rule (4.212) enables us to derive the magnitude $g_1^A(0)$, it does not predict its sign. Indeed the structure of the commutation relations we have been using is such that they cannot predict the sign of $g_1^A(0)$. This is just the ambiguity we noted in (4.86) for the quantity R_A. To date the sign of $g_1^A(0)$ remains unexplained, and it is difficult to think of any reasonable dynamical assumption which would 'explain' it.

The curious thing about the sign ambiguity is that the two cases describe two very different dynamical situations. The sign of $g_1^A(0)$ is negative, so that the matrix element of J_λ^+ between neutron and proton states has a form quite similar to the matrix element of the leptonic current, which also has opposite signs for its vector and axial vector parts $[\gamma_\lambda(1-\gamma_5)]$. Thus it is often argued that $g_1^A(0)$ has undergone a 'small' renormalisation from its 'natural' value -1, to its actual value -1.23. However, as far as the Adler-Weisberger Sum Rule is concerned, this small renormalisation is unexplained and might equally have been the large renormalisation from -1 to $+1.23$.

Neutron Decay (Continued)

Aside from this sign ambiguity, we have now completely determined the form of the hadronic matrix element (4.144) in the limit that we set $q = 0$. We may therefore proceed, as before, to evaluate the transition probability. We may simplify this calculation by continuing with the approximation of neglecting q for the hadronic matrix element.

Evidently in this approximation both neutron and proton have the same momenta, and using (2.145) it follows that

$$<p, s_p|J_\lambda^+|p, s_n> = 2\,m_N[g_1^V(0)m_N^{-1}\,p_\lambda\,I_2 + g_1^A(0)\sigma^i n_\lambda^{(i)}]_{s_p s_n}, \qquad (4.216)$$

where s_p, s_n are the spin vectors of the neutron and proton, respectively. Since both particles have the same momenta, $s_p = \pm\, s_n$. Taking the complex conjugate and summing over s_p, but not s_n, gives

$$H_{\mu\lambda} \equiv \sum_{\pm s_p} \langle p, s_p | J_\mu^+ | p, s_n \rangle^* \langle p, s_p | J_\lambda^+ | p, s_n \rangle$$

$$= 4 \left\{ [g_1^V(0)^* p_\mu I_2 + g_1^A(0)^* m_N \sigma^j n_\mu^{(j)}][g_1^V(0) p_\lambda I_2 + g_1^A(0) m_N \sigma^i n_\lambda^{(i)}] \right\}_{s_n s_n}$$

$$= 4 \left\{ |g_1^V(0)|^2 p_\mu p_\lambda + g_1^V(0) g_1^A(0)^* m_N p_\lambda s_{n\mu} + g_1^V(0)^* g_1^A(0) m_N p_\mu s_{n\lambda} \right.$$

$$\left. + |g_1^A(0)|^2 m_N^2 [n_\mu^{(i)} n_\lambda^{(i)} + i\epsilon^{jik} \sigma^k_{s_n s_n} n_\mu^{(j)} n_\lambda^{(i)}] \right\} , \qquad (4.217a)$$

since

$$\sigma^j \sigma^i = \delta^{ji} + i\epsilon^{jik} \sigma^k \qquad (4.217b)$$

and only σ^3 has non-zero diagonal matrix elements. Since the vectors $n^{(i)}$, p form a complete orthogonal set, it is easy to see that

$$\epsilon^{jik} n_\mu^{(j)} n_\lambda^{(i)} = - m_N^{-1} \epsilon_{\mu\lambda}{}^{\rho\sigma} n_\rho^{(k)} p_\sigma \qquad (4.218a)$$

and

$$n_\mu^{(i)} n_\lambda^{(i)} = - g_{\mu\lambda} + m_N^{-2} p_\mu p_\lambda \ . \qquad (4.218b)$$

Thus finally we have

$$\tfrac{1}{4} H_{\mu\lambda} = |g_1^V(0)|^2 p_\mu p_\lambda + g_1^V(0) g_1^A(0)^* m_N p_\lambda s_{n\mu} + g_1^V(0)^* g_1^A(0) m_N p_\mu s_{n\lambda}$$

$$+ |g_1^A(0)|^2 m_N^2 [-g_{\mu\lambda} + m_N^{-2} p_\mu p_\lambda - i m_N^{-1} \epsilon_{\mu\lambda}{}^{\rho\sigma} s_{n\rho} p_\sigma] \ . \qquad (4.219)$$

The calculation of the transition now proceeds straightforwardly by performing the phase space integration of the square of the invariant matrix element

$$|M|^2 = \frac{G^2}{2} \cos^2 \theta_c H_{\mu\lambda} E^{\lambda\mu} , \qquad (4.220)$$

where $E_{\mu\lambda}$ is given in (3.20).

In the rest frame of the neutron this gives the transition probability

$$d\Gamma = G^2 \cos^2 \theta_c (2\pi)^{-5} |\underline{p}_e|^2 d|\underline{p}_e| d\Omega_e (E_o - E_e)^2 d\Omega_\nu \left\{ |g_1^V(0)|^2 + \right.$$

$$3|g_1^A(0)|^2 + \underline{v}_e \cdot \underline{v}_\nu (|g_1^V(0)|^2 - |g_1^A(0)|^2) - 2\underline{s}_n \cdot \underline{v}_e (|g_1^A(0)|^2 + \mathrm{Re}\, g_1^V(0) g_1^A(0)^*)$$

$$\left. + \underline{s}_n \cdot \underline{v}_\nu (|g_1^A(0)|^2 - \mathrm{Re}\, g_1^V(0) g_1^A(0)^*) - 2\underline{s}_n \cdot (\underline{v}_e \times \underline{v}_\nu) \,\mathrm{Im}\, g_1^A(0) g_1^A(0)^* \right\} , (4.221)$$

where $s_n = (0, \underline{s}_n)$ is the neutron's spin vector in its rest frame, and

$\underline{v}_e = \underline{p}_e/E_e$, $\underline{v}_\nu = \underline{p}_\nu/E_\nu$ are the velocities of the electron and antineutrino. E_o is the maximum possible energy of the electron in the decay. Neglecting the recoil kinetic energy of the proton we have, to excellent approximation, $E_o = m_n - m_p$.

It is apparent from (4.221) that a study of the various correlations with the neutron's spin $\underline{s}_n$ enables one in principle to determine both the magnitude and phase of $g_1^A(0)/g_1^V(0)$. In particular, if $g_1^V(0)$ and $g_1^A(0)$ are relatively real (as we predict), the triple correlation term $\underline{s}_n \cdot (\underline{v}_e \times \underline{v}_\nu)$ is absent, and this is confirmed by experiment. Further if $g_1^V(0) \simeq g_1^A(0)$, the correlation $\underline{s}_n \cdot \underline{v}_e$ is present, but if $g_1^A(0) \simeq -g_1^A(0)$ it is not. Experimentally the latter situation is what is found. The experimental data on free neutron decay are not very accurate, but, if we assume (for the present) that $g_1^V(0)$ and $g_1^A(0)$ are relatively real, then there is no doubt from the asymmetry data that they have opposite sign, and we may conclude[24] that

$$g_1^A(0)/g_1^V(0) = -1.25 \pm 0.03 \quad . \tag{4.222}$$

It is possible to obtain more detailed and precise information from a study of nuclear beta decay, as we shall see. However we see that (4.222) agrees spectacularly with the theoretical prediction from (4.215) and (4.148). In addition to the polarization experiments, already discussed, the lifetime of a free unpolarized neutron is also measured experimentally. Integrating over the phase space in (4.221) for an unpolarized neutron gives

$$\Gamma(n \to pe^- \bar{\nu}_e) = \frac{G^2}{2\pi^3} \cos^2\theta_c \, m_e^5 \, f_o(\beta_o) \left(|g_1^V(0)|^2 + 3|g_1^A(0)|^2 \right), \tag{4.223a}$$

where

$$f_o(\beta_o) = \int_0^{\beta_o} d\beta \, \mathrm{sh}^2\beta \, \mathrm{ch}\beta \, (\mathrm{ch}\beta - \mathrm{ch}\beta_o)^2 \quad , \tag{4.223b}$$

with

$$\mathrm{ch}\beta_o = E_o/m_e \quad .$$

We may write $\Gamma = \tau^{-1}$, where τ is the lifetime of the decay. Then $\tau \ln 2 = t$ is the "half-life", and it is customary to rewrite (4.223) in the form

$$f_o(\beta_o)t = 2\pi^3 \ln 2\,(G\cos\theta_c)^{-2} m_e^{-5} \left(|g_1^V(0)|^2 + 3\,|g_1^A(0)|^2\right)^{-1} \quad . \tag{4.224}$$

The left-hand side is called the ft-value and can be determined directly from the experimental values of t and E_o. If we have independent information on $G\cos\theta_c\,|g_1^V(0)|$, the ratio $|g_1^A(0)/g_1^V(0)|$ can be determined more accurately from (4.224) than is possible using just the asymmetry data. The required independent information is obtained from a study of nuclear beta decay, as we shall now show.

4.2.4 ALLOWED NUCLEAR BETA DECAY

The decay of the neutron may also occur when the neutron is bound in a nucleus, provided that the binding energy of the daughter nucleus does not prohibit the decay kinematically. There are many examples of nuclear beta decay

$$(A, Z-1) \to (A, Z)\, e^- \bar{\nu}_e \quad , \tag{4.225}$$

and their study has played an important historical role in the development of the theory as well as providing a considerable body of precise data. The decay of a bound proton is also permitted although that of the free proton is not. This gives rise to nuclear decays in which a positron is emitted together with a neutrino

$$(A, Z+1) \to (A, Z)\, e^+ \nu_e \quad . \tag{4.226}$$

The theoretical treatment of these processes assumes that the nucleons in the nucleus behave like free nucleons as far as their weak interactions are concerned (independent particle approximation). Nuclear effects are included by the use of the appropriate nuclear wave functions. It is also assumed that the leptons' wave functions do not vary over the nuclear region ('allowed' approximation). Together with the previous assumption that we may neglect recoil effects, these assumptions enable us to write the hadronic part of the matrix element in the following form in the rest frame of the nuclei

$$\langle Z\,|\,J_o^+\,|\,Z-1\rangle = g_1^V(0)\, M_F \quad , \tag{4.227a}$$

$$<Z|\underline{J}^{+}|Z-1> = g_1^A(0)\,\underline{M}_{GT} \quad , \tag{4.227b}$$

where

$$M_F = \int d^3x_1 \ldots d^3x_A\, \varphi_Z^{\dagger}(\underline{x}_1, \ldots \underline{x}_A) \sum_{a=1}^{A} \tfrac{1}{2}(\tau_a^1 + i\tau_a^2)\, \varphi_{Z-1}(\underline{x}_1 \ldots \underline{x}_A) \tag{4.227c}$$

$$\underline{M}_{GT} = \int d^3x_1 \ldots d^3x_A\, \varphi_Z^{\dagger}(\underline{x}_1, \ldots \underline{x}_A) \sum_{a=1}^{A} \tfrac{1}{2}(\tau_a^1 + i\tau_a^2)\, \underline{\sigma}_a \varphi_{Z-1}(\underline{x}_1, \ldots \underline{x}_A) \; , \tag{4.227d}$$

with φ_{Z-1}, φ_Z the wave functions of the decaying and daughter nuclei, and τ_a^i, σ_a^i are the isospin and spin matrices for the ath nucleon in the nucleus. The above form follows directly from (4.216) if we quantize spin along the third spatial axis (i.e $n_j^{(i)} = \delta_j^i$). The τ-matrices just look after the isospin book-keeping to ensure that non-zero matrix elements arise only as a neutron in the nucleus decays leaving a proton. Then, proceeding as before, we may derive an expression for the ft-value for the nuclear decay. The precision of the measurements necessitates the inclusion of electromagnetic corrections and the following form results

$$ft\,(1+\delta_R') = 2\pi^3 \ln 2\, m_e^{-5} (G \cos\theta_c)^{-2} \{ |g_V'|^2 |M_F|^2 + |g_A'|^2 |\underline{M}_{GT}|^2 \}^{-1} \; . \tag{4.228}$$

The constant f now includes the effect of Coulomb scattering of the electron in the field of the daughter nucleus. Thus

$$m_e^5 f = \int_{m_e}^{E_o} d\,E_e\, |\underline{p}_e|\, E_e (E_o - E_e)^2\, F(Z, E_e) \quad , \tag{4.229}$$

where $F(Z, E_e)$ is the Fermi function. The remaining differences arise from the truly radiative corrections. δ_R' is determined from a known finite universal energy-dependent function[(25)] $g(E_e, E_o)$

$$\delta_R'(E_o) = \frac{\alpha}{2\pi} \left\{ \int_{m_e}^{E_o} dE_e |\underline{p}_e| E_e (E_o - E_e)^2\, g(E_e, E_o) \right\} \left\{ \int_{m_e}^{E_o} dE_e |\underline{p}_e| E_e (E_o - E_e)^2 \right\}^{-1} . \tag{4.230}$$

The constants g_V' and g_A' are given by

$$g_V' = (1 + \frac{\alpha}{4\pi} C)\, g_1^V(0) \quad , \quad g_A' = (1 + \frac{\alpha}{4\pi} D)\, g_1^A(0) \ , \qquad (4.231)$$

where C and D are constants which are usually divergent and in any case 'model-dependent'. The complications of strong interactions make the calculation of the electromagnetic corrections to hadron decays considerably less straightforward than those for the purely leptonic decays. One is forced to make approximations and assumptions. In general different approximations lead to different values of C and D but not of $g(E_e, E_o)$. This is why we say C and D are model-dependent, while $g(E_e, E_o)$ is not. We shall return at a later stage to discuss the theoretical importance of the radiative corrections.

For nuclear transitions of the type $0^+ \to 0^+$ it is apparent from the definition (4.227) that $\underline{M}_{GT}$ is zero. Also, if we consider superallowed transitions - that is, allowed transitions between nuclear states <u>belonging to the same isomultiplet</u> - M_F is determined by CVC independently of the approximation (4.227c). For example, suppose the initial and final states are the $I_3 = -1, 0$ components of a triplet of states with $I = 1$, then in precisely the way that we deduced (4.134), we find

$$M_F = \sqrt{2}\,(1 - \delta_c)^{\frac{1}{2}} \ , \qquad (4.232)$$

where δ_c represents the unknown $O(\alpha)$ terms arising from isospin non-conservation.

For such decays, therefore, everything in (4.228) is known except the product $G \cos\theta_c\, g_V'$ which is universal. Thus by studying these super-allowed decays we may check the self-consistency of the theory and determine the above quantity. Over the years the self-consistency of the theory has been open to some doubt, since different decays yielded different values for the above constant. However recent and more accurate experiments on nuclei with mass number $A < 40$ are now all consistent and yield a best value[(26)] of

$$|G \cos\theta_c\, g_V'| = (1.4131 \pm 0.0005) \times 10^{-49}\ \text{erg cm}^3 \ . \qquad (4.233)$$

If we again <u>ignore</u> the model-dependent electromagnetic corrections contained in g_V' it may be replaced by $g_1^V(0) = 1$. Then using the value of G given in (3.62) we can estimate $\cos\theta_c$

$$\cos\theta_c = 0.9845 \quad . \tag{4.234}$$

Larger values of g'_V are associated with larger values of θ_c. Given the accuracy of the present data the neglect of the electromagnetic corrections is plainly unjustified. We shall discuss later the problem of calculating these corrections and the conclusions concerning $\cos\theta_c$ which can be derived from the assumption of various models.

Now, using the value (4.233) we may determine the other constant g'_A from any other transition having $\underline{M}_{GT}$ non-zero. It is customary to use the free neutron data for this purpose, since it is free of the complications of nuclear physics. Comparing (4.224) and (4.228) we see that for neutron decay we set

$$|M_F|^2 = 1 \quad , \quad |\underline{M}_{GT}|^2 = 3$$

in (4.228). Then using the experimental value[27] of the neutron's lifetime

$$\tau_n\big|_{exp} = (0.918 \pm 0.014)\,10^3 \text{ s} \,, \tag{4.235}$$

we find[24]

$$|g'_A/g'_V| = 1.248 \pm 0.010 \quad . \tag{4.236}$$

If we ignore the electromagnetic corrections temporarily we may replace g'_A and g'_V by $g_1^A(0)$ and $g_1^V(0)$. In that case (4.236) again agrees spectacularly with the value obtained from (4.148) and Adler's value of $|g_1^A(0)|$ given in (4.215).

4.2.5 CONSERVED VECTOR CURRENT HYPOTHESIS (CVC)

We have seen how CVC predicts the weak vector form factors in terms of the electromagnetic form factors, and we have used in particular the prediction (4.148) arising from the charge form factors. In principle it is possible to test all of the predictions (4.146) using high energy neutrino experiments, which, however, are extremely difficult to perform at present. In beta decay the presence of the form factors implies some q-dependence of the hadronic matrix element, which in turn implies that they contribute some shape correction factor to the allowed energy spectrum. Obviously to maximize such effects it is advisable to study decays with a large energy release. A famous experiment which does just this is the ^{12}B - ^{12}N - ^{12}C experiment suggested by Gell-Mann[28]. The nuclei ^{12}B, ^{12}N are the $J^P = 1^+$, $I_3 = \pm 1$ components of an isotriplet which decay to the ground state of ^{12}C having $J^P = 0^+$, $I = 0$. The $I_3 = 0$ component of the isotriplet, $^{12}C^*$, decays electromagnetically to the ground state. Plainly the beta decays are both allowed Gamow-Teller transitions. The experiment consists of measuring the deviations from the allowed spectrum arising from interference between the dominant g_1^A term and the 'weak magnetic' term g_2^V in (4.144). Let us try and evaluate this correction.

As before we approximate the hadronic matrix element, but now retain terms linear in q. With the aid of (2.145) we obtain

$$\langle p', s_p | J_\lambda^+ | p, s_n \rangle = \Big\{ g_1^V(0) \left[(p+p')_\lambda - \frac{i}{m_N} (\lambda q p n^{(i)}) \sigma^i \right] + 2 g_2^V(0)\, i (\lambda q p n^{(i)}) \sigma^i + 2m_N\, g_1^A(0) \sigma^i n_\lambda^{(i)} \Big\}_{s_p s_n} ,$$

where $$q = p - p' . \tag{4.237}$$

In deriving this we have dropped the contribution from all terms proportional to q_λ (e.g. those from g_3^V, g_2^A), since they yield terms proportional to m_e when coupled to the leptonic current, by use of the Dirac equation. We have also dropped the contribution from g_3^A, since the Cabibbo hypothesis and T-invariance imply its absence, as we shall see. In the Gamow-Teller transition under consideration the term proportional to $(p + p')_\lambda$ cannot contribute. Proceeding as before we may calculate the contribution to $H_{\mu\lambda}$ arising from the terms proportional to σ. Then averaging over the neutron spin configurations gives :

$$\tfrac{1}{4} H_{\mu\lambda}^{GT} = |g_1^A(0)|^2\, m_N^2\, n_\mu^{(i)} n_\lambda^{(i)} - g_1^A(0)\, \Delta\, i\, m_N (\lambda q p n^{(i)})\, n_\mu^{(i)} + g_1^A(0)\, \Delta\, i\, m_N (\mu q p n^{(i)})\, n_\lambda^{(i)} + O(q^2) , \tag{4.238a}$$

where

$$\Delta \equiv (2m_N)^{-1} g_1^V(0) - g_2^V(0) , \tag{4.238b}$$

and we have assumed $g_1^A(0)$ is <u>real</u>, as before. Using orthonormality, as in (4.218b), gives

$$\tfrac{1}{4} H_{\mu\lambda}^{GT} = |g_1^A(0)|^2 (p_\mu p_\lambda - g_{\mu\lambda} m_N^2) - 2i\Delta g_1^A(0) m_N (\lambda q p \mu) \quad .$$

It is now straightforward to evaluate the electron spin-averaged decay spectrum. Taking $E^{\lambda\mu}$ from (3.20) gives in the rest frame

$$|M|_{GT}^2 \propto |g_1^A(0)|^2 (3 - \underline{v}_e \cdot \underline{v}_\nu) + 4\Delta g_1^A(0) [2E_e - E_o - m_e^2 E_e^{-1} - \underline{v}_e \cdot \underline{v}_\nu (2E_e - E_o)] \quad . \tag{4.239}$$

The terms proportional to $\underline{v}_e \cdot \underline{v}_\nu$ integrate out of the spectrum, if we do not observe the direction of the neutrino. We also drop the term proportional to $m_e^2 E_e^{-1}$, having previously dropped similar terms. Thus finally

$$d\Gamma^{GT} \propto [3|g_1^A(0)|^2 - 4\Delta g_1^A(0) E_o] (1 + AE_e) + O(\Delta^2) \quad , \tag{4.240a}$$

where

$$A = \frac{8}{3} \frac{\Delta}{g_1^A(0)} \quad . \tag{4.240b}$$

So the energy spectrum for the β^- decay is predicted to have the allowed Gamow-Teller spectrum corrected by the factor $1 + AE_e$. In the same way the β^+ decay of ^{12}N is corrected by the factor $1 - AE_e$. A more accurate prediction can be made for the <u>ratio</u> of the ^{12}B and ^{12}N spectra. Contributions from the induced pseudoscalar form factor g_2^A, which we neglected, in fact cancel out of this ratio, which is thus predicted to be $1 + 2AE_e$. The coefficient A is completely determined by CVC. From (4.146) we have

$$g_1^V(0) = 1 + O(\alpha) \tag{4.241a}$$

$$-g_2^V(0) = \frac{1}{2m_N} (\varkappa_p - \varkappa_n) + O(\alpha) \quad , \tag{4.241b}$$

where $\varkappa_p$, $\varkappa_n$ are the anomalous magnetic moments of the proton and neutron. Experimentally

$$1 + \varkappa_p - \varkappa_n = 4.7 \quad , \tag{4.242a}$$

so

$$A = 0.53\% \ / \mathrm{MeV} . \tag{4.242b}$$

Of course the accuracy of this prediction depends on the validity of the independent particle approximation. In fact CVC allows us to predict the magnitude of A independently of this approximation. Since ^{12}N, $^{12}C^*$, ^{12}B form an isotriplet

$$(v^1 + i\,v^2)\,|^{12}B> = \surd 2\,|^{12}C^*> \quad . \tag{4.243}$$

Thus using (4.129) it follows that

$$<^{12}C\,|j_\lambda|\,^{12}C^*> = \surd 2 <^{12}C\,|\,V^1_\lambda + i\,V^2_\lambda|^{12}B> \quad , \tag{4.244}$$

since ^{12}C is an isoscalar. This relates the required vector current matrix element to the matrix element for the electromagnetic decay of $^{12}C^*$, and this determines the magnitude of the required quantity but not its sign. If we assume that the sign at least of A is correctly predicted by (4.242b), CVC predicts that

$$A = (0.55 \pm 0.12)\% \ / \mathrm{MeV} \quad , \tag{4.245}$$

which differs only marginally from our cruder estimate. Experimentally the results are as follows:

$$A = (0.55 \pm 0.12)\% \ / \ \mathrm{MeV} \qquad (^{12}B) \tag{4.246a}$$

$$A = (0.53 \pm 0.06)\% \ / \ \mathrm{MeV} \qquad (^{12}N) \tag{4.246b}$$

$$2A = (1.07 \pm 0.24)\% \ / \ \mathrm{MeV} \qquad (^{12}B : {}^{12}N) . \tag{4.246c}$$

All of these are in excellent agreement with the theoretical prediction (4.245) derived using CVC.

4.2.6 GENERAL CONSIDERATIONS

The operator v^2 is the generator of rotations about the second axis in isospin space. Thus $e^{i\pi v^2}$ represents a rotation of 180° about the second axis. When this "charge symmetry" operation is applied to an element of any representation of SU(2) it plainly gives another element of the same representation. In the case of the regular (vector) representation, to which the strangeness conserving part of the Cabibbo current $J^\pm_\lambda$ belongs, the component in the direction of the second axis is unaffected, while those in the direction of the first and third axes are reversed in sign; this is just the effect of such a rotation upon the components of a vector.

Thus

$$e^{-i\pi v^2} J_\lambda^- e^{+i\pi v^2} = - J_\lambda^+ \quad , \tag{4.247}$$

and this relation holds for both vector and axial parts separately. The above relation is important, since it says that J_λ^+ and its <u>hermitian conjugate</u> are connected by the charge symmetry operation $e^{i\pi v^2}$. Of course, this is only possible because the Cabibbo hypothesis places both J_λ^+ and its hermitian conjugate in the <u>same</u> isospin representation. No such relationship exists, for example, for the strangeness changing current S_λ^+ which belongs to an isodoublet; its hermitian conjugate S_λ^- belongs to a different isodoublet and is not therefore connected to S_λ^+ by the charge symmetry operation. Similarly, if we apply the charge symmetry operation to a proton state, for example, we will obtain some multiple of the neutron state, since this is the state in the nucleon isodoublet which has opposite I_3 to the proton:

$$e^{i\pi v^2} |p> = \omega |n> \quad . \tag{4.248}$$

Since $e^{i\pi v^2}$ is a unitary operator, $|\omega| = 1$. It is easy to calculate ω. In the two-dimensional representation of the isospin group SU(2)

$$v^i = \tfrac{1}{2} \tau^i \qquad (i = 1,2,3) \quad ,$$

where τ^i are the 2×2 Pauli matrices. Thus

$$e^{i\pi v^2} = e^{\frac{1}{2} i\pi\tau_2} = \cos \frac{\pi}{2} I_2 + i \sin \frac{\pi}{2} \tau_2 \quad ,$$

summing up even and odd powers of τ_2 separately. Hence

$$e^{i\pi v^2} = i\tau_2$$

and

$$e^{i\pi v^2} | \alpha > = (i\tau_2)_{\beta\alpha} | \beta > \quad ,$$

where α, β run over the two basis vectors (p and n) of the representation. Now

$$i\tau_2 = \begin{pmatrix} 0 & 1 \\ -1 & 0 \end{pmatrix} \quad ,$$

so

$$e^{i\pi v^2}|p\rangle = -|n\rangle \qquad (4.249a)$$

$$e^{i\pi v^2}|n\rangle = +|p\rangle \quad , \qquad (4.249b)$$

and

$$e^{-i\pi v^2}|n\rangle = -|p\rangle \qquad (4.249c)$$

$$e^{-i\pi v^2}|p\rangle = +|n\rangle \quad . \qquad (4.249d)$$

This verifies the form (4.248) and evaluates the constant ω. Now suppose we take the matrix element of (4.247) between proton and neutron states - it is advisable to exhibit the momenta explicitly. We find

$$\begin{aligned} -\langle p(k')|J_\lambda^+|n(k)\rangle &= \langle p(k')|e^{-i\pi v^2} J_\lambda^- \, e^{i\pi v^2}|n(k)\rangle \\ &= -\langle n(k')|J_\lambda^-|p(k)\rangle + O(\alpha) \\ &= -\langle p(k)|J_\lambda^+|n(k')\rangle^* + O(\alpha) \quad , \end{aligned} \qquad (4.250)$$

using (2.249). (The $O(\alpha)$ terms arise because isospin is not exactly conserved - as a result (2.249) are only accurate if we neglect $O(\alpha)$.) The general form of the matrix element on the <u>left</u> of (4.250) is given in (4.144), where now $q = k - k'$. So interchanging k and k' and taking the complex conjugate gives the matrix element on the right of (4.250) :

$$\langle p(k)|J_\lambda^+|n(k')\rangle^* = \bar{u}(k')\Big\{ g_1^V(q^2)^* \gamma_\lambda + g_2^V(q^2)^* i\sigma_{\lambda\mu} q^\mu$$

$$- g_3^V(q^2)^* q_\lambda + g_1^A(q^2)^* \gamma_\lambda\gamma_5 + g_2^A(q^2)^* q_\lambda\gamma_5 - g_3^A(q^2)^* i\sigma_{\lambda\mu} q^\mu \gamma_5\Big\} u(k) \, .$$

(4.251)

Then substituting into (4.250) and neglecting $O(\alpha)$ we find that

$$g_1^V, \; g_2^V, \; g_1^A, \; g_2^A \qquad \text{are real,} \qquad (4.252a)$$

and

$$g_3^V, \; g_3^A \qquad \text{are imaginary.} \qquad (4.252b)$$

Of course we have already observed that CVC (which is predicted by the Cabibbo hypothesis) implies that $g_{1,2}^V$ are real (and known) quantities and that g_3^V is zero, and these results are consistent with (4.252). If we were to abandon the Cabibbo hypothesis, there would be no particular reason to believe (4.247), since in general there is no a priori reason for placing the strangeness conserving current with $\Delta Q = +1$ in the same isomultiplet as its hermitian conjugate current with $\Delta Q = -1$. Thus in general some or all of the relations (4.252) might be violated.

Additional constraints are placed upon the form factors g_i^V, g_i^A if we assume T-invariance. The properties of the leptonic current L_α given in (2.264) are easily ascertained using (2.218), (2.262):

$$\mathcal{T}\bar{\psi}_{(e)}(x^o,\underline{x})\gamma_\lambda a\psi_{(\nu_e)}(x^o,\underline{x})\mathcal{T}^{-1} = \mathcal{T}\psi_{(e)}(x^o,\underline{x})\mathcal{T}^{-1}\gamma_\lambda^* \mathcal{T} a\psi_{(\nu_e)}(x^o,\underline{x})\mathcal{T}^{-1}$$

$$= \eta_{(e)}\bar{\psi}_{(e)}(-x^o,\underline{x})B^{-1}\gamma^{0T}A^*\gamma_\lambda^*\eta^*_{(\nu_e)}A^{T-1}\gamma^{0T}Ba\psi_{(\nu_e)}(-x^o,\underline{x})$$

$$= \eta_{(e)}\eta^*_{(\nu_e)}\bar{\psi}_{(e)}(-x^o,\underline{x})\gamma^\lambda a\psi_{(\nu_e)}(-x^o,\underline{x}) \quad ,$$

where $a \equiv \frac{1}{2}(1-\gamma_5)$ and we have used (2.78) and (2.114). Hence if we take $\eta_{(e)} = \eta_{(\nu_e)}$,

$$\mathcal{T}L_\lambda(x^o,\underline{x})\mathcal{T}^{-1} = L^\lambda(-x^o,\underline{x}) \quad .$$

It follows that if we assume T-invariance of the Hamiltonian (4.1) then the hadronic current $\mathcal{J}_\lambda^\pm$ must satisfy a similar equation:

$$\mathcal{T}\mathcal{J}_\lambda^\pm(x^o,\underline{x})\mathcal{T}^{-1} = \mathcal{J}_\lambda^\pm(-x^o,\underline{x}) \quad . \tag{4.253}$$

Plainly this equation is satisfied separately by $J_\lambda^\pm$ and $S_\lambda^\pm$. Now taking $x = 0$ and remembering that $\mathcal{T}$ is an anti-unitary operator gives

$$<p|\mathcal{T}^{-1}J_\lambda^+\mathcal{T}|n> = <p^t|J_\lambda^+|n^t>^* = <p|J^{+\lambda}|n> \quad .$$

Thus

$$<p^t|J_\lambda^+|n^t> = <p|J^{+\lambda}|n>^* \quad . \tag{4.254}$$

To evaluate the expression on the left we use (4.144) and the expression (2.154) for the time reversed spinors; writing $q^t = n^t - p^t$, so $(q^t)^2 = q^2$,

$$<p^t|J^+_\lambda|n^t> = p^T\gamma^{0T} B \Big\{ g_1^V(q^2)\,\gamma_\lambda + g_2^V(q^2)\, i\sigma_{\lambda\mu}\, q^{t\mu} + g_3^V(q^2)\, q^t_\lambda$$

$$+ g_1^A(q^2)\,\gamma_\lambda\gamma_5 + g_2^A(q^2)\, q^t_\lambda\gamma_5 + g_3^A(q^2)\, i\sigma_{\lambda\mu}\, q^{t\mu}\gamma_5 \Big\} B^{-1}\gamma^{0T}\bar{n}^T .$$

Now we use the defining property of B given in (3.114) and, since $q^t_\lambda = q^\lambda$, we find

$$<p^t|J^+_\lambda|n^t> = p^T \Big\{ g_1^V(q^2)\gamma^{\lambda T} - g_2^V(q^2)\, i\sigma^{\lambda\mu T} q_\mu + g_3^V(q^2)\, q^\lambda$$

$$+ g_1^A(q^2)\,(\gamma^\lambda\gamma_5)^T - g_2^A(q^2)\, q^\lambda\gamma_5^T + g_3^A(q^2)\, i(\sigma^{\lambda\mu}\gamma_5)^T q_\mu \Big\} \bar{n}^T . \tag{4.255}$$

The right-hand side of (4.254) may be calculated directly from (4.144)

$$<p|J^{+\lambda}|n>^* = \bar{n} \Big\{ g_1^V(q^2)^*\gamma^\lambda - g_2^V(q^2)\, i\sigma^{\lambda\mu} q_\mu + g_3^V(q^2)^* q^\lambda$$

$$+ g_1^A(q^2)^*\gamma^\lambda\gamma_5 - g_2^A(q^2)^* q^\lambda\gamma_5 + g_3^A(q^2)^*\, i\sigma^{\lambda\mu} q_\mu\gamma_5 \Big\} p . \tag{4.256}$$

Comparing (4.255) and (4.256) we see that T-invariance implies

$$g_1^V,\ g_2^V,\ g_3^V,\ g_1^A,\ g_2^A,\ g_3^A \quad \text{are \underline{all} real .} \tag{4.257}$$

Thus (4.252a) is consistent with (4.257). However (4.252b) and (4.257) together imply that g_3^V and g_3^A are <u>zero</u>.

Clearly we may multiply the entire hadronic matrix element by an arbitrary phase factor without observable consequences. As a result only the phase <u>differences</u> between form factors are observable. Now, if we make the 'allowed' approximation of dropping all q-dependence,the hadronic matrix element is entirely determined by $g_1^V(0)$ and $g_1^A(0)$, and the only way that 'unusual' effects could manifest themselves in allowed decays is by $g_1^V(0)$ and $g_1^A(0)$ differing in phase. In fact the experimental evidence does not indicate any such phase difference. Interference between g_1^V and g_1^A affects the asymmetry coefficient in the decay of polarized neutrons, as we noted earlier. Ignoring electromagnetic corrections, which have been shown to be very small[29], experiment gives[30]

$$\arg\,[\, g_1^A(0)/g_1^V(0)\,] = (180.2 \pm 0.4)^\circ . \tag{4.258}$$

Since allowed decays provide no evidence of unusual effects, the next place to look is at quantities sensitive to the q-dependent terms in the hadronic matrix element. The 'best' terms to look for are g_3^V and/or g_3^A, since the existence of either of these would indicate that one or other of our hypotheses is false. If we accept CVC (and we have seen that experiment supports this), or even just that the vector current is conserved, then g_3^V is <u>zero</u>. We see from (4.250) and (4.251) that if g_3^A is <u>real</u> then its sign in the $n \to p$ transition is opposite to that for the $p \to n$ transition. Proceeding as we did in the previous section it is easy to see that the g_3^A term does not appreciably affect the <u>shape</u> of the allowed Gamow-Teller spectrum for the decay of an unpolarized nucleus. However, it does affect the total decay rate and thereby the ft-value. Thus we may double the effect of such a term by measuring the difference between the ft-values for the decays of mirror nuclei. We define

$$\delta \equiv \frac{(ft)^+}{(ft)^-} - 1 \quad ,$$

where $(ft)^{\pm}$ are the ft-values for $\beta^{\pm}$ decays. A calculation similar to that in the previous section gives δ in terms of g_3^A :

$$\delta \simeq \frac{4}{3} \frac{g_3^A(0)}{g_1^A(0)} (E_o^+ + E_o^-) \ ,$$

where $E_o^{\pm}$ are the maximum energies available to the $\beta^{\pm}$.

Recent measurements of this quantity show no evidence[26] that it is non-zero:

$$\delta(26) = -\,0.012 \pm 0.016$$

$$\delta(30) = -\,0.002 \pm 0.006 \ ,$$

where $\delta(A)$ is the value of δ for the mirror nuclei with mass number A. In deriving the above expression for δ it is assumed, as we have assumed hitherto, that as far as the weak interactions are concerned the nucleons in the nucleus may be treated as free particles. This assumption is not strictly justified of course, because the nucleons are bound by the nuclear potential. But asymmetries can also arise owing to 'trivial' differences in the binding energies on each side of the mirror. Further, if the nucleons are off-shell, as they are because of their binding, the current matrix element (4.144) is not the most general - we no longer have the Dirac equation to reduce the number of independent terms. In addition the exchanged mesons, which give rise to the binding, also participate in the weak interactions, and the resultant exchange current may have charge symmetry violating terms which contribute to δ. Thus if charge symmetry violating currents exist their manifestation in nuclei is drastically influenced by binding effects[31]. We must, therefore, be extremely cautious of deducing from the small measured values of δ, quoted above, that such currents are not present. In fact, a better test is likely to be provided by the original correlation experiments which established the existence of parity violation in the first place. Because these are spectrum shape measurements the uncertainties in nuclear wave functions do not contribute. Of course one still has to be sure that other contributing form factors, in particular the weak magnetism term, are independently known, since both g_3^A and g_2^V contribute oppositely in $n \rightarrow p$ and $p \rightarrow n$ transitions.

4.2.7 MUON CAPTURE

The Hamiltonian (4.1) predicts the existence of the process

$$\mu^- p \rightarrow n\nu_\mu \quad .$$

This process is observed experimentally when a beam of muons is directed into hydrogen, for example. The muon is captured in one of the excited states, cascades down to the ground state, and finally, if it does not itself decay, is captured via the above process. Capture has been observed in a liquid hydrogen bubble chamber, in complex nuclei, and, more recently, in gaseous hydrogen. In the gaseous hydrogen capture takes place chiefly from the atomic state $(p\mu^-)$; from the theoretical point of view this makes the process relatively clean, since it is freed of any complications arising from nuclear physics or molecular effects. Unfortunately, for low Z nuclei, and hydrogen in particular, the muon is much more likely to decay than to be captured. Thus the simplest process from a theoretical view is the hardest to study experimentally.

In all cases it is assumed that capture occurs when the muon is at rest (the binding energy is, of course, negligible). Thus it follows that

$$q^2 \equiv (p-n)^2 = (\nu - \mu)^2 = m_\mu (m_\mu - 2\nu_o) .$$

Also

$$m_n^2 = n^2 = (p - \nu + \mu)^2 = (m_p + m_\mu)^2 - 2\nu_o (m_\mu + m_p) .$$

Hence

$$q^2 = m_\mu (m_n^2 - m_p^2 - m_\mu m_p)(m_\mu + m_p)^{-1}$$

$$= - 0.877\, m_\mu^2 . \tag{4.259}$$

Since $q \sim 100$ MeV, it is impossible to neglect all q-dependence as we were were able to do in beta decay. The vector form factors $g^V_{1,2}$ are related to the electromagnetic form factors, as we have seen in (4.146), and these are measured up to $q^2 = - 25\ (\mathrm{GeV})^2$. So from the data directly, or using the conventional parametrization given in (4.334), we obtain, for $q^2 = - 0.877\, m_\mu^2$,

$$g_1^V(q^2) = 0.95 \tag{4.260a}$$

$$g_2^V(q^2) = - 1.75\, m_p^{-1} . \tag{4.260b}$$

Similarly, $g_1^A(q^2)$ can in principle be measured directly by neutrino scattering experiments, as we shall see in 4.2.9. Using the assumed double pole form given in (4.345) and (4.346) we obtain for $q^2 = - 0.877\, m_\mu^2$

$$g_2^A(q^2) = - 1.21 . \tag{4.261}$$

The Cabibbo hypothesis and T-invariance imply

$$g_3^V(q^2) = g_3^A(q^2) = 0 , \tag{4.262}$$

as we have seen, so it only remains to consider the "induced pseudoscalar" form factor $g_2^A(q^2)$. This is <u>not</u> expected to be slowly varying. Consider the contribution to $\langle n|J_\lambda^-|p\rangle$ shown in Fig. 4.2 arising from pion exchange.

The matrix element for this diagram is proportional to

$$i q_\lambda \frac{1}{q^2 - m_\pi^2} (\bar{n}\, i\gamma_5 p) ,$$

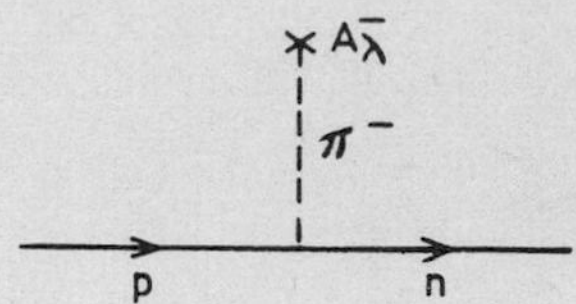

Fig. 4.2 Pion pole contribution to $\langle n|J_\lambda^-|p\rangle$

so $g_2^A(q^2)$ certainly has a pion pole, and we cannot ignore its q^2-dependence. The only information we have on g_2^A is that contained in the Goldberger-Treiman formula (4.161a). This gives

$$g_2^A(q^2) = (q^2)^{-1}\left\{-f_\pi\sqrt{2}\, g_{\pi NN}K(q^2)q^2(m_\pi^2-q^2)^{-1} - f_\pi\sqrt{2}\, g_{\pi NN}K(q^2) + 2m_n g_1^A(q^2)\right\}. \tag{4.263}$$

Thus, if we assume

$$g_1^A(q^2) \simeq g_1^A(0) \text{ and } K(q^2) \simeq K(0)$$

and use (4.161b), we have

$$g_2^A(q^2) = -f_\pi\sqrt{2}\, g_{\pi NN}K(0)(m_\pi^2 - q^2)^{-1} \tag{4.264a}$$

$$\approx -2m_n\, g_1^A(0)\,(m_\pi^2 - q^2)^{-1}\ . \tag{4.264b}$$

As before we may set $K(0) \simeq K(m_\pi^2) \approx 1$, in which case we obtain two values for $g_2^A(q^2)$. Using (4.124) and (4.234) we deduce

$$|f_\pi| = 130.4\ \text{MeV}\ . \tag{4.265}$$

The πN-coupling constant is given by

$$|g_{\pi NN}| = 13.66 \pm 0.5\ , \tag{4.266}$$

so, substituting (4.259) for q^2, (4.264a) gives

$$m_\mu\, g_2^A(q^2) = -7.4\, g_1^A(0)\ , \tag{4.267a}$$

where $g_1^A(0) \approx 1.248$ as indicated by (4.236). On the other hand, (4.264b) gives directly

$$m_\mu\, g_2^A(q^2) = -6.8\, g_1^A(0)\ . \tag{4.267b}$$

The difference between these two values for $g_2^A(q^2)$ arises from the neglect of the unknown form factor dependence of $K(q^2)$:

$$K(m_\pi^2) - K(0) \equiv \Delta = 1 - (m_p + m_n)\, g_1^A(0) \Big/ f_\pi \sqrt{2}\, g_{\pi NN}$$

$$= 0.070 \pm 0.026 \quad . \tag{4.268}$$

The magnitude of this variation is in accord with our most naive expectation that it derives from the 3π continuum states, as we have already remarked. However, more realistic attempts(32) to estimate Δ show that the 3π continuum contributions provide only 1% of the desired variation and are therefore completely negligible. The $\rho\pi$ and $\sigma\pi$ intermediate states contribute something like 0.03 to Δ and the high energy states (those of mass greater than $2m_N$) make a contribution of less than 0.01. Thus at the present time the magnitude of Δ is not completely understood. However, recent shifts in the experimental data have had the effect of reducing Δ and lend the hope that theory and experiment will ultimately agree.

Returning to the problem in hand, we have now completely determined the hadronic matrix element:

$$<n|J_\lambda^-|p> = \bar{n} \Big\{ g_1^V(q^2)\,\gamma_\lambda + g_2^V(q^2)\, i\sigma_{\lambda\mu} q^\mu$$

$$+ g_1^A(q^2)\,\gamma_\lambda\gamma_5 + g_2^A(q^2)\, q_\lambda \gamma_5 \Big\} p \quad , \tag{4.269}$$

where

$$q = p - n$$

and the quantities g_i^V, g_i^A are given in (4.260), (4.261), (4.267). We therefore have all of the information needed to calculate the required capture rate. However there is a feature of this calculation which we have not hitherto encountered. The forms (4.269) and (3.20) involve free particle spinors for the muon and proton. In reality, of course, these particles are not free but are bound by the electrostatic Coulomb potential. In coordinate space this has the effect of multiplying the free particle spinors by the wave function $\psi(\underline{r})$, $\underline{r} = \underline{r}_\mu - \underline{r}_p$, describing their relative motion. Now the weak interaction is a point interaction and occurs only when $\underline{r} = 0$. Thus we must multiply the free particle matrix element by $\psi(\underline{0})$ and the free particle capture rate by $|\psi(\underline{0})|^2$. In other words, we simply multiply the free capture rate by the probability that the muon and the proton are in the same place (so that the weak interaction can occur), as we might have guessed at the outset. In the case under discussion the muon is captured from the ground state of the atom, so the required wave function is

$$\psi(\underline{r}) = N e^{-r/a_o} ,$$

where a_o is the Bohr radius, given by

$$a_o^{-1} = \alpha\mu = \frac{\alpha m_\mu m_p}{m_\mu + m_p} ,$$

and $N = \psi(0)$ is the normalization factor fixed by

$$\int d^3r \, |\psi(\underline{r})|^2 = 1 .$$

Thus

$$|\psi(0)|^2 = \frac{(\alpha\mu)^3}{\pi} , \tag{4.270}$$

and we find the total capture rate is given by

$$\Gamma = (2\pi)^{-2} \frac{\nu_o \, (\alpha\mu)^3}{(m_\mu + m_p) \, 16 m_\mu m_p} \sum_{(s_\mu, s_p, s_n)} |M(\mu p \to n\nu)|^2 . \tag{4.271}$$

We can get some idea of the numbers involved, if we make the crude (non-relativistic) approximation of dropping all momentum dependence in (4.269) and, using (4.219) and (3.18), we find

$$\sum_{s_n} |M(\mu p \to n\nu)|^2 = \tfrac{1}{2} G^2 \cos^2\theta_c \, \{A + B \, \underline{s}_\mu \cdot \underline{s}_p\} \, 16 m_n m_\mu m_p \nu_o , \tag{4.271a}$$

where

$$A = |g_1^V(0)|^2 + 3\,|g_1^A(0)|^2 \tag{4.271b}$$

$$B = 2\, g_1^A(0) \, [\, g_1^V(0) - g_1^A(0)\,] . \tag{4.271c}$$

Thus the total capture rate is

$$\Gamma = \frac{G^2 \alpha^3}{2\pi^2} \cos^2\theta_c \, \nu_o^2 \, m_\mu^{-1} \, \mu^4 A = (29.81 \; s^{-1}) \, A , \tag{4.272}$$

using the value (4.233) for $G \cos\theta_c$.

Now the $(p\,\mu^-)$ atom in its ground state has two hyperfine states, $F = 0, 1$ corresponding to total angular momentum of 0 and 1. In the $F = 0$ (singlet) state the elementary rules of angular momentum show that the average value of $\underline{s}_\mu \cdot \underline{s}_p = -3$, while in the $F = 1$ (triplet) state $\underline{s}_\mu \cdot \underline{s}_p = +1$. Thus we may write down the rates Γ_F for capture from the hyperfine state F:

$$\Gamma_o = (29.81\ s^{-1})\ (A - 3B) = (29.81\ s^{-1})\ [g_1^V(0) - 3g_1^A(0)]^2 \qquad (4.273a)$$

$$\Gamma_1 = (29.81\ s^{-1})\ (A + 3B) = (29.81\ s^{-1})\ [g_1^V(0) + g_1^A(0)]^2 . \qquad (4.273b)$$

As a check we note that

$$\Gamma = \tfrac{1}{4}\Gamma_o + \tfrac{3}{4}\Gamma_1 .$$

Since $g_1^V(0) + g_1^A(0) \sim 0$ we see that

$$\Gamma_o >> \Gamma_1 ,$$

so that capture occurs almost entirely from the singlet state. The calculation using all the information in (4.269) is straight forward but tedious and we simply quote the result:

$$\Gamma_o = (29.81\ s^{-1})\,|1.1578\ g_1^V - 0.2046\ m_N\ g_2^V - 3.0526\ g_1^A - 0.0526\ m_\mu\ g_2^A|^2 \qquad (4.274a)$$

$$\Gamma_1 = (29.81\ s^{-1})\tfrac{1}{3}\,|0.974\ (g_1^V + g_1^A) + 0.2165\ g_2^V - 0.0526\ m_\mu\ g_2^A|^2 + (29.81\ s^{-1})\tfrac{2}{3}\,|1.0526\ (g_1^V + g_1^A) + 0.006\, m_N\ g_2^V - 0.0526\, m_\mu\, g_2^A|^2 . \qquad (4.274b)$$

Using the values of g_i^V, g_i^A derived previously these expressions give

$$\Gamma_o = 647\ s^{-1} \qquad (4.275a)$$

$$\Gamma_1 = 12.1\ s^{-1} . \qquad (4.275b)$$

The electromagnetic corrections to these numbers have also been calculated. Of course we have already included some electromagnetic effects by using the Coulomb bound state wave function. In addition we must consider order α corrections arising from the 'small' components of the muon and proton spinors, since these particles are not at rest; it turns out, however, that the neglect of these components is justified to $O(\alpha^2)$. We must also include the effects of radiative corrections, being careful not to double count the Coulomb contribution already included. Goldman[33] finds the net effect of the corrections to be

$$\Gamma_o = \Gamma_o^{(0)}[1 - \frac{\alpha}{\pi} 6.3] = 638 \text{ s}^{-1} \tag{4.276a}$$

$$\Gamma_1 = \Gamma_1^{(0)}[1 + \frac{\alpha}{\pi} 5.5] = 12.3 \text{ s}^{-1} . \tag{4.276b}$$

These then are the theoretical predictions and we turn now to the data. The "cleanest" experiment from a theoretical viewpoint is that in which the capture rate is measured in gaseous hydrogen[34]. This essentially measures Γ_o, since at the gas density used atomic collisions rapidly induce transitions from the $F = 1$ state to the $F = 0$ state. The result then is

$$\Gamma_o \big|_{exp} = (651 \pm 57) \text{ s}^{-1} ,$$

which agrees well with the above prediction. With liquid hydrogen as a target the process is complicated by the formation of mesic molecules:

$$(\mu^- p) + (e^- p) \rightarrow (p\mu^- p) + e^- .$$

The mesic molecule $(p\mu^- p)$ can be formed in a para-molecular state $(\ell = 0)$ or an ortho-molecular state $(\ell = 1)$; however it can be shown that production of the ortho-state exceeds production of the para-state by five orders of magnitude. Thus absorption takes place essentially from the ortho-state for which the absorption rate is[35]

$$\begin{aligned} \Gamma_{ortho} &= \tfrac{3}{4} \Gamma_o + \tfrac{1}{4} \Gamma_1 \\ &= 481 \text{ s}^{-1} , \end{aligned} \tag{4.277}$$

using the figures in (4.276). The experiments in liquid hydrogen are done both with counters and bubble chambers. If counters are used, only delayed neutrons can be seen and it can be shown that only Γ_{ortho} is measured. The capture rates measured with counters are

$$\Gamma = (515 \pm 85) \text{ s}^{-1} \quad \text{(Bleser et al. [36])}$$

$$\Gamma = (464 \pm 42) \text{ s}^{-1} \quad \text{(Rothberg et al. [37])}$$

and both of these are consistent with the prediction (4.277). In a bubble chamber experiment all recoil neutrons are seen, and capture occurs from both the mesic atom and the ortho-molecule with relative probabilities of 17% and 83% respectively. Thus the prediction in this case is

$$\begin{aligned}\Gamma &= 0.17\,\Gamma_o + 0.83\,\Gamma_{ortho}\\ &= 508\ s^{-1}\ ,\end{aligned} \tag{4.278}$$

using (4.276) and (4.277). The capture rates measured with bubble chambers are

$$\Gamma = (428 \pm 85) s^{-1} \qquad \text{(Hildebrand}^{(38)}\text{)}$$

$$\Gamma = (450 \pm 50) s^{-1} \qquad \text{(Bertolini et al}^{(39)}\text{)} .$$

These too are consistent with the prediction (4.278).

All of the quantities we have discussed are essentially determined by Γ_o, which in turn is largely determined by $g_1^V(q^2)$ and $g_1^A(q^2)$; in particular the results are not sensitive to the value of $m_\mu g_2^A(q^2)$, so that the uncertainty in its value, which we discussed, turns out to be immaterial. One process which is sensitive to its value is radiative muon capture, in which the capture is accompanied by the emission of a real photon. This has been observed in nuclei but the complications of nuclear physics make it difficult to draw positive conclusions. Such complications are beyond the scope of this book and the interested reader is referred to other texts[(40)].

4.2.8 $K \to Ke\nu$ AND $\Sigma - \Lambda$ BETA DECAY

The quantum numbers of $J_\lambda^{\pm}$ predict the existence of a number of hadronic transitions besides those so far discussed. In the first place, there are direct analogues of pion and nucleon beta decays for other isomultiplets:

$$\begin{aligned}&K^o \to K^+ e^- \bar{\nu}_e\\ &\bar{K}^o \to K^- e^+ \nu_e\\ &\Sigma^o \to \Sigma^+ e^- \bar{\nu}_e\\ &\Sigma^- \to \Sigma^o e^- \bar{\nu}_e\\ &\Xi^- \to \Xi^o e^- \bar{\nu}_e \ .\end{aligned}$$

The decay rates for these processes can immediately be calculated in a way precisely analogous to those already used. For example, the kaon decays above, like π_{e3}, are both pure vector transitions whose hadronic matrix elements follow from CVC. We find

$$< K^+ | J_\lambda^+ | K^o > = - (K_\lambda^+ + K_\lambda^o) + O(\alpha) .$$

The only difference is that π_{e3} has a Clebsch Gordan factor of $\sqrt{2}$. Thus, aside from the phase space difference, $K^o \to K^+ e^- \bar{\nu}_e$ has a decay rate just $\frac{1}{2}$ that of π_{e3}. So from (4.139) we have

$$\Gamma (K^o \to K^+ e^- \bar{\nu}_e) \simeq \frac{G^2 \cos^2 \theta_c}{60 \pi^3} \Delta^5 , \tag{4.279}$$

where now $\Delta \equiv m_{K^o} - m_{K^+}$. This gives a decay rate for this mode of about 0.1 s^{-1}. We have already noted that hypercharge is <u>not</u> conserved by the weak interactions. As a result the K^o state does not have a well defined lifetime. The relevant states for weak processes are the linear combinations of $|K^o>$ and $|\bar{K}^o>$ which diagonalise the effective weak Hamiltonian. Assuming only TCP-invariance, these states are

$$|K^o_{\substack{S \\ L}}> = (1 + |r|^2)^{-\frac{1}{2}} (|K^o> \mp r |\bar{K}^o>) , \tag{4.280}$$

where r is a complex number[41]. The suffixes S and L refer to the experimental fact that the two states actually occurring have short and long lifetimes ($\tau_{K^o_L} \sim 600\ \tau_{K^o_S}$). To calculate the decay rates of these states into $K^+ e^- \bar{\nu}_e$ we need only observe that

$$< K^+ | J_\lambda^+ | \bar{K}^o > = 0 ,$$

since $\bar{K}^o$ and K^+ have the <u>same</u> value of I_3. Thus

$$\Gamma (K^o_S \to K^+ e^- \bar{\nu}_e > = \Gamma (K^o_L \to K^+ e^- \bar{\nu}_e)$$

$$\simeq (1 + |r|^2)^{-1} \frac{G^2 \cos^2 \theta_c}{60 \pi^3} \Delta^5 . \tag{4.281}$$

Likewise,

$$\Gamma(K_S^o \to K^- e^+ \nu_e) = \Gamma(K_L^o \to K^- e^+ \nu_e)$$

$$= |r|^2 \, \Gamma(K_S^o \to K^+ e^- \bar{\nu}_e) \ . \qquad (4.282)$$

We shall see later that

$$|r| \sim 1 \ ,$$

so that all of these decay rates are of the order of $10^{-1}\ s^{-1}$. The total decay rates of these states are both considerably faster

$$\Gamma(K_S^o) \sim 10^{10}\ s^{-1}$$

$$\Gamma(K_L^o) \sim 10^{7}\ s^{-1} \ .$$

Thus the decay modes we have been considering constitute at most 10^{-8} of the total decay rate. They are thus completely unobservable. The physical reason for this is that the other decay modes, into states like $\pi e \nu$ and 2π, have considerably more phase space. The same is true of the above mentioned hyperon decays, which will typically have decay rates of the order of the neutron's. The kinematically favoured decay modes of Σ and Ξ give these particles total decay rates of about $10^{10}\ s^{-1}$.

There is, however, one strangeness conserving hyperon decay which is less afflicted by this phase space inhibition and which has a novel feature not possessed by the processes so far considered. The processes

$$\Sigma^+ \to \Lambda e^+ \nu_e \qquad (4.283a)$$

$$\Sigma^- \to \Lambda e^- \bar{\nu}_e \qquad (4.283b)$$

are both predicted by the Hamiltonian (4.1). The novel feature is that the hadrons do not belong to the same isomultiplet. As a result the energy available ($m_\Sigma - m_\Lambda \simeq 70$ MeV) is considerably larger than that available in ordinary beta decay.

As for neutron decay, Lorentz covariance requires that

$$\langle \Lambda | J_\lambda^+ | \Sigma^- \rangle = \bar{\Lambda} \Big\{ f_1^V(q^2)\,\gamma_\lambda + f_2^V(q^2)\, i\sigma_{\lambda\mu} q^\mu + f_3^V(q^2)\, q_\lambda +$$

$$+ \ f_1^A(q^2)\,\gamma_\lambda\gamma_5 + f_2^A(q^2)\, q_\lambda \gamma_5 + f_3^A(q^2)\, i\sigma_{\lambda\mu}\, q^\mu \gamma_5 \Big\} \Sigma \ , \qquad (4.284a)$$

where

$$q = \Sigma - \Lambda , \tag{4.284b}$$

and the form factors $f_i^V(q^2)$, $f_i^A(q^2)$ $(i = 1,2,3)$ arise respectively from the vector and axial vector parts of J_λ^+. As before, we may use CVC to relate f_i^V to the form factors arising in the electromagnetic decay of the Σ^o,

$$\Sigma^o \to \Lambda\gamma . \tag{4.285}$$

$$\begin{aligned} < \Lambda| V_\lambda^1 + i V_\lambda^2 | \Sigma^- > &= < \Lambda| [j_\lambda , v^1 + i v^2] | \Sigma^- > \\ &= \sqrt{2} < \Lambda| j_\lambda |\Sigma^o > + O(\alpha) , \end{aligned} \tag{4.286}$$

since

$$< \Lambda | (v^1 + i v^2) = 0$$

as Λ is an isoscalar. The matrix element of the electromagnetic current j_λ has the general form

$$< \Lambda|j_\lambda |\Sigma^o> = \bar{\Lambda}\left\{F_1^{\Sigma\Lambda}(q^2)\, \gamma_\lambda + F_2^{\Sigma\Lambda}(q^2)\, i\sigma_{\lambda\mu} q^\mu + F_3^{\Sigma\Lambda}(q^2)\, q_\lambda\right\}\Sigma , \tag{4.287}$$

and using (4.286) we deduce

$$f_i^V(q^2) = \sqrt{2}\; F_i^{\Sigma\Lambda}(q^2) \quad (i = 1,2,3) . \tag{4.288}$$

Current conservation gives

$$< \Lambda | \partial^\lambda j_\lambda | \Sigma^o > = - iq^\lambda < \Lambda| j_\lambda | \Sigma^o > = 0 ,$$

so that

$$(m_\Sigma - m_\Lambda)\, F_1^{\Sigma\Lambda}(q^2) + q^2\, F_3^{\Sigma\Lambda}(q^2) = 0 . \tag{4.289}$$

Hence

$$F_1^{\Sigma\Lambda}(0) = f_1^V(0) = 0 . \tag{4.290}$$

As a first approximation we neglect <u>all</u> q-dependence in the matrix element, so that it is completely determined by $f_1^A(0)$. Then the calculation of the decay rate follows immediately from that for neutron decay

$$\Gamma(\Sigma^- \to \Lambda e^- \bar{\nu}_e) \simeq \frac{3G^2}{2\pi^3} \cos^2\theta_c\, m_e^5 f_o(\beta_o) \,|f_1^A(0)|^2 ,$$

where f_o is given by (4.223b) with

$$m_e \operatorname{ch}\beta_o = m_{\Sigma^-} - m_\Lambda \equiv \Delta_- . \tag{4.291}$$

Given the large value of β_o we may as well neglect the electron's mass, so that

$$m_e^5 f_o(\beta_o) = \int_0^{\Delta_-} dE\, E^2 (E - \Delta_-)^2 = \frac{1}{30} \Delta_-^5 .$$

Thus,

$$\Gamma(\Sigma^- \to \Lambda e^- \bar{\nu}_e) \simeq \frac{G^2}{20\pi^3} \cos^2\theta_c\, \Delta_-^5 \,|f_1^A(0)|^2$$

$$= (0.235 \times 10^{12}\, s^{-1})\, (\Delta_-/m_p)^5 |f_1^A(0)|^2 . \tag{4.292}$$

Now the charge symmetry property (4.247) of $J_\lambda^\pm$ gives

$$<\Lambda| J_\lambda^+ |\Sigma^-> = - <\Lambda| e^{-i\pi v^2} J_\lambda^- e^{i\pi v^2} |\Sigma^->$$

$$= <\Lambda| J_\lambda^- |\Sigma^+> + O(\alpha) , \tag{4.293}$$

since Λ is an isoscalar and Σ an isovector. Thus, as long as we make the approximation of neglecting the q-dependence of the matrix elements, the ratio of the decay rates of the processes (4.283) is determined by phase space alone[32].

$$\rho \equiv \frac{\Gamma(\Sigma^+ \to \Lambda e^+ \nu_e)}{\Gamma(\Sigma^- \to \Lambda e^- \bar{\nu}_e)} = (\Delta_+/\Delta_-)^5 = 0.61 , \tag{4.294}$$

where

$$\Delta_\pm = m_{\Sigma^\pm} - m_\Lambda .$$

Of course, to derive this result we have neglected the $O(\alpha)$ terms in (4.293). But, in the limit of exact isospin symmetry the Σ^+ and Σ^- have the same mass and the phase space available for each decay is the same. This is an abiding problem in applying symmetry considerations to particle physics. Here and elsewhere we adopt the minimal procedure of applying the symmetry constraints to the unknown form factors, while retaining the actual (physically different) spinors and kinematic factors. Plainly the result is valid even if $f_1^V(0)$ does not have the value (4.290) we have deduced. The prediction agrees well with the experimental value

$$\rho_{\text{exp.}} = 0.67 \pm 0.15 \quad , \tag{4.295}$$

although the large experimental errors make a more accurate test desirable.

To determine the absolute decay rates for the processes under consideration we need an estimate of f_1^A. As a start we can generalize the Goldberger-Treiman formula (4.163) to the decays we are considering. This gives

$$f_1^A(0) = (m_\Sigma + m_\Lambda)^{-1} f_\pi g_{\pi\Sigma\Lambda} . \tag{4.296}$$

As before, we have the choice of using the physical value (4.265) of f_π, or the value derived from (4.163) using the physical value (4.236) of $g_1^A(0)$, which gives

$$f_1^A(0) = \frac{\sqrt{2}\, m_N}{m_\Sigma + m_\Lambda} \quad g_1^A(0) \; \frac{g_{\pi\Sigma\Lambda}}{g_{\pi NN}} \quad . \tag{4.297}$$

Both of these expressions require knowledge of the strong interaction coupling constant $g_{\pi\Sigma\Lambda}$, which unfortunately has not been measured. The best we can do is to assume SU(3) symmetry which enables us to express all of the baryon-baryon-meson coupling constants in terms of two parameters. Then using the known values of two of the coupling constants we may evaluate all of the others.

SU(3) Symmetry And The Meson-Baryon Coupling Constants

The general baryon-baryon-meson coupling constant is defined, analogously to (4.159), by

$$< B^j | j_p^i | B^k > = g_{ijk} K(q^2) \bar{B}^j i\gamma_5 B^k \quad , \qquad \text{(no summation)} \tag{4.298}$$

where $i, j, k = 1, \ldots 8$, and linear combinations of these matrix elements specify the physical couplings. The currents j_p^i are the sources of the pseudoscalar meson octet of fields φ_p^i. As before $K(q^2)$ is normalized to unity when $q^2 = m_p^2$. Since we are assuming that φ_p^i belong to an octet representation of SU(3), as defined in (4.34), we have

$$<B^\ell|\mathbb{F}^i\varphi_p^j|B^m> - <B^\ell|\varphi_p^j\,\mathbb{F}^i|B^m> = i f^{ijk} <B^\ell|\varphi_p^k|B^m> . \qquad (4.299)$$

Further, since we are assuming that the baryons also belong to an octet, their field operators B^j transform like the φ^j in (4.34). Thus the state vectors satisfy

$$\mathbb{F}^i\,|B^m> = if^{imk}\,|B^k>$$

$$<B^\ell|\,\mathbb{F}^i = -\,if^{i\ell k} <B^k| \quad ,$$

and if we substitute into (4.299) we obtain the following restriction upon the matrix elements under consideration

$$f^{i\ell k}<B^k|\varphi_p^j|B^m> + f^{imk}<B^\ell|\varphi_p^j|B^k> + f^{ijk}<B^\ell|\varphi_p^k|B^m> = 0 \ .$$

Interpreting SU(3) symmetry in the minimal sense discussed earlier, we impose this condition upon the coupling constants g_{ijk}, rather than the full matrix element. Thus the g_{ijk} satisfy

$$f^{i\ell k}\,g_{jkm} + f^{imk}\,g_{j\ell k} + f^{ijk}\,g_{k\ell m} = 0 \ . \qquad (4.300)$$

By comparing this equation with the Jacobi identity (4.20) we can immediately find one solution to this equation, namely

$$g_{ijk} \propto f^{ijk} \ . \qquad (4.301)$$

In fact there is another Jacobi identity satisfied by the λ^i

$$[\{\lambda^i, \lambda^j\}, \lambda^a] + [\{\lambda^j, \lambda^a\}, \lambda^i] + [\{\lambda^a, \lambda^i\}, \lambda^j] \equiv 0 \ ,$$

and using (4.10) and (4.12) this gives

$$d^{ijk} f^{kam} + d^{jak} f^{kim} + d^{aik} f^{kjm} = 0 \ . \qquad (4.302)$$

Thus (4.300) has a second solution namely

$$g_{ijk} \propto d^{ijk} \ . \qquad (4.303)$$

In fact (4.301) and (4.303) are the only independent solutions to (4.300). This is most easily seen by using the tensor notation used at the beginning of section 4.1 . Thus we denote the baryon octet of fields by B^α_β ($\alpha, \beta = 1,2,3$), and the pseudoscalar meson octet by P^α_β. Both are traceless ($B^\alpha_\alpha = 0$) so that there are eight independent elements as required. Then we are trying to find how many independent SU(3) invariant couplings of the octets $\bar{B}^\alpha_\beta$, B^α_β and P^α_β can be made. Plainly there are just two, as we have asserted,

$$\bar{B}^{\alpha}_{\beta}\, B^{\beta}_{\gamma}\, P^{\gamma}_{\alpha}\,,\quad \bar{B}^{\alpha}_{\beta}\, B^{\gamma}_{\alpha}\, P^{\beta}_{\gamma}\,, \tag{4.304}$$

since there are no other combinations saturating all the indices, given that each tensor is traceless. The two couplings in (4.304) may be combined to form symmetric and antisymmetric couplings corresponding to (4.303) and (4.301).

Let us try to put this result in perspective by considering an analogous problem using the isospin group SU(2). Suppose we wanted to know how many isospin invariant couplings could be made of the Σ hyperons to a pion. The Σ is an isovector denoted $\underline{\Sigma}$, as is the pion $\underline{\pi}$, and the only isoscalar which can be constructed is

$$\underline{\bar{\Sigma}} \times \underline{\Sigma} \,.\, \underline{\pi} = \epsilon_{ijk}\, \bar{\Sigma}^{i}\, \Sigma^{j}\, \pi^{k} \,. \tag{4.305}$$

This is the direct analogue of the antisymmetric coupling (4.301). The product of two isovectors (e.g. $\underline{\bar{\Sigma}}, \underline{\Sigma}$) is not itself an irreducible representation of SU(2) but it can be decomposed into the sum of irreducible representations. In fact the laws for addition of angular momentum or isospin are well known and

$$\underline{1} \times \underline{1} = \underline{0} + \underline{1} + \underline{2} \,.$$

Writing this in terms of the dimension of each representation gives

$$\underline{3} \times \underline{3} = \underline{1} + \underline{3} + \underline{5} \,. \tag{4.306}$$

Now the only representation which can be coupled to the isovector $\underline{\pi}$ to give an isoscalar is the isovector $\underline{3}$, which occurs just once in the product (4.306). This is why there is just <u>one</u> isoscalar $\bar{\Sigma}\Sigma\pi$ coupling given by (4.305).

In the analogous problem in SU(3), we are concerned with the product $\underline{8} \times \underline{8}$ of two octet representations (e.g. B, $\bar{B}$). What we have shown is that when we decompose this product into irreducible representations of SU(3) the representation $\underline{8}$ occurs <u>twice</u>. In fact it can be shown that the analogue of (4.306) is

$$\underline{8} \times \underline{8} = \underline{1} + \underline{8}_s + \underline{8}_a + \underline{10} + \underline{\overline{10}} + \underline{27} \,. \tag{4.307}$$

The antisymmetric combination $\underline{8}_a$ is the analogue of $\underline{3}$ in (4.306), but the occurrence of the symmetric combination $\underline{8}_s$ is a novel feature of SU(3).

Returning to the problem in hand, we have shown that SU(3) invariance requires the coupling constants g_{ijk} in (4.298) to have the form

$$g_{ijk} = -\, i f^{ijk} F + d^{ijk} D \; . \tag{4.308}$$

Insertion of the factor i allows us to take F and D real, as required by T-invariance. The pion nucleon coupling constant defined in (4.160) gives

$$\sqrt{2}\, g_{\pi NN} \sim \frac{1}{\sqrt{2}} < p | j^1 + i j^2 | n > \; . \tag{4.309}$$

The quantum numbers of the proton and neutron require us to take (cf. (4.26))

$$| p > = \frac{1}{\sqrt{2}} (| B^4 > + \; i | B^5 >) \tag{4.310a}$$

$$| n > = \frac{1}{\sqrt{2}} (| B^6 > + \; i | B^7 >) \; . \tag{4.310b}$$

Thus using (4.309), the definition (4.298), (4.308) and the values (4.11) and (4.13), we find

$$\sqrt{2}\, g_{\pi NN} = \frac{1}{\sqrt{2}} (D + F) \; . \tag{4.311}$$

In the same way, since

$$| \Sigma^- > = \frac{1}{\sqrt{2}} (| B^1 > - \, i | B^2 >)$$

and

$$| \Lambda > = | B^8 > \; ,$$

$$g_{\pi\Sigma\Lambda} = \frac{1}{\sqrt{3}} D \; . \tag{4.312}$$

Thus

$$\frac{g_{\pi\Sigma\Lambda}}{g_{\pi NN}} = \frac{2D}{\sqrt{3}\,(F + D)} \; . \tag{4.313}$$

The overall analysis of the low energy strong interaction data[(43)] gives general consistency with SU(3) symmetry and a value

$$\frac{F}{F + D} = 0.41 \pm 0.07 \; . \tag{4.314}$$

Thus using (4.313) and the observed value (4.266) of $g_{\pi NN}$ we can estimate $g_{\pi\Sigma\Lambda}$, which in turn determines $f_1^A(0)$. Using (4.296) and

(4.265) we find

$$f_1^A(0) = -0.52 \quad , \tag{4.315}$$

so that, using (4.292),

$$\Gamma(\Sigma^+ \to \Lambda e^+ \nu_e) = 0.19 \times 10^6 \ s^{-1} \tag{4.316a}$$

$$\Gamma(\Sigma^- \to \Lambda e^- \bar{\nu}_e) = 0.32 \times 10^6 \ s^{-1} \quad . \tag{4.316b}$$

To compare with experiment we convert these into branching ratios using the measured lifetimes[10] of $\Sigma^{\pm}$

$$\rho^+ \equiv \Gamma(\Sigma^+ \to \Lambda e^+ \nu_e)\,\tau\,(\Sigma^+) = 0.16 \times 10^{-4} \tag{4.317a}$$

$$\rho^- \equiv \Gamma(\Sigma^- \to \Lambda e^- \bar{\nu}_e)\,\tau\,(\Sigma^-) = 0.48 \times 10^{-4} \quad . \tag{4.317b}$$

The observed branching ratios are[10]

$$\rho^+ \big|_{\text{exp.}} = (0.202 \pm 0.047)\, 10^{-4} \tag{4.318a}$$

$$\rho^- \big|_{\text{exp.}} = (0.604 \pm 0.060)\, 10^{-4} \quad . \tag{4.318b}$$

The agreement between the two is quite good, bearing in mind the theoretical approximations made. However the predictions (4.317) are 14% larger than those following if we use (4.297) rather than (4.296), so that we are presenting the figures in their most favourable form.

It would be possible to eliminate the assumption of SU(3) invariance for the strong coupling constants $g_{\pi BB}$ if a direct measurement of $g_{\pi\Sigma\Lambda}$ were available, but we would still be using the Goldberger-Treiman formula with the concomitant 14% uncertainty in the predictions. We can dispense with this formula at the expense of assuming SU(3) invariance for the matrix element of the axial vector current, as we shall see in section 4.3.3 .

Aside from these potential improvements, it might be thought that inclusion of the q-dependent terms in (4.284) would improve our theoretical estimates. The terms proportional to q_λ, namely the f_3^V and f_2^A contributions, can be dropped immediately. When coupled to the lepton spinors they yield a factor of m_e using the Dirac equation, so, unless the form factors are immense, they are completely negligible.

The weak magnetism term $f_2^V(0)$ may be determined (in principle) directly from the decay (4.285). Since the photon is on the mass shell, (4.290) shows that $F_1^{\Sigma\Lambda}$ cannot contribute and transversality ensures that $F_3^{\Sigma\Lambda}(0)$ cannot either. Thus the decay rate for (4.285) is entirely determined by the magnetic form factor

$$F_2^{\Sigma\Lambda}(0) \equiv \frac{1}{2m_N} \mu_{\Sigma\Lambda} . \qquad (4.319)$$

A straightforward calculation then gives[44]

$$\Gamma(\Sigma^o \rightarrow \Lambda\gamma) = \alpha \left(\frac{m_\Sigma^2 - m_\Lambda^2}{2m_\Sigma}\right)^3 \left(\frac{\mu_{\Sigma\Lambda}}{m_N}\right)^2$$

$$= \mu_{\Sigma\Lambda}^2 \; (5.0 \times 10^{18} \; s^{-1}) . \qquad (4.320)$$

Unfortunately the decay rate has not been measured, so $\mu_{\Sigma\Lambda}$ is unknown. Our only recourse is to use SU(3) symmetry for the matrix elements of the vector current octet V_μ^i between the states of the baryon octet. This enables us to express $\mu_{\Sigma\Lambda}$ in terms of the neutron's magnetic moment[45] μ_n

$$\mu_{\Sigma\Lambda} = \frac{\sqrt{3}}{2} \mu_n . \qquad (4.321)$$

The inclusion of the weak magnetism term modifies the decay rate (4.292)

$$\Gamma(\Sigma^- \rightarrow \Lambda e^- \bar{\nu}_e) = \frac{G^2 \cos^2\theta_C}{20\pi^3} |f_1^A(0)|^2 \Delta_-^5 \left[1 + \frac{2}{21}\left(\frac{\mu_{\Sigma\Lambda}\Delta_-}{f_1^A(0)m_p}\right)^2\right] . \qquad (4.322)$$

Using the above estimates of $f_1^A(0)$ and $\mu_{\Sigma\Lambda}$ the extra term is quite negligible and increases our estimates by less than 1%. Finally we should consider the term proportional to f_3^A: the axial analogue of weak magnetism. In view of the estimated size of the f_2^V contributions it is most unlikely that this term makes any significant correction to the decay rate, particularly since in the limit of SU(3) symmetry for the matrix elements of the axial vector current octet f_3^A is <u>zero</u>.

4.2.9 QUASI-ELASTIC NEUTRINO SCATTERING

In principle, experiments involving the scattering of neutrinos by various targets offer the best opportunity yet devised of testing the hypotheses which have been advanced to explain the weak interactions. The weak processes so far considered are characterized by a small momentum transfer; the decay processes have q^2 positive, and the largest value yet encountered is in $\Sigma - \Lambda$ decay with $\sqrt{q^2}$ less than about 70 MeV, while muon capture has a negative q^2 and $\sqrt{-q^2}$ about 100 MeV. Thus it has been largely impossible to test for the existence of all the form factors which might exist, let alone to detect the q^2 dependence of those which are predicted to exist. The neutrino experiments have q^2 negative and allow the investigation of weak interaction theory in the region where $\sqrt{-q^2}$ is measured in GeV rather than MeV. Thus, increasing experimental effort is being devoted to this field. The neutrinos used in these experiments are produced by the decay in flight of accelerator produced pions and kaons. We have seen already that the vast majority of pion decays are via the muonic mode, and the same is true of kaon decay. Thus the beam consists chiefly of muonic neutrinos, although there is some admixture of the electronic variety which arises mainly from another kaon decay mode and also from muon decay. Of course the theory for electronic neutrinos precisely parallels that for muonic neutrinos, so we shall consider both cases simultaneously. But it is worth bearing in mind that most of the data relates to the muonic variety.

The target of the neutrino beam is the nucleons in a bubble chamber or spark chamber, and the simplest processes to analyse are the quasi-elastic processes

$$\nu_\ell n \to \ell^- p \tag{4.323a}$$

$$\bar{\nu}_\ell p \to \ell^+ n \ . \tag{4.323b}$$

The hadronic part of the matrix elements has the same form as that in nuclear beta decays. It is therefore straightforward to calculate the differential cross sections, and we merely give an outline. We denote by k, k' respectively the momenta of the initial and final leptons. Likewise p, p' are the momenta of the initial and final hadrons. Thus

$$k + p = k' + p' \ .$$

We define

$$q \equiv k - k' = p' - p \tag{4.324a}$$

and

$$m_N \nu \equiv p.q \quad . \tag{4.324b}$$

Then,

$$k'^2 = m_\ell^2 = (k-q)^2 = q^2 - 2k.q \tag{4.325a}$$

$$p'^2 = m_N^2 = (p+q)^2 = m_N^2 + 2m_N \nu + q^2 \; . \tag{4.325b}$$

Thus in the cases under consideration q^2 and ν are related. As before, the square of the invariant element M is given as in (3.21) by

$$|M|^2 = \tfrac{1}{2} G^2 \cos^2\theta_c \, L_{\alpha\beta} H^{\alpha\beta} , \tag{4.326}$$

where $L_{\alpha\beta}$ is the tensor constructed from the lepton spinors as in (3.20) and $H^{\alpha\beta}$ is the hadronic analogue. If the spin s of the final lepton is not observed we are concerned only with the tensor $\sum_{\pm s} L_{\alpha\beta}$, which can be written in terms of k and q

$$\sum_{\pm s} L_{\alpha\beta} = 8 \{ [\alpha k \beta k\text{-}q] - i(\alpha k \beta q) \} \; . \tag{4.327}$$

This contains terms at most quadratic in the momentum k. In the same way, if we sum over the spin configurations of the final nucleon and average over those of the target nucleon, we may write the hadronic tensor in terms of p and q and like $L_{\alpha\beta}$ it contains terms at most quadratic in p. Of course in this case the tensor is much more complicated than (4.327) because of the extra q-dependence of the hadronic matrix element (4.144). The various terms will be multiplied by functions of q^2 arising from the form factors. From what we have said it is apparent that the spin summed/averaged scalar quantity $\Sigma|M|^2$ is a quadratic function of p.k with coefficients which are functions of q^2 alone; any dependence on k.q or $p.q = m_N \nu$ can be rewritten in terms of q^2 using (4.325). Thus

$$\Sigma|M|^2 = a(q^2) + b(q^2)\, p.k + c(q^2)\, (p.k)^2 \; . \tag{4.328}$$

Thus only <u>three</u> combinations of the six form factors in (4.144) can be measured experimentally if no spins are observed. As in (3.68) the differential cross section measured in the laboratory is related to

$\Sigma|M|^2$ by a relation of the form

$$d\sigma_L \propto \frac{1}{k_o} \Sigma|M|^2 \delta(p+k-p'-k') \frac{d^3p'}{p_o'} \frac{d^3k'}{k_o'} \quad . \tag{4.329}$$

Since

$$\frac{d^3p'}{2p_o'} = d^4p' \, \delta\left(p'^2 - m_N^2\right) ,$$

we may do the p' integration and, since $k_o = E_\nu$ is energy of the incident neutrino, we obtain

$$E_\nu d\sigma_L \propto \Sigma|M|^2 \delta(q^2 + 2m_N\nu) \frac{d^3k'}{k_o'} \quad . \tag{4.330}$$

Now as in (3.73), with $z = \cos\theta$,

$$\frac{d^3k'}{k_o'} = 2\pi|\underline{k}'|dk_o' \, dz \quad . \tag{4.331}$$

Also

$$q^2 = (k-k')^2 = m_\ell^2 - 2E_\nu(k_o' - |\underline{k}'|z)$$

and

$$m_N\nu = p.q = m_N(E_\nu - k_o') \quad .$$

Thus

$$d|q^2|d\nu = 2E_\nu|\underline{k}|dk_o' \, dz \quad , \tag{4.332}$$

and using (4.331) we have

$$\frac{d^3k'}{k_o'} = \frac{\pi}{E_\nu} d|q^2|d\nu \quad . \tag{4.333}$$

So finally we have

$$E_\nu^2 d\sigma_L \propto \Sigma|M|^2 \delta(q^2 + 2m_N\nu) \, d|q^2|d\nu \quad . \tag{4.334}$$

Thus doing the trivial ν integration we find that $E_\nu^2 \dfrac{d\sigma_L}{d|q^2|}$ is a quadratic function[(46)] of $p.k = m_N E_\nu$. In fact this result is true even if the lepton tensor has a structure as complicated as the hadronic tensor's. It derives from the 'vector-vector' form of the matrix element, which ensures that only zero or one unit of angular momentum is exchanged between leptons and hadrons[47].

To write down the explicit form of $\frac{d\sigma_L}{d|q^2|}$ it is convenient to rewrite E_ν in terms of $s - u$, where

$$s \equiv (p+k)^2 = m_N^2 + 2m_N E_\nu \quad , \tag{4.335a}$$

$$u \equiv (p-k')^2 = m_N^2 + m_\ell^2 - q^2 - 2m_N E_\nu \quad . \tag{4.335b}$$

Thus

$$s - u = 4m_N E_\nu + q^2 - m_\ell^2 \tag{4.335c}$$

is linear in E_ν, and a quadratic in E_ν may be rewritten as a quadratic in $(s - u)$. The advantage of using this variable is that the relationship between the cross sections for the two processes under consideration is much clearer. To see this we first note that the charge symmetry condition (4.250) shows that the hadronic tensor is the same for both processes. However, the lepton matrix elements differ; we have

$$\bar{u}_\ell(k')\gamma_\alpha(1-\gamma_5)u_{\nu_\ell}(k) \text{ for } \nu_\ell \to \ell^-$$

and

$$\bar{v}_{\nu_\ell}(k)\gamma_\alpha(1-\gamma_5)v_\ell(k') \text{ for } \bar{\nu}_\ell \to \ell^+ \; .$$

Thus

$$L_{\alpha\beta}^{\nu_\ell \to \ell^-}(k,k') = L_{\alpha\beta}^{\bar{\nu}_\ell \to \ell^+}(-k',-k) \; , \tag{4.336}$$

since use of the v-spinors amounts to changing the sign of the momenta. Thus

$$d\sigma_L^{\nu_\ell \to \ell}(s,q^2,u) = d\sigma_L^{\bar{\nu}_\ell \to \ell^+}(u,q^2,s) \; , \tag{4.337}$$

so finally we have the general result

$$\frac{d\sigma_L}{d|q^2|}\begin{pmatrix}\nu_\ell n \to \ell^- p \\ \nu_\ell p \to \ell^+ n\end{pmatrix} = \frac{G^2\cos^2\theta_c}{8\pi m_N^2 E_\nu^2}\left[A(q^2) \mp B(q^2)(s-u) + C(q^2)(s-u)^2\right] . \tag{4.338}$$

The calculation of A, B, C is straightforward but tedious, and we merely quote the result[48]. The expressions for A, B and C can be simplified, if we neglect the mass m_ℓ of the lepton, which is a good approximation at high energies. With this approximation g_3^V and g_2^A cannot contribute, since they yield terms proportional to m_ℓ when contracted with the lepton current by the Dirac equation. Then

$$A(q^2) = \tfrac{1}{4} q^2 \left(q^2 - 4m_N^2\right)\left(|g_1^A(q^2)|^2 + q^2 |g_3^A(q^2)|^2\right) + 2m_N (q^2)^2 \operatorname{Re} g_1^V(q^2) g_2^V(q^2)^*$$
$$+ \tfrac{1}{4} q^2 \left(q^2 + 4m_N^2\right)\left(|g_1^V(q^2)|^2 + q^2 |g_2^V(q^2)|^2\right) \quad (4.339a)$$

$$B(q^2) = -q^2 \operatorname{Re} g_1^A(q^2)^* \left[g_1^V(q^2) + 2m_N g_2^V(q^2)\right] \quad (4.339b)$$

$$C(q^2) = \tfrac{1}{4}|g_1^A(q^2)|^2 - \tfrac{1}{4} q^2 |g_3^A(q^2)|^2 + \tfrac{1}{4}|g_1^V(q^2)|^2 - \tfrac{1}{4} q^2 |g_2^V(q^2)|^2 \ . \quad (4.339c)$$

Finally we should note the kinematic restrictions on the values of q^2 which are accessible for a given value of E_ν. In the approximation we are using $(m_\ell = 0)$

$$-q^2 = 2 E_\nu k_o' (1-z) \ ,$$

so that

$$0 \le -q^2 \le 4 E_\nu k_o' \ ,$$

with the lower bound occurring at $z = 1$ and the upper at $z = -1$. Now using (4.325b) we have when $z = -1$

$$-q^2 = 4 E_\nu k_o' = 2 m_N \nu = 2m_N (E_\nu - k_o') \ ,$$

so that

$$k_o' = m_N E_\nu^{-1} (2 E_\nu + m_N)^{-1} \ .$$

Thus the kinematically accessible values of q^2 are

$$0 \le |q^2| \le 4m_N E_\nu^2 (2 E_\nu + m_N)^{-1} \ . \quad (4.340)$$

As a check on our result let us evaluate the cross sections in the (fictitious) case that the nucleons are <u>point</u> particles like the leptons. That is to say we take

$$g_1^V(q^2) = -g_1^A(q^2) = 1 \ ,$$

and all other form factors zero. Then

$$A - B(s-u) + C(s-u)^2 = 8\,m_N^2 E_\nu^2$$

and

$$\frac{d\sigma_L}{d|q^2|}(\nu_\ell n \to \ell^- p) = \frac{1}{\pi} G^2 \cos^2\theta_c \quad . \tag{4.341}$$

Hence integrating over the allowed kinematic region (4.340) we find the total cross section

$$\sigma_L(\nu_\ell n \to \ell^- p) = \frac{G^2}{\pi} \cos^2\theta_c\, 4m_N E_\nu^2\, (2E_\nu + m_N)^{-1} \quad . \tag{4.342}$$

Aside from the $\cos^2\theta_c$ and m_N instead of m_ℓ, this is precisely the result already obtained for the leptonic case in (3.78). In the same way we may derive the analogue of (3.79), and as before in the high energy limit [cf. (3.80)]

$$\sigma_L(\nu_\ell n \to \ell^- p) = 3\sigma_L(\bar{\nu}_\ell p \to \ell^+ n) = \frac{2G^2}{\pi} \cos^2\theta_c\, m_N E_\nu \quad . \tag{4.343}$$

In the approximation we are using only four of the possible six hadronic form factors can contribute, and only three linear combinations of these can be observed, as we have seen. Thus to extract any information about any one of these surviving form factors we shall be compelled to make further assumptions. Of course, if our hypotheses are correct, the charge symmetry property of $J_\lambda^\pm$ and T-invariance predict g_3^A is zero, as already noted following (4.257). Even if g_3^A is non-zero its (presumably small) value will be difficult to detect, since it enters the cross sections only in the form $|g_3^A|^2$ - there are no interference terms with other 'large' form factors. In fact the accuracy of the experiments available <u>at present</u> is such that the best we can do is to use <u>all</u> of the hypotheses so far advanced and then use the data to fix the q^2-dependence of the remainder.

We have seen that the vector form factors $g_{1,2}^V$ are specified by CVC in terms of the nucleon electromagnetic form factors, which are known up to $-q^2 = 25\ (\text{GeV})^2$ from electron scattering experiments. It turns out that, if one uses Sachs form factors G_E and G_M, rather than Dirac form factors F_1 and F_2, then both electric and magnetic form factors may be parametrized by a double-pole fit with the same mass m_V. Thus we have from (4.146)

$$g_1^V(q^2) = F_1^p(q^2) - F_1^n(q^2) \equiv \left(1 - \frac{q^2}{4m_N^2}\right)^{-1} \left[G_E^V(q^2) - \frac{q^2}{4m_N^2} G_M^V(q^2)\right] ,$$

(4.344a)

$$-g_2^V(q^2) = F_2^p(q^2) - F_2^n(q^2) \equiv \left(1 - \frac{q^2}{4m_N^2}\right)^{-1} \left[G_E^V(q^2) - G_M^V(q^2)\right] \frac{1}{2m_N} ,$$

(4.344b)

where

$$G_E^V(q^2) = \left(1 - \frac{q^2}{m_V^2}\right)^{-2} , \quad G_M^V(q^2) = (1 + \varkappa_p - \varkappa_n) \left(1 - \frac{q^2}{m_V^2}\right)^{-2} \qquad (4.344c)$$

and

$$m_V = (0.84 \pm 0.03) \text{ GeV} . \qquad (4.344d)$$

If we now drop g_3^A which is predicted to be zero, this leaves only the form factor $g_1^A(q^2)$ to be determined by experiment. Even so the data is not good enough to fix g_1^A, without specifying its normalization and shape. In fact, as we have seen, the normalization is known, and it is customary to <u>assume</u> a double pole shape for g_1^A. Thus we have

$$g_1^A(q^2) = -1.25 \left(1 - \frac{q^2}{m_A^2}\right)^{-2} , \qquad (4.345)$$

and the experiments consist of determining the single remaining unknown m_A. With this parametrization the experiments give

$$m_A = (0.87 \pm 0.12) \text{ GeV} , \qquad (4.346)$$

a value remarkably close to m_V given in (4.344d). However it must be emphasised that the shape assumption (4.345) is purely ad hoc. Plainly it is an attractive hypothesis that the vector and axial vector form factors have the same shape, but there is, as yet, no known physical reason why this should be so. It is therefore most important to improve the experiments and just measure $g_1^A(q^2)$ directly.

The best way of doing this <u>in principle</u> is to use both neutrino and anti-neutrino experiments. Then from (4.338) and (4.339b)

$$\frac{d\sigma_L}{d|q^2|}(\nu_\ell n \to \ell^- p) - \frac{d\sigma_L}{d|q^2|}(\bar{\nu}_\ell p \to \ell^+ n) = \frac{G^2 \cos^2\theta_c}{8\pi m_N^2 E_\nu^2}(q^2 + 4m_N E_\nu) 2q^2 g_1^A(q^2) G_M^V(q^2) , \tag{4.347}$$

since (4.344) shows that

$$G_M^V = g_1^V(q^2) + 2m_N g_2^V(q^2) . \tag{4.348}$$

However, this method is not feasible until the neutrino and anti-neutrino fluxes are better known.

In any case, if we wish to measure the form factor at large values of $-q^2$, it is essential to use high energies because of the kinematic constraint (4.340). But at high energies

$$s - u \sim 4m_N E_\nu ,$$

using (4.335), so that the most important term in the cross sections is $C(q^2)(s-u)^2$ which contributes equally to both ν and $\bar{\nu}$ processes. It follows that measurement of the difference of the unpolarized cross sections is not a good way of determining the high $-q^2$ behaviour of $g_1^A(q^2)$. The polarization of the recoil nucleon is proportional[49] to $g_1^A(q^2)$, and measurement of it might be a worthwhile experiment at high energies.

In any event it would be of great interest to check the internal self-consistency of the theory by comparing two determinations of $g_1^A(q^2)$ from the same process, as it would be to compare these with the form factor as determined from some other process such as electro-production[50].

Finally, for future reference, we note the following simple properties[51] of the quasi-elastic differential cross sections

$$\frac{d\sigma}{d|q^2|}(\nu_\ell n \to \ell^- p)\Big|_{q^2=0} = \frac{d\sigma}{d|q^2|}(\bar{\nu}_\ell p \to \ell^+ n)\Big|_{q^2=0}$$

$$= \frac{G^2 \cos^2\theta_c}{2\pi}\left(|g_1^V(0)|^2 + |g_1^A(0)|^2\right) \tag{4.349a}$$

$$\lim_{E_\nu \to \infty} \frac{d\sigma}{d|q^2|}(\nu_\ell n \to \ell^- p) = \lim_{E_\nu \to \infty} \frac{d\sigma}{d|q^2|}(\bar{\nu}_\ell p \to \ell^+ n)$$

$$= \frac{G^2 \cos^2\theta_c}{2\pi}\left(|g_1^V(q^2)|^2 - q^2|g_2^V(q^2)|^2 + |g_1^A(q^2)|^2 - q^2|g_2^A(q^2)|^2\right). \tag{4.349b}$$

Both of these follow directly from (4.338) and (4.339).

4.2.10 EXCLUSIVE INELASTIC NEUTRINO SCATTERING

In addition to the quasi-elastic scattering processes discussed in the previous section the Hamiltonian (4.1) also predicts the existence of innumerable inelastic processes when a neutrino or anti-neutrino scatters from a nucleon target. The adjective "exclusive" is used to describe those processes in which a particular final hadronic state is identified. At moderate energies and momentum transfers the kinematically allowed final states are limited. In particular the first inelastic processes which may be studied are those when a pion is produced with the nucleon or when the excited nucleon resonances Δ are produced. The detailed dynamical (and kinematical) description of these processes is extremely complicated[52] and beyond the scope of this book. We shall confine our remarks to the selection rules which follow from the assumed quantum numbers of $J_\lambda^\pm$. These all follow from its isovector character as expressed by the commutation relation

$$[v^1 + i\,v^2, J_\lambda^+] = 0 \ . \tag{4.350}$$

We now take matrix elements between $<\Delta^{++}|$ and $|n>$. The Δ^{++} has isospin quantum numbers $(I, I^3) = (\frac{3}{2}, \frac{3}{2})$. Thus using (4.27) we have

$$(v^1 - i\,v^2)|\Delta^{++}> = \sqrt{3}\,|\Delta^+>$$

and, since

$$(v^1 + i\,v^2)|n> = |p> \ ,$$

we have from (4.350)

$$<\Delta^{++}|J_\lambda^+|p> = \sqrt{3}\ <\Delta^+|J_\lambda^+|n> \ . \tag{4.351}$$

Thus we may relate the invariant matrix elements

$$M\,(\nu_\ell p \to \ell^- \Delta^{++}) = \sqrt{3}\ M\,(\nu_\ell n \to \ell^- \Delta^+) \ .$$

At high energies, where phase space differences are negligible, we have then for the cross sections

$$\sigma\,(\nu_\ell p \to \ell^- \Delta^{++}) = 3\,\sigma\,(\nu_\ell n \to \ell^- \Delta^+) \ , \tag{4.352}$$

and in the same way

$$\sigma(\bar{\nu}_\ell n \to \ell^+ \Delta^-) = 3\sigma(\bar{\nu}_\ell p \to \ell^+ \Delta^0) \quad . \tag{4.353}$$

In fact, the charge symmetry condition (4.247) enables us to relate the matrix elements of J^+_λ to those of J^-_λ. As we have already noted, the symmetry condition expresses the fact that J^+_λ and J^-_λ belong to the same (isovector) multiplet. We may express this fact another way using the isospin raising operator $v^1 + i\,v^2$; if it is applied twice to J^-_λ, plainly we obtain a multiple of J^+_λ, and the commutation relations (4.53) show

$$\left[v^1 + i v^2 , [\, v^1 + i v^2 , J^-_\lambda \,] \right] = -2 J^+_\lambda \; . \tag{4.354}$$

Taking matrix elements between $< \Delta^{++} |$ and $|p>$ and proceeding as above we obtain

$$\begin{aligned} \sqrt{3} < \Delta^0 | J^-_\lambda | p > &= - < \Delta^{++} | J^+_\lambda | p > \\ &= - \sqrt{3} < \Delta^+ | J^+_\lambda | n > . \end{aligned}$$

Thus we have the 'mirror' relations

$$< \Delta^0 | J^-_\lambda | p > = - < \Delta^+ | J^+_\lambda | n > \tag{4.355a}$$

$$< \Delta^- | J^-_\lambda | n > = - < \Delta^{++} | J^+_\lambda | p > . \tag{4.355b}$$

However, it is important to realize that these do <u>not</u> imply that the cross sections for $\bar{\nu}_\ell p \to \ell^+ \Delta^0$ and $\nu_\ell n \to \ell^- \Delta^+$ are equal, since the lepton spinors differ in the two cases. Tests of the charge symmetry condition have been discussed by Llewellyn-Smith and Pais[(53)].

Analogous relations can be derived for the processes

$$\nu_\ell N \to \ell^- N \pi \tag{4.356a}$$

and

$$\bar{\nu}_\ell N \to \ell^+ N\pi \; . \tag{4.356b}$$

The commutation relation (4.350) yields

$$< n\pi^+ | J^+_\lambda | n> - \sqrt{2} < p\pi^0 | J^+_\lambda | n> \;=\; < p\,\pi^+ | J^+_\lambda | p> \tag{4.357a}$$

$$< p\pi^- | J^-_\lambda | p> + \sqrt{2} < n\pi^0 | J^-_\lambda | p> \;=\; < n\,\pi^- | J^-_\lambda | n> \; , \tag{4.357b}$$

with corresponding relationships between the invariant scattering amplitudes. In principle we may test these sum rules as they stand, although they do not convert into corresponding relationships between cross sections. Such triangular relationships between complex numbers do however imply triangular inequalities for their magnitudes and therefore for the square roots of the cross sections. Thus

$$\sqrt{\sigma}(\nu_\ell n \to \ell^- n \pi^+) + \sqrt{2}\sqrt{\sigma}(\nu_\ell n \to \ell^- p \pi^o) \leq \sqrt{\sigma}(\nu_\ell p \to \ell^- p \pi^+) \tag{4.358a}$$

$$\sqrt{\sigma}(\nu_\ell n \to \ell^- n \pi^+) + \sqrt{\sigma}(\nu_\ell p \to \ell^- p \pi^+) \leq \sqrt{2}\sqrt{\sigma}(\nu_\ell n \to \ell^- p \pi^o) \tag{4.358b}$$

$$\sqrt{\sigma}(\nu_\ell p \to \ell^- p \pi^+) + \sqrt{2}\sqrt{\sigma}(\nu_\ell n \to \ell^- p \pi^o) \leq \sqrt{\sigma}(\nu_\ell n \to \ell^- n \pi^+) , \tag{4.358c}$$

with similar inequalities for the processes (4.356b). The commutator (4.354) yields charge symmetry relations as before

$$< p\pi^+ | J_\lambda^+ | p > \; = \; < n\pi^- | J_\lambda^- | n > \tag{4.359a}$$

$$< n\pi^+ | J_\lambda^+ | n > \; = \; < p\pi^- | J_\lambda^- | p > \tag{4.359b}$$

$$< p\pi^o | J_\lambda^+ | n > \; = \; - < n\pi^o | J_\lambda^- | p > . \tag{4.359c}$$

4.2.11 INCLUSIVE INELASTIC NEUTRINO PROCESSES

Inclusive processes are defined as those in which only the final lepton is observed. Thus the inclusive inelastic processes are

$$\nu_\ell N \to \ell^- X \tag{4.360a}$$

$$\bar{\nu}_\ell N \to \ell^+ X , \tag{4.360b}$$

where X is any hadronic state other than the nucleon. Since we are concerned in this section only with hypercharge conserving processes, we further restrict X to have the same hypercharge ($Y = 1$) as the target nucleon. This of course requires some minimal observation of X, but the generalization to hypercharge nonconserving processes is straightforward and will be dealt with in 4.3.5 . As for the elastic processes, we denote by k, k' the momenta of the initial and final leptons and by p, p' the momenta of the initial and final hadronic states. Measurement of the momenta k, k' means that

$$q = k - k' \tag{4.361}$$

is completely determined. Thus another way of looking at these processes is that we are measuring the total cross sections for scattering a fictitious particle of momentum q on a nucleon target. This is illustrated in Fig. 4.3.

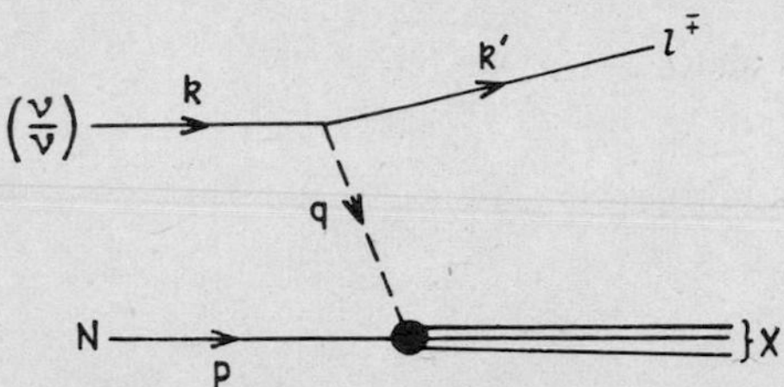

Fig. 4.3. Inclusive inelastic neutrino scattering.

Of course, if the W-boson exists as discussed in Chapter 3, q is the momentum of the virtual W in this process; it plainly is not a real W, since $q^2 \le 0$ in these processes, and it is largely immaterial to the considerations of this section whether or not the W actually exists.

Adler's Theorems

We consider first a theorem due to Adler[54], which shows how we may use these processes to test two fundamental hypotheses of weak interactions: CVC and PCAC. As before, we assume that the energies involved are high enough for us to neglect the mass m_ℓ of the final state lepton. The "parallel" configuration is defined as that in which the final lepton energes parallel to the incident neutrino $\underline{k} \,||\, \underline{k}'$. Because of the masslessness of the neutrino and lepton it follows that their momenta are proportional

$$k \propto k' \propto q \quad , \tag{4.362}$$

where $q(= k - k')$ is non-zero because we are considering an inelastic process. The invariant matrix element for the process (4.360a), for example, is given by

$$M = \frac{G}{\sqrt{2}} \cos\theta_c \, \bar{u}(k') \, \gamma^\lambda (1-\gamma_5) \, u(k) \; \langle X | J_\lambda^+ | N \rangle \quad , \tag{4.363}$$

where u is the zero mass neutrino spinor discussed in section 2.4.2 .

It follows from (2.245) that u(k) has the dimensions of $[M^{\frac{1}{2}}]$. Thus

$$u(k) = \left(\frac{k_o}{q_o}\right)^{\frac{1}{2}} u(q)$$

and

$$\bar{u}(k') = \left(\frac{k'_o}{q_o}\right)^{\frac{1}{2}} \bar{u}(q) \; .$$

Hence using (2.245) and (2.246a) we have

$$\bar{u}(k')\,\gamma_\lambda(1-\gamma_5)\,u(k) = 4\,(k_o k'_o)^{\frac{1}{2}}\,\frac{q_\lambda}{q_o} \; , \qquad (4.364)$$

so that

$$(k_o k'_o)^{-\frac{1}{2}}\,M = -\,i\;G2\sqrt{2}\,\cos\theta_c\,q_o^{-1} < X|\partial^\lambda J^+_\lambda|N> \; . \qquad (4.365)$$

Now CVC says that the vector part of J^+_λ is conserved, so using PCAC, as expressed in (4.152) and (4.156), we have

$$\partial^\lambda J^+_\lambda = \partial^\lambda (A^1_\lambda + i\,A^2_\lambda) = f_\pi m^2_\pi\,\varphi_{(\pi^-)} .$$

This enables us to relate the matrix element M to that for pion-nucleon scattering; using (4.158), since $q^2 = 0$, we have

$$(k_o k'_o)^{-\frac{1}{2}}\,M = -\,i2\sqrt{2}\,f_\pi\,G\cos\theta_c\,q_o^{-1} < X|J_{(\pi^-)}|N>$$

$$= 2\sqrt{2}\,f_\pi q_o^{-1}\,G\cos\theta_c < X|\pi^+N;\,in> \; .$$

Thus, proceeding as before, it is straightforward to relate the cross section for the process (4.360a) to that for πN-scattering, with the result

$$\left.\frac{d^2\sigma}{d|q^2|\,d\nu}(\nu_\ell N\to\ell^- X)\right|_{\theta=0} = \left(\frac{f_\pi G\cos\theta_c}{2\pi}\right)^2 \frac{2(E_\nu-\nu)}{\nu E_\nu}\,\sigma^o(\pi^+N\to X;\nu) \; , \qquad (4.366a)$$

where

$$W^2 \equiv (p+q)^2 = m^2_N + 2m_N\nu + q^2 = m^2_N + 2\,m_N\nu \; , \qquad (4.366b)$$

$$m_N\nu \equiv p.q \; , \qquad (4.366c)$$

and $\sigma^o(\pi^+N\to X;\nu)$ is the cross section for scattering (zero-massed) pions on nucleons to the state $|X>$ at centre of mass energy W given by (4.366b).

Unfortunately it is rather difficult to test the prediction (4.366), since there are no events at $\theta = 0$. This means that we must accept events with small finite values of q^2 and θ and attempt to extrapolate. But rapid variations may occur in this neighbourhood, which would make firm conclusions hard to draw. One possibility might be to use a phenomenological model of a particular process, such as Δ production, and extract the required value from that. However most such phenomenological descriptions use the Born term, and the Goldberger-Treiman formula guarantees[55] that this term satisfies Adler's theorem (4.366). A crude test of the theorem has been made[56] which is consistent with (4.366) but a stringent test is still awaited.

The inelastic neutrino processes also provide a test of the current algebra assumptions independently of PCAC[57]. We have seen how current algebra plus PCAC leads to the Adler-Weisberger formula (4.211). PCAC is used to relate D^{ij}, defined in (4.170b), to the πN-scattering amplitude. We have also seen in (4.366) how PCAC may be used to relate πN-scattering to the neutrino scattering processes (4.360). It is plain, therefore, that without using PCAC we may derive a sum rule for $|g_1^A(0)|^2$ in terms of the differential cross sections for the processes (4.360). The result of such a derivation is obtained by substituting (4.366) into (4.211), having summed over all states $|X>$:

$$|g_1^A(0)|^2 - 1 = \frac{2\pi}{G^2\cos^2\theta_c} \int_{m_\pi + m_\pi^2/2m_N}^{\infty} d\nu \left(1 - \frac{\nu}{E_\nu}\right)^{-1} \left[\frac{d^2\sigma}{d|q^2|\,d\nu}(\nu_\ell p) - \frac{d^2\sigma}{d|q^2|\,d\nu}(\bar{\nu}_\ell p)\right] . \tag{4.367}$$

If we now take the limit $E_\nu \to \infty$, we may rewrite this as

$$\lim_{E_\nu \to \infty} \left[\frac{d\sigma}{d|q^2|}(\bar{\nu}_\ell p) - \frac{d\sigma}{d|q^2|}(\nu_\ell p)\right]_{q^2=0} = \frac{G^2\cos^2\theta_c}{2\pi}\left[2 - |g_1^A(0)|^2 - |g_1^V(0)|^2\right],$$

since $g_1^V(0) = 1$, or more compactly using (4.349a)

$$\lim_{E_\nu \to \infty} \left[\frac{d\sigma}{d|q^2|} \left(\bar{\nu}_\ell p \to \ell^+ X(Y=1) \right) - \frac{d\sigma}{d|q^2|} \left(\nu_\ell p \to \ell^- X(Y=1) \right) \right]_{q^2=0} = \frac{G^2}{\pi} \cos^2\theta_c , \tag{4.368}$$

where now the cross sections include the contribution from the quasi-elastic process (4.323b) as well as the inelastic processes (4.360). For scattering by neutrons a similar formula may be obtained in which the right-hand side of (4.368) has its sign reversed.

In fact (4.368) is a special case of a more general relation proved by Adler[58] for general q^2, not just $q^2 = 0$. To derive this general relation we consider first the hadronic tensor $W^{(\nu)}_{\mu\lambda}$ which occurs in any process of the type

$$\nu_\ell N \to \ell^- X(Y=1) , \tag{4.369}$$

where N is a nucleon and X is any hadronic state with the same hypercharge (Y = 1). To compute the total cross section for such a process we need to know

$$W^{(\nu)}_{\mu\lambda} \equiv (2\pi)^3 (2m_N)^{-1} \tfrac{1}{2} \sum_{s_N} \sum_{|X>} <N| J^-_\mu |X><X| J^+_\lambda |N> \delta(p+k-p'-k') , \tag{4.370}$$

where, as before, k, k' are the momenta of the initial and final leptons and p, p' are the momenta of the initial and final hadron states; the factor $\frac{1}{2} \sum_{s_N}$ means that we are concerned with the scattering from an unpolarized target nucleon. Since the leptons' momenta k, k' enter only in the combination

$$q = k - k'$$

in the delta function, Lorentz covariance alone shows that $W^{(\nu)}_{\mu\lambda}$ has the general form

$$W^{(\nu)}_{\mu\lambda} = -g_{\mu\lambda} W^{(\nu)}_1 + \frac{p_\mu p_\lambda}{m_N^2} W^{(\nu)}_2 - \tfrac{1}{2} i m_N^{-2} (\mu\lambda pq) W^{(\nu)}_3 + \\ + \frac{q_\mu q_\lambda}{m_N^2} W^{(\nu)}_4 + \frac{1}{2m_N^2} \left(p_\mu q_\lambda + p_\lambda q_\mu \right) W^{(\nu)}_5 + \frac{i}{2m_N^2} \left(p_\mu q_\lambda - p_\lambda q_\mu \right) W^{(\nu)}_6 , \tag{4.371}$$

where the $W^{(\nu)}_i$ are scalar functions which therefore can only depend on

q^2 and $\nu = p.q/m_N$:

$$W_i^{(\nu)} = W_i^{(\nu)}(q^2, \nu) \ .$$

It follows from the definition (4.370) that

$$W_{\mu\lambda}^{*} = W_{\lambda\mu} \ , \qquad (4.372)$$

from which we deduce that the $W_i^{(\nu)}$ are <u>real</u> functions.

Positivity Constraints On W_i

The form (4.370) of $W_{\mu\lambda}^{(\nu)}$ means that it is positive semi-definite, i.e.

$$\epsilon^{\mu *} W_{\mu\lambda} \epsilon^{\lambda} \geq 0 \qquad (4.373)$$

for any complex vector ϵ. This leads to a number of inequalities restricting the possible values of the functions W_i. For example, we may take ϵ to be the polarization vector of the (fictitious) W-boson absorbed by the target nucleon. Then (4.373) expresses the fact that the absorption cross section must be positive. If we work in the laboratory frame in which

$$p = (m_N, 0, 0, 0) \ , \quad q = (\nu, 0, 0, q_3) \ , \qquad (4.374)$$

then the polarization vectors representing left and right-handed states of the W are

$$\epsilon_{\substack{L \\ R}} = \frac{1}{\sqrt{2}} (0, 1, \mp i, 0) \ , \qquad (4.375a)$$

and the third polarization vector is

$$\epsilon_S = \frac{1}{\sqrt{-q^2}} (q_3, 0, 0, \nu) \qquad (q_3 \geq 0) \ . \qquad (4.375b)$$

Substituting (4.375a) into (4.373) gives

$$\tfrac{1}{2} (W_{11} + W_{22} \mp i W_{12} \pm i W_{21}) \geq 0 \ . \qquad (4.376)$$

In the laboratory frame (4.374) gives

$$W_{11} = W_{22} = W_1$$

$$W_{12} = -W_{21} = -\tfrac{1}{2} i \frac{q_3}{m_N} W_3 \ ,$$

so that (4.376) yields

$$W_1 \mp \frac{q_3}{2m_N} W_3 \geq 0 . \tag{4.377}$$

In the same way substituting (4.375) for ϵ_S gives

$$(-q^2)^{-1} \left[q_3^2 W_{00} + \nu^2 W_{33} - \nu q_3 (W_{03} + W_{30}) \right] \geq 0 ,$$

which yields

$$-W_1 + \left(1 - \frac{\nu^2}{q^2}\right) W_2 \geq 0 . \tag{4.378}$$

So altogether, since $q_3 = (\nu^2 - q^2)^{\frac{1}{2}}$, we have[59]

$$0 \leq \frac{1}{2m_N}\left(\nu^2 - q^2\right)^{\frac{1}{2}} |W_3| \leq W_1 \leq \left(1 - \frac{\nu^2}{q^2}\right) W_2 . \tag{4.379}$$

There are a number of other inequalities which may be deduced from (4.373) involving the 'structure functions' $W_{4,5,6}$. However we see from the definition (4.371) that these terms are always multiplied by q_μ or q_λ. When multiplied by the lepton tensor $L^{\mu\lambda}$ these terms give contribution proportional to m_ℓ, by the Dirac equation. They are therefore extremely difficult to detect at high energies, and henceforth we make the approximation $m_\ell = 0$; in fact the term W_6 is not present in any case if, as hitherto, we assume T-invariance. For a full discussion of all of the inequalities following from positivity the interested reader is referred to Doncel and de Rafael[60].

Adler's Generalized Sum Rule

To derive the sum rule under discussion we rewrite (4.370)

$$\begin{aligned}(2m_N) W^{(\nu)}_{\mu\lambda} &= \frac{1}{4\pi} \sum_{s_N} \sum_{|X>} \int dx\, e^{iqx} <N(p)|J^-_\mu(x)|X(p')><X(p')|J^+_\lambda|N(p)> \\ &= \frac{1}{4\pi} \sum_{s_N} \int e^{iqx} <N(p)|J^-_\mu(x) J^+_\lambda|N(p)>\, dx \end{aligned} \tag{4.380}$$

using translation invariance and $\sum_{|X>} |X><X| = 1$. Next, consider

$$\int dx\, e^{iqx} < N(p) | J^{+}_{\lambda}\, J^{-}_{\mu}(x) | N(p) >$$

$$= \sum_{|X>} < N(p) | J^{+}_{\lambda} | X(p') > < X(p') | J^{-}_{\mu} | N(p) > (2\pi)^4 \delta(q + p' - p)$$

using the same properties. Plainly this quantity is non-zero only when

$$W^2 = p'^2 = (p-q)^2 = m_N^2 - 2m_N \nu + q^2 \geq m_N^2 ,$$

and this inequality cannot be satisfied with $q^2 \leq 0$ and $\nu \geq 0$. It follows that we may replace the current product $J^{-}_{\mu}(x)\, J^{+}_{\lambda}$ in (4.380) by the commutator. Thus

$$(2m_N) \int dq_o\, W^{(\nu)}_{oo} = \int dq_o (2\pi)^{-1} \tfrac{1}{2} \sum_{s_N} \int dx\, e^{iqx} < N(p) | [J^{-}_{o}(x), J^{+}_{o}] | N(p) >$$

$$= \tfrac{1}{2} \sum_{s_N} \int dx\, e^{-i\underline{q}.\underline{x}}\, \delta(x^o) < N(p) | [J^{-}_{o}(x), J^{+}_{o}] | N(p) > ,$$

which is determined by the current commutators given in (4.69). These give

$$\delta(x^o)\, [J^{-}_{o}(x), J^{+}_{o}] = - 4\, \delta(x)\, (V^3_o + A^3_o) .$$

Now V^3_o is the isovector current, so

$$< N(p) | V^3_o | N(p) > = \bar{u}(p)\, \gamma_o\, u(p) < I_3 > = 2 p_o < I_3 > ,$$

using (2.145a). $< I_3 >$ is the third component of isospin of the nucleon N. By virtue of (2.145d) the matrix element of A^3_o is proportional to $(s_N)_o$, so its spin averaged value is zero

$$\sum_{s_N} < N, s_N | A^3_o | N, s_N > = 0 .$$

Putting all this together we have

$$(2m_N) \int dq_o\, W^{(\nu)}_{oo} = - 8\, p_o < I_3 > . \qquad (4.381)$$

Next we change the integral over q_o to one over ν and work in a frame where $\underline{p}.\underline{q} = 0$. Then

$$m_N \nu = p_o q_o - \underline{p}.\underline{q} = p_o q_o$$

and

$$dq_o = \frac{m_N}{p_o} d\nu .$$

We divide both sides of (4.381) by p_o and let $|\underline{p}| \to \infty$, then since

$$p_o^{-2} W_{oo}^{(\nu)} \to m_N^{-2} W_2^{(\nu)} ,$$

we deduce from (4.381) that

$$\int_{-\infty}^{\infty} d\nu\, W_2^{(\nu)} (q^2, \nu) = -4 < I_3 > . \tag{4.382}$$

To change this into an integral over the physically accessible values of ν we use a crossing property of $W_{\mu\lambda}$. This is most easily deduced from (4.380) with the current product replaced by the commutator.

$$W_{\mu\lambda}^{(\nu)} (p, q) \sim \int dx\, e^{iqx} < N | [J_\mu^-(x), J_\lambda^+] | N > .$$

Likewise,

$$W_{\mu\lambda}^{(\bar{\nu})} (p, q) \sim \int dx\, e^{iqx} < N | [J_\mu^+(x), J_\lambda^-] | N > .$$

Hence

$$W_{\mu\lambda}^{(\bar{\nu})} (p, -q) \sim \int dx\, e^{iqx} <N| [J_\mu^+ (-x), J_\lambda^-] |N>$$

$$= \int dx\, e^{iqx} <N| [J_\mu^+, J_\lambda^- (x)] | N> .$$

Thus

$$W_{\mu\lambda}^{(\bar{\nu})} (p, -q) = - W_{\lambda\mu}^{(\nu)} (p, q) . \tag{4.383}$$

If we write $W_{\mu\lambda}^{(\bar{\nu})}$ in the same form as $W_{\mu\lambda}^{(\nu)}$ in (4.371), the crossing property (4.383) implies

$$W_i^{(\nu)} (q^2, \nu) = - W_i^{(\bar{\nu})} (q^2, -\nu) \quad (i \neq 5) \tag{4.384a}$$

$$W_5^{(\nu)} (q^2, \nu) = + W_5^{(\bar{\nu})} (q^2, -\nu) . \tag{4.384b}$$

With the aid of this relation we may rewrite the integral (4.382)

$$\int_0^{\infty} d\nu \left[W_2^{(\nu)} (q^2, \nu) - W_2^{(\bar{\nu})} (q^2, \nu) \right] = -4 < I_3 > . \tag{4.385}$$

It is straightforward to relate $W_2^{(\bar{\nu})}$ to the differential cross section for the process (4.369)

$$d\sigma(\nu_\ell N \to \ell^- X) = \frac{G^2}{2}\cos^2\theta_c (2\pi)^{-2} \frac{m_N}{4E_\nu p_o} \frac{d^3k'}{k'_o} \sum_{s_\ell} L^{\lambda\mu} W^{(\nu)}_{\mu\lambda} , \tag{4.386}$$

where $L^{\lambda\mu}$ is given by (3.20) with m_e replaced by m_ℓ. Thus with the form (4.371) we find

$$\frac{d^2\sigma}{d|q^2|d\nu}(\nu_\ell N \to \ell^- X) = \frac{G^2}{4\pi}\cos^2\theta_c E_\nu^{-2}\left\{-q^2 W_1^{(\nu)} + W_2^{(\nu)}\left[2E_\nu(E_\nu - \nu) + \tfrac{1}{2}q^2\right]\right.$$
$$\left. + W_3^{(\nu)} \frac{q^2}{2m_N}(2E_\nu - \nu)\right\} , \tag{4.387}$$

so that

$$\lim_{E_\nu \to \infty} \frac{d^2\sigma}{d|q^2|d\nu}(\nu_\ell N \to \ell^- X) = \frac{G^2}{2\pi}\cos^2\theta_c W_2^{(\nu)}(q^2, \nu) . \tag{4.388}$$

Substituting in (4.385) we obtain

$$\lim_{E_\nu \to \infty}\left[\frac{d\sigma}{d|q^2|}\left(\bar{\nu}_\ell N \to \ell^+ X(Y=1)\right) - \frac{d\sigma}{d|q^2|}\left(\nu_\ell N \to \ell^- X(Y=1)\right)\right] = \frac{G^2}{\pi}\cos^2\theta_c \langle 2I_3 \rangle. \tag{4.389}$$

When N is a proton $\langle 2I_3 \rangle = +1$, and, as promised, the sum rule is a generalization of (4.368) but with q^2 <u>not</u> restricted to zero. This property, that the right-hand side of (4.389) is a non-zero constant independent of q^2, has been encountered previously. As shown in (4.341) it is a characteristic of the scattering from point-like objects, like the leptons; it is apparent from (4.349b) that the contribution $\frac{d\sigma}{d|q^2|}$ from any individual hadronic state X, the single nucleon, for example, will <u>not</u> have this property. Rather, the weak form factors ensure that at large q^2 this contribution falls off like $(q^2)^{-3}$, if we suppose that the form factors continue to possess a double pole shape.

Bjorken's Scaling Hypothesis

Another way to see the point-like behaviour implied by Adler's sum rules is to go back to (4.385). We define a new variable

$$x = \frac{-q^2}{2m_N \nu} \tag{4.390}$$

Then, since q^2 is fixed,

$$\frac{dx}{x} = -\frac{d\nu}{\nu} .$$

The functions $W_2(q^2, \nu)$ have dimension $[M^{-1}]$, so without loss of generality we may write

$$\nu W_2^{(\nu)}(q^2, \nu) \equiv G_2^{(\nu)}\left(x, \frac{q^2}{m_N^2}\right). \tag{4.391}$$

We may now rewrite the sum rule (4.385) as an integral over x

$$4 < I_3 > = \int_0^\infty d\nu \left[W_2^{(\bar{\nu})}(q^2, \nu) - W_2^{(\nu)}(q^2, \nu) \right]$$

$$= \int_{-q^2/2m_N}^\infty d\nu \, \frac{1}{\nu} \left[\nu W_2^{(\bar{\nu})}(q^2, \nu) - \nu W_2^{(\nu)}(q^2, \nu) \right]$$

$$= \int_0^1 \frac{dx}{x} \left[G_2^{(\bar{\nu})}\left(x, \frac{q^2}{m_N^2}\right) - G_2^{(\nu)}\left(x, \frac{q^2}{m_N^2}\right) \right] . \tag{4.392}$$

Note that the change of the lower limit of integration in the second equation is justified since, by definition, $W_2 = 0$ unless $(p+q)^2 \geq m_N^2$. Written in this form the sum rule suggests that, as $-q^2 \to \infty$ with x fixed, the individual functions approach non-zero limits

$$\lim_{\substack{-q^2 \to \infty \\ x \text{ fixed}}} \nu W_2^{(\nu, \bar{\nu})}(q^2, \nu) \equiv \lim_{-q^2 \to \infty} G_2^{(\nu, \bar{\nu})}\left(x, \frac{q^2}{m_N^2}\right) = F_2^{(\nu, \bar{\nu})}(x) \neq 0 .$$

Plainly this is a sufficient condition for (4.392) but not necessary.

It is true of course for point-like particles, as is apparent from (4.328) and (4.334). Using (4.388) we see that

$$W_2(q^2, \nu) \propto m_N C(q^2) \, \delta(q^2 + 2m_N \nu)$$

with $C(q^2)$ given by (4.339c). For a point-like particle $C(q^2)$ is a constant, so, since $x \equiv -q^2/2m_N \nu$, we have

$$\nu W_2 \propto \nu m_N \, \delta\left[2m_N \nu (1-x)\right] = \tfrac{1}{2} \delta(1-x) .$$

Thus by definition (4.391),

$$G_2\left(x, \frac{q^2}{m_N^2}\right) \propto \delta\,(1-x) \quad , \tag{4.393}$$

which is indeed independent of q^2. In the same way, by considering terms linear and independent of E_ν in (4.388), we find for a point particle

$$\nu W_3\,(q^2, \nu) \propto m_N \nu\, \delta(q^2 + 2m_N\nu) \propto \delta(1-x) \tag{4.394a}$$

and

$$m_N W_1\,(q^2, \nu) \propto \nu m_N\, \delta\,(q^2 + 2m_N\nu) \propto \delta\,(1-x) \quad . \tag{4.393b}$$

This is the germ of Bjorken's Scaling Hypothesis[61], which states that with the definitions of the dimensionless functions $G_i\,(x, q^2/m_N^2)$

$$m_N W_1\,(q^2, \nu) \equiv G_1\left(x, \frac{q^2}{m_N^2}\right) \tag{4.395a}$$

$$\nu W_2\,(q^2, \nu) \equiv G_2\left(x, \frac{q^2}{m_N^2}\right) \tag{4.395b}$$

$$\nu W_3\,(q^2, \nu) \equiv G_3\left(x, \frac{q^2}{m_N^2}\right) \quad , \tag{4.395c}$$

$$\lim_{\substack{-q^2\to\infty \\ x\ \text{fixed}}} G_i\left(x, \frac{q^2}{m_N^2}\right) = F_i(x) \neq 0 \quad . \tag{4.395d}$$

We may rewrite the differential cross section (4.387) in terms of the G_i. Defining

$$x = -q^2/(2m_N\nu) \qquad \text{and} \qquad y = \nu/E_\nu \quad , \tag{4.396}$$

the Jacobian gives

$$d|q^2|d\nu = 2m_N E_\nu^2\, y\, dx\, dy\, .$$

Then (4.387) has the form

$$\frac{d^2\sigma}{dx\,dy}\,(\nu N) = \frac{G^2}{\pi}\cos^2\theta_c\, m_N E_\nu \left\{ G_2^{(\nu)}\left(1-y-\frac{m_N}{E_\nu}\,x\,y\right) + x y^2\, G_1^{(\nu)} - \right.$$
$$\left. - x y \left(1 - \frac{y}{2}\right) G_3^{(\nu)} \right\} \quad . \tag{4.397}$$

Thus if the scaling hypothesis (4.395) is correct,

$$\lim \begin{Bmatrix} -q^2 \to \infty \\ E_\nu \to \infty \\ x, y \text{ fixed} \end{Bmatrix} \frac{d^2\sigma}{dx\,dy}(\nu N) = \frac{G^2}{\pi} \cos^2\theta_c \, m_N E_\nu \Big\{ F_2^{(\nu)}(x)(1-y) + $$

$$+ xy^2 F_1^{(\nu)}(x) - xy(1-\frac{y}{2}) F_3^{(\nu)}(x) \Big\} , \qquad (4.398)$$

A similar form is of course obtained if we consider the $\bar{\nu}_\ell$ process (4.360b)

$$\lim \begin{Bmatrix} -q^2 \to \infty \\ E_\nu \to \infty \\ x, y \text{ fixed} \end{Bmatrix} \frac{d^2\sigma}{dx\,dy}(\bar{\nu} N) = \frac{G^2}{\pi} \cos^2\theta_c \, m_N E_\nu \Big\{ F_2^{(\bar{\nu})}(x)(1-y) + $$

$$+ xy^2 F_1^{(\bar{\nu})}(x) + xy(1-\frac{y}{2}) F_3^{(\bar{\nu})}(x) \Big\} . \qquad (4.399)$$

It is easy to understand how the F_3 term has opposite sign in this case. The term in question arises from the antisymmetric part $\sim (\alpha, k, \beta, k')$ of $L_{\alpha\beta}$. The relationship between the lepton tensors for $\nu_\ell \to \ell^-$ and $\bar{\nu}_\ell \to \ell^+$ is given in (4.336), and

$$(\alpha, -k', \beta, -k) = -(\alpha, k, \beta, k')$$

$$[\alpha, -k', \beta, -k] = +[\alpha, k, \beta, k'] .$$

This explains the sign flip between (4.398) and (4.399).

Kinematic Restrictions

To obtain the total cross section we must integrate over the kinematically allowed values of x and y. To find this region we note

$$\nu = E_\nu - k_o' \;:\; k_o' \geq 0$$

$$Q^2 \equiv -q^2 = 2E_\nu k_o'(1-z) \;:\; z = \cos\theta , \; |z| \leq 1 .$$

Also

$$m_x^{2\,\prime} = (p+q)^2 = m_N^2 + 2m_N\nu - Q^2 \geq m_N^2 .$$

Thus

$$Q^2 \leq 2m_N\nu , \qquad (4.400a)$$

$$z \leq 1 \Rightarrow Q^2 \geq 0 \qquad (4.400b)$$

and

$$z \geq -1 \Rightarrow Q^2 \leq 4E_\nu k'_o = 4E_\nu (E_\nu - \nu) . \tag{4.400c}$$

The physical region in the $(Q^2, 2m_N \nu)$ plane is shown in Fig. 4.4.

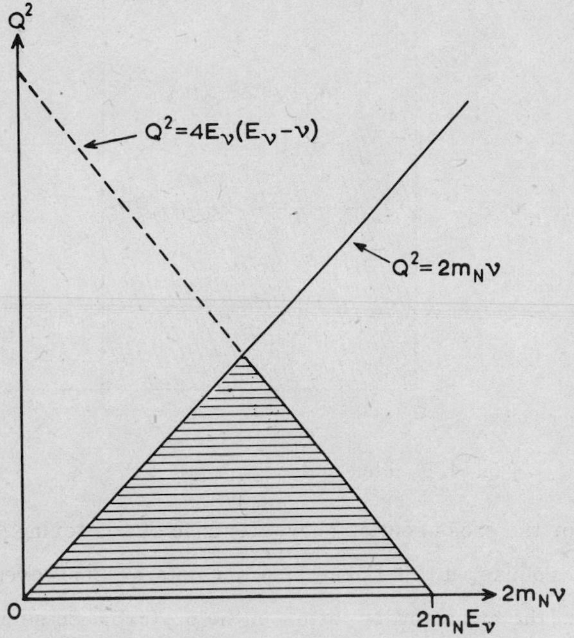

Fig. 4.4 Physical region for $\nu N \to \ell X$

In terms of the variables x and y (4.400) yields the restrictions

$$0 \leq Q^2 \leq 2m_N \nu \Rightarrow 0 \leq xy \leq y \tag{4.401a}$$

$$Q^2 \leq 4E_\nu (E_\nu - \nu) \Rightarrow m_N xy \leq 2E_\nu (1-y) \tag{4.401b}$$

and the physical region in the (x, y) plane is shown in Fig. 4.5.

Total Cross Sections

In the limit $E_\nu \to \infty$ we see that the physical region occupies the whole of the square

$$0 \leq x \leq 1 \quad , \quad 0 \leq y \leq 1 \quad . \tag{4.402}$$

Thus, the computation of the total cross section is trivial. We find

$$\sigma\begin{pmatrix}\nu \\ \bar{\nu}\end{pmatrix} = \frac{G^2}{\pi} \cos^2\theta_c \, m_N E_\nu \int_0^1 dx \left[\tfrac{1}{2} F_2(x) + \tfrac{1}{3} x F_1(x) \mp \tfrac{1}{3} x F_3(x)\right] , \tag{4.403}$$

where the superfixes $^{(\nu)}$ or $^{(\bar{\nu})}$ are left understood. As might have been anticipated the scaling hypothesis has led to the linear energy

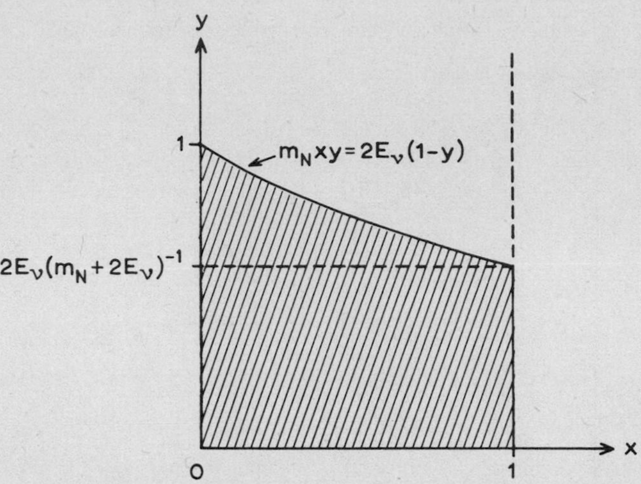

Fig. 4.5 Physical region for $\nu N \to \ell X$

dependence for the cross section characteristic of scattering from a point particle. Of course, if this linear rise persists, we shall eventually have to worry about the higher order terms in the perturbation series, as discussed in Chapter 1. So we are assuming that the scaling hypothesis is valid at energies where it is still all right to use first order perturbation theory. In fact this linear energy dependence is precisely what is seen in the high energy neutrino experiments, as we shall see shortly. To go further we need more information on the structure functions $F_i(x)$. At the moment all we know is that they are constrained by the positivity conditions (4.379). Rewriting these in terms of G_i, defined in (4.395), we find

$$0 \le \tfrac{1}{2}\left(1 - x^2 \frac{4m_N^2}{q^2}\right) | G_3 | \le G_1 \le \frac{1}{2x} \left(1 - x^2 \frac{4m_N^2}{q^2}\right) G_2 \ .$$

Then taking the "Bjorken limit" ($-q^2 \to \infty$ with x fixed) we obtain

$$0 \le x \mid F_3(x) \mid \le 2xF_1(x) \le F_2(x) \ . \tag{4.404}$$

The Adler theorem (4.392) tells us that

$$\int_0^1 \frac{dx}{x}\left[F_2^{(\bar{\nu})}(x) - F_2^{(\nu)}(x)\right] = 4 < I_3 > . \qquad (4.405)$$

The derivation of the actual form of the structure functions requires the assumption of a specific model. In any case the consideration of a specific model has considerable pedagogic value in clarifying and crystallizing the ideas we have introduced.

The Quark-Parton Model[62]

We have seen that Bjorken's scaling hypothesis leads to behaviour of the cross sections which is characteristic of scattering from point-like particles. So the simplest way to "derive" the scaling behaviour is to postulate that in the kinematic region under consideration the nucleon behaves as if it were composed of point-like constitutents known as "partons". What could these partons be ? Evidently the simplest hypothesis is that they are "quarks". By this we mean that they belong to the fundamental triplet representation of SU(3) and that they are fermions with spin $\frac{1}{2}$. After all, we have already postulated that the octet of vector and axial vector currents transforms in the same way as the octets of currents (4.65) constructed from the quark fields q. The quantum numbers of this fundamental representation are given in (4.64)

$$(Y, I, I_3) = (\tfrac{1}{3}, \tfrac{1}{2}, \tfrac{1}{2}), (\tfrac{1}{3}, \tfrac{1}{2}, -\tfrac{1}{2}), (-\tfrac{2}{3}, 0, 0) \qquad (4.64a)$$

$$Q = \tfrac{2}{3} \quad , \quad -\tfrac{1}{3} \quad , \quad -\tfrac{1}{3} . \qquad (4.64b)$$

We denote these three states by

$$"q" = "p" \quad , \quad "n" \quad , \quad "\lambda" \quad , \qquad (4.406)$$

the notation reflecting that the isospin, but not the hypercharge, is the same as the proton, neutron and lambda hyperon. Obviously we need at least three of these constituents to construct another fermion, like the nucleon, which has integral hypercharge; and in fact with three quarks we can arrange that the nucleons belong to an octet representation of SU(3), as is experimentally well-verified. With just three quarks the only way to construct the nucleons is

$$\text{proton} \sim \text{"p" "p" "n"} \tag{4.407}$$

$$\text{neutron} \sim \text{"p" "n" "n"} \,. \tag{4.408}$$

Let us now compute the structure functions for neutrino scattering from one of these constituents assuming, as postulated, that they are point-like. By definition (4.370)

$$W_{\mu\lambda}^{(\nu\,\text{"n"})} = (2\pi)^3 (2\mu)^{-1} \tfrac{1}{2} \sum_{s_n} \sum_{|x>} <\text{"n"}|J_\mu^-|X><X|J_\lambda^+|\text{"n"}> \delta(p+k-p'-k') \,,$$

where μ is the mass of "n". Now J_λ^+ has $(Y, I, I^3) = (0, 1, +1)$ and, since we are assuming a point-like interaction, the only possibility for the state $|X>$ is that it is $|\text{"p"}>$. Furthermore

$$<\text{"p"}|J_\lambda^+|\text{"n"}> = \bar{p}\,\gamma_\lambda\,(g_V + g_A\,\gamma_5)\; n \;, \tag{4.409}$$

where g_V and g_A are constants, since any momentum dependence of the matrix element would be at variance with our assumption of a point-like interaction. Thus

$$W_{\mu\lambda}^{(\nu\,\text{"n"})} = \int \frac{d^3p'}{2p_o'} (2\mu)^{-1} \tfrac{1}{2} \,\text{Tr} \left[\gamma_\mu(g_V+g_A\gamma_5)(\not{p}'+\mu)\gamma_\lambda(g_V+g_A\gamma_5)(\not{p}+\mu)\right] \delta(p+k-p'-k')$$

$$= \int d^4p'\delta(p'^2-\mu^2)\delta(p+k-p'-k')\mu^{-1} \Big\{ g_{\mu\lambda}\mu^2(g_V^2-g_A^2)+(g_V^2+g_A^2)[\mu p'\lambda p] - $$

$$- 2i\,g_V\,g_A\,(\mu p'\lambda p) \Big\}$$

$$= \mu^{-1}\delta(q^2+2p.q) \Big\{ g_{\mu\lambda}\left[-2g_A^2\mu^2-(g_V^2+g_A^2)p.q\right] + p_\mu p_\lambda 2(g_V^2+g_A^2) - $$

$$- \frac{i}{2}\,(\mu\lambda p q)\,4 g_V g_A + \cdots\cdots \Big\} \,. \tag{4.410}$$

We observe that the W_i for scattering from a single parton has the characteristic δ-function form already noted. In the same way we consider the scattering of ν_ℓ off a "p" target. Since there is no quark having $I^3 = \frac{3}{2}$, we deduce immediately that

$$W_{\mu\lambda}^{(\nu\,\text{"p"})} = 0 \,. \tag{4.411}$$

Now let us turn to the problem which really concerns us. We want to calculate the structure functions for ν_ℓ scattering off of a proton target, for example. We assume the scattering is incoherent, so that the cross section is simply the sum of the cross sections for scattering from the partons inside the proton. The hadrons enter the expression for the cross section only in the combination

$$\frac{1}{2P_o}(2\pi)^3 \tfrac{1}{2}\sum_{s_N}\sum_{|X>} <N|J^-_\mu|X><X|J^+_\lambda|N>\delta(p+k-p'-k') \equiv \frac{m_N}{P_o} W^{(\nu N)}_{\mu\lambda} \quad . \tag{4.370}$$

Thus it is the quantities $\frac{\mu}{p_o} W_{\mu\lambda}(p,q)$ which are additive. Let $u_{n,p}(f)$ be the probability of a "n", "p" quark in the proton having momentum $p = fP\,(0 \le f \le 1)$, where now P is the momentum of the target proton. Thus we are assuming that each parton carries a fraction f of the proton's momentum and are ignoring the possibility of the partons having momenta $\underline{p}$ perpendicular (or transverse) to $\underline{P}$. The above statement about additivity gives

$$\frac{m_N}{P_o} W^{(\nu p)}_{\mu\lambda}(P,q) = \int_0^1 df \left\{ u_n(f)\, \frac{\mu}{p_o} W^{(\nu''n'')}_{\mu\lambda}(p,q) + u_p(f)\, \frac{\mu}{p_o} W^{(\nu''p'')}_{\mu\lambda}(p,q) \right\} \quad , \tag{4.412}$$

and using (4.410) and (4.411) we may immediately identify the various structure functions $W^{(\nu p)}_i$ which we want. For example, identifying the terms proportional to $p_\mu p_\lambda$ gives

$$\frac{P_\mu P_\lambda}{m_N P_o} W^{(\nu p)}_2 = \int_0^1 df\, u_n(f)\, \frac{\mu}{p_o}\, \mu^{-1}\, \delta(q^2 + 2p.q)\; 2(g_V^2 + g_A^2)\, p_\mu p_\lambda$$

$$= 2(g_V^2 + g_A^2)\, \frac{P_\mu P_\lambda}{P_o} \int_0^1 df\, u_n(f)\, f\, \delta(q^2 + 2fP.q) \ .$$

Thus, recalling $m_N\nu = P.q$ and $x = -q^2/2m_N\nu$, we have

$$m_N^{-1} W^{(\nu p)}_2 = 2(g_V^2 + g_A^2) \int_0^1 df\, u_n(f)\, f\, \delta\left[2m_N\nu(f-x)\right]$$

$$= \frac{g_V^2 + g_A^2}{m_N\nu}\, u_n(x)\, x \quad .$$

Hence

$$\nu W_2^{(\nu p)} \equiv G_2^{(\nu p)}\left(x, \frac{q^2}{m_N^2}\right) = \left(g_V^2 + g_A^2\right) x\, u_n(x) = F_2^{(\nu p)}(x) \quad . \tag{4.413a}$$

In the same way we find

$$m_N W_1^{(\nu p)} \equiv G_1^{(\nu p)}\left(x, \frac{q^2}{m_N^2}\right) = \left[\tfrac{1}{2}\left(g_V^2 + g_A^2\right) - 2 g_A^2 \frac{\mu^2}{q^2}\right] u_n(x)$$

$$\overset{Bj}{\rightarrow} \tfrac{1}{2}\left(g_V^2 + g_A^2\right) u_n(x) = F_1^{(\nu p)}(x) \tag{4.413b}$$

$$\nu W_3^{(\nu p)} = G_3^{(\nu p)}\left(x, \frac{q^2}{m_N^2}\right) = 2 g_V g_A u_n(x) = F_3^{(\nu p)}(x) \quad . \tag{4.413c}$$

Thus we see that the structure functions $F_i^{(\nu p)}$ are simply related to the longitudinal momentum distribution of the "n" parton inside the proton. Plainly this is a feature not just of the simple model we have been considering but of any model in which the nucleon is composed of point-like constituents. Notice that the structure functions we have derived satisfy the positivity constraints (4.404), as they must. In fact in this model we saturate one of the inequalities, since

$$2 x F_1^{(\nu p)}(x) = F_2^{(\nu p)}(x) \quad . \tag{4.414}$$

This is the Callan-Gross relation. Again, it is not a specific feature of this model. It derives from (4.410) and would remain true even if we had more quarks and/or anti-quarks inside the nucleon. It depends specifically only on the assumption that the constituents have spin ½ rather than spin 0[63]. Further, it is not specific to the particular process which we have considered; it is true both for ν and $\bar{\nu}$ scattering on n or p. Of course, if we are really taking this quark-parton model seriously, we should take

$$g_V = - g_A = 1 \tag{4.415}$$

in (4.409), since these are the couplings required to ensure that our current algebra assumptions are satisfied. With these values all of the positivity inequalities are saturated and

$$-x F_3^{(\nu p)}(x) = 2x F_1^{(\nu p)}(x) = F_2^{(\nu p)}(x) = 2x u_n(x) . \tag{4.416}$$

In the same way we may consider the scattering of anti-neutrinos $\bar{\nu}_\ell$ by a proton target with the result

$$-x F_3^{(\bar{\nu} p)}(x) = 2x F_1^{(\bar{\nu} p)}(x) = F_2^{(\bar{\nu} p)}(x) = 2x u_p(x) . \tag{4.417}$$

We might also consider the scatterings from a neutron target. The structure functions would be determined by $u'_{n,p}$, which specify the longitudinal momentum distributions inside a neutron. But by isospin symmetry

$$u'_n(x) = u_p(x) , \quad u'_p(x) = u_n(x) . \tag{4.418}$$

Thus,

$$F_i^{(\nu n)}(x) = F_i^{(\bar{\nu} p)}(x) , \quad F_i^{(\bar{\nu} n)}(x) = F_i^{(\nu p)}(x) . \tag{4.419}$$

In fact these relations follow directly, and independently of the model, from the charge symmetry property of the currents $J_\mu^\pm$ given in (4.247). This yields

$$<p| [J_\mu^-(x) , J_\lambda^+] |p> = <n| [J_\mu^+(x) , J_\lambda^-] |n> ,$$

$$W_{\mu\lambda}^{(\nu p)} = W_{\mu\lambda}^{(\bar{\nu} n)} ,$$

and similarly

$$W_{\mu\lambda}^{(\bar{\nu} p)} = W_{\mu\lambda}^{(\nu n)} .$$

Thus all of the weak structure functions are determined by the unknown distributions $u_n(x)$ and $u_p(x)$. The same distributions also determine the electromagnetic structure functions which are measured in deep inelastic electron scattering experiments

$$eN \to eX . \tag{4.420}$$

The electromagnetic structure functions are defined as in (4.370) with the electromagnetic current j_μ, j_λ replacing J_μ^-, J_λ^+. As before, we may calculate $W_1^{(\gamma N)}$ and $W_2^{(\gamma N)}$ ($W_3^{(\gamma N)}$ is zero since j_μ is a pure vector current) on the assumption of a point electromagnetic structure for the partons. That is to say we take

$$< "q" | j_\lambda | "q" > = \bar{q}\, \gamma_\lambda\, Q_q\, q \ , \tag{4.421}$$

where "q" = "p", "n" and Q_q is given by (4.64b). We leave this calculation as an exercise and simply quote the result

$$2\,x\, F_1^{(\gamma p)}(x) = F_2^{(\gamma p)}(x) = x \left[\frac{4}{9}\, u_p(x) + \frac{1}{9}\, u_n(x)\right] \tag{4.422a}$$

$$2\,x\, F_1^{(\gamma n)}(x) = F_2^{(\gamma n)}(x) = x \left[\frac{4}{9}\, n_n(x) + \frac{1}{9}\, u_p(x)\right] \ . \tag{4.422b}$$

It follows that the weak structure functions are completely determined by the electromagnetic structure functions. In terms of the F_2 structure functions we have

$$F_2^{(\nu p)} + F_2^{(\nu n)} = F_2^{(\bar{\nu} n)} + F_2^{(\bar{\nu} p)} = 2x\,(u_p + u_n) = \frac{18}{5}\left(F_2^{(\gamma p)} + F_2^{(\gamma n)}\right) \tag{4.423a}$$

$$F_2^{(\nu p)} - F_2^{(\nu n)} = F_2^{(\bar{\nu} n)} - F_2^{(\bar{\nu} p)} = 2x\,(u_n - u_p) = 6\left(F_2^{(\gamma n)} - F_2^{(\gamma p)}\right) \tag{4.423b}$$

Now the Adler Sum rule (4.405) tells us something about the isovector combination of structure functions. On a proton target, for example, we have, using (4.419) and (4.423b),

$$\int_o^1 \frac{dx}{x}\left[F_2^{(\bar{\nu} p)} - F_2^{(\nu p)}\right] = \int_o^1 \frac{dx}{x}\left[F_2^{(\bar{\nu} p)} - F_2^{(\bar{\nu} n)}\right]$$

$$= \int_o^1 dx\; 2\,(u_p - u_n)$$

$$= 2 \ ,$$

since $< I^3 >_p = \frac{1}{2}$. Thus we deduce

$$\int_o^1 dx \left[\tfrac{1}{2} u_p - \tfrac{1}{2} u_n\right] = \tfrac{1}{2} \ . \tag{4.424}$$

In fact we could have written this down a priori, since it simply states that I^3 for the proton is the sum of the I^3 carried by "p" quarks and "n" quarks. Indeed we might conjecture that the probability distributions for any individual quark inside the proton were all equal. Then, since a proton contains two "p" quarks and one "n" quark, we would deduce

$$\tfrac{1}{2}\, u_p(x) = u_n(x) \equiv u(x) \ . \tag{4.425}$$

The isospin sum rule (4.424) then follows simply from the fact that the probability that an individual quark carries <u>some</u> fraction of the proton's momentum is unity

$$\int_0^1 dx\, u(x) = 1 \quad . \tag{4.426}$$

Under this <u>strongest</u> hypothesis we have finally for the F_2's

$$F_2^{(\nu p)} = F_2^{(\bar{\nu} n)} = 2x\, u(x) \tag{4.427a}$$

$$F_2^{(\bar{\nu} p)} = F_2^{(\nu n)} = 4x\, u(x) \tag{4.427b}$$

$$F_2^{(\gamma p)} = x\, u(x) \tag{4.427c}$$

$$F_2^{(\gamma n)} = \tfrac{2}{3}\, x\, u(x) \quad . \tag{4.427d}$$

The data on the neutrino scattering processes is not accurate enough to permit the determination of the actual structure functions. However the electromagnetic structure functions $F_2^{(\gamma p)}$ and $F_2^{(\gamma n)}$ are well measured and do <u>not</u> have the same shape. The ratio $F_2^{(\gamma n)}/F_2^{(\gamma p)}$, which according to (4.427c,d) would be $\frac{2}{3}$, in fact varies from 1 to about $\frac{1}{4}$. It therefore seems likely that the strongest hypothesis (4.425) is FALSE, and we shall study the experimental implications of the somewhat weaker conclusions contained in (4.423). Nevertheless simple integrals

$$\int_0^1 dx\, [\alpha\, u_p(x) + \beta\, u_n(x)] \quad (\alpha, \beta \text{ constant})$$

have the values they would have if (4.425) were true. It is therefore useful as a mnemonic. For example, since the proton has unit charge,

$$\int_0^1 dx\, [\tfrac{2}{3} u_p - \tfrac{1}{3} u_n] = 1 \quad .$$

So using (4.424) we deduce

$$\int_0^1 dx\, \tfrac{1}{2}\, u_p = \int_0^1 dx\, u_n = 1 \quad , \tag{4.429}$$

as implied by (4.425, 6) (but <u>not</u> conversely).

Comparison With The Data

This then is the naive quark parton model. Obviously the treatment we have followed could be generalized to allow the nucleon to contain arbitrary mixtures of quarks and anti-quarks consistent with the overall nucleon quantum numbers. And a number of the sum rules have a more general validity than our treatment would indicate. The astonishing thing is that this simple model, in so far as it has been checked experimentally, is reasonably consistent with the data.

Consider first the total cross sections. At present these are measured only for "isospin averaged" targets. That is to say only the quantities

$$\sigma(\nu N) \equiv \tfrac{1}{2}[\sigma(\nu p) + \sigma(\nu n)] \tag{4.430a}$$

and

$$\sigma(\bar{\nu} N) \equiv \tfrac{1}{2}[\sigma(\bar{\nu} p) + \sigma(\bar{\nu} n)] \tag{4.430b}$$

have been measured. Of course the measured cross sections include final states in which hypercharge is conserved but in addition those characterized by $|\Delta Y| = 1$. These latter are expected to be damped by a factor $\tan^2\theta_c$ relative to the former. With the presently achieved experimental accuracy it is a good approximation to neglect these and set $\theta_c \simeq 0$. Substituting (4.416), (4.417) and (4.419) into (4.403) gives

$$\sigma(\nu N) = \sigma_o\, \omega \int_o^1 dx\, x\, (u_p + u_n) \;, \tag{4.431a}$$

$$\sigma(\bar{\nu} N) = \sigma_o\, \omega\, \tfrac{1}{3} \int_o^1 dx\, x\, (u_p + u_n) \;, \tag{4.431b}$$

where

$$\sigma_o \equiv \frac{G^2}{\pi} \cos^2\theta_c\, m_N^2 \simeq 1.5 \times 10^{-38}\ \mathrm{cm}^2 \tag{4.431c}$$

$$\omega = E_\nu / m_N \;. \tag{4.431d}$$

Both of the measured quantities exhibit the linear energy dependence we have discussed even at relatively low neutrino energies[64], $(0 < E_\nu < 10\,\mathrm{GeV})$. The latest experiments were performed at NAL with neutrinos in the 100 GeV range. Writing the cross sections in the form $\sigma = \alpha E_\nu$, these experiments give[65]

$$\alpha_\nu = (0.83 \pm 0.11)\, 10^{-38}\ \mathrm{cm}^2/\mathrm{GeV} \tag{4.432a}$$

$$\alpha_{\bar{\nu}} = (0.28 \pm 0.055)\, 10^{-38}\ \mathrm{cm}^2/\mathrm{GeV} \ . \tag{4.432b}$$

The best fit to the cross section ratio is

$$R \equiv \frac{\sigma(\bar{\nu}N)}{\sigma(\nu N)} = 0.33 \pm 0.08 \ , \tag{4.433}$$

which is remarkably close to the value $\frac{1}{3}$ predicted by the naive quark parton model (4.431). Another way to get at R is by measuring the y-distribution of ν or $\bar{\nu}$ events. Using the Callan-Gross relation (4.414), we see, from (4.398), that at $y = 1$

$$\frac{d\sigma}{dy}(\nu) = \sigma_o\, \omega \int_o^1 dx\ F_2^{(\nu)}(x)\ \ \tfrac{1}{2}(1 + B) \ , \tag{4.434a}$$

where

$$B \equiv -\int_o^1 x\ F_3^{(\nu)}(x)\ dx \ \ \left[\int_o^1 F_2^{(\nu)}(x)\ dx\right]^{-1} . \tag{4.434b}$$

Using the same notation

$$\sigma(\nu) = \sigma_o\, \omega \int_o^1 dx\ F_2(x)\ \ \tfrac{1}{3}(2 + B) \ . \tag{4.435a}$$

Thus

$$\frac{1}{\sigma}\frac{d\sigma}{dy}(\nu N) = \frac{3}{2}\,\frac{1 + B}{2 + B} \quad \text{at} \ \ y = 1 \ . \tag{4.435b}$$

In the same way

$$\frac{1}{\sigma}\frac{d\sigma}{dy}(\bar{\nu} N) = \frac{3}{2}\,\frac{1 - B}{2 - B} \quad \text{at} \ \ y = 1 \ , \tag{4.436a}$$

and

$$R = \frac{2 - B}{2 + B} \ . \tag{4.436b}$$

Thus a measurement of $\frac{1}{\sigma}\frac{d\sigma}{dy}$ at $y = 1$ determines B and predicts R. Using this method a value

$$R = 0.30 \pm 0.04$$

is obtained which again agrees well with our prediction.

The relationship to the electromagnetic structure functions, contained in (4.423a), provides us with another prediction[66] which can be tested. On the one hand we have (from (4.423a))

$$\frac{5}{18}\int_o^1 dx\, x\,(u_p + u_n) = \int_o^1 dx\, \tfrac{1}{2}\left[F_2^{(\gamma p)} + F_2^{(\gamma n)}\right] = 0.14 \pm 0.02 \quad , \qquad (4.437)$$

using the SLAC/MIT electroproduction data[67]. On the other hand from (4.431) we have

$$\sigma(\nu N) + \sigma(\bar{\nu} N) = \frac{4}{3}\sigma_o\,\omega\int_o^1 dx\, x\,(u_p + u_n) = (\alpha_\nu + \alpha_{\bar{\nu}})\, E_\nu \, .$$

Thus

$$\int_o^1 dx\, x\,(u_p + u_n) = \frac{3}{4}\,\frac{M_N}{\sigma_o}\,(\alpha_\nu + \alpha_{\bar{\nu}}) = 0.53 \quad , \qquad (4.438)$$

using (4.432). Then comparing (4.437) and (4.438) we see that[65]

$$\frac{5}{18} = \tfrac{1}{2}\left[Q^2_{"p"} + Q^2_{"n"}\right] = 0.27 \pm 0.05 \, . \qquad (4.439)$$

This last prediction, as we have indicated, depends crucially on the one third integral charges (4.64) which we have assumed for the "p" and "n" quarks, and the consistency with the experimentally measured value on the right of (4.439) is the more remarkable.

By definition of $u_q(x)$ ($q =$ "p", "n", "λ"), the average longitudinal momentum carried by the quarks q in the proton[68] is

$$\int_o^1 dx\; x\; P_\mu\; u_q(x) \quad .$$

Thus the integral on the left of (4.438) measures the fraction of longitudinal momentum carried by all the quarks in the proton. Since this is seen experimentally to be about 50%, it follows that something else is carrying the remaining 50% of the proton's momentum P_μ. This "something else" is presumably the "glue" which binds the constituents together to form the nucleon. It must be neutral to both electromagnetic and weak interactions, since the consistency of the model with experiment was achieved by ignoring the presence of the glue. Indeed the glue cannot be binding the constituents very tightly <u>in the kinematic region under consideration</u>, since the derivation of the model's predictions assumed scattering from <u>free</u> constituents. So we are left with a picture of the nucleon which raises almost as many questions as those it has so elegantly answered. In the deep inelastic region we have been considering the nucleon has three quarks and some neutral glue all just sitting there essentially unbound. Evidently this picture does not persist in other kinematic domains, since no quark has ever been produced singly. Presumably the glue binds more strongly in some regions than in others. But why should this be so ? If we believe in field theory, what kind of a (strong-interaction) field theory is it that can "turn itself off" at the high energies and momentum transfers characteristic of the deep inelastic region ? At the time of writing some tentative answers to these questions are beginning to emerge. We postpone our discussion of these ideas until after Chapter 6. However it is worthwhile to make one remark whose pertinence will become more apparent when we return to the subject.

The Scaling Region And The Light Cone

Consider the hadronic tensor $W^{(\nu)}_{\mu\lambda}$ as written in (4.380). It is the Fourier transform of the matrix element of the product (or commutator) of two current operators $J^{-}_{\mu}(x)$ and $J^{+}_{\lambda}(0)$ at different space time points (x and 0). In the laboratory frame p and q are given in (4.374). We define

$$q_{\pm} = q_o \pm q_3 \quad .$$

Then,

$$q^2 = q_o^2 - q_3^2 = q_+ q_-$$

and

$$\nu = q_o = \tfrac{1}{2}(q_+ + q_-) \; .$$

In the limit $q_+ \to \infty$ with q_- fixed and negative we see

$$-q^2 \to \infty$$

$$\nu \to \infty$$

$$\frac{-q^2}{2m_N\nu} \to \frac{-q_-}{m_N} \quad .$$

This is just the Bjorken limit which interests us. In the same notation the argument of the exponential e^{iqx} in (4.380) is

$$q.x = q_o x_o - q_3 x_3$$
$$= \tfrac{1}{2} (q_+ x_- + q_- x_+) ,$$

where

$$x_\pm = x_o \pm x_3 .$$

Now the exponential $e^{iq_+ x_-}$ will, for general x, oscillate rapidly as $q_+ \to \infty$ thereby rendering negligible contributions to the x-integration. The exception is the region of space-time specified by

$$x_- = x_o - x_3 = 0 ,$$

which implies

$$x^2 \equiv x_+ x_- - x_1^2 - x_2^2$$
$$= - x_1^2 - x_2^2$$
$$\leq 0 .$$

Thus the behaviour of the hadronic tensor in the Bjorken limit is controlled by the commutator of two currents in the region $x^2 \leq 0$. But the commutator vanishes when $x^2 < 0$, because of causality. Hence the investigation of deep inelastic scattering amounts to the study of the product of two current operators on the light cone[(69)] ($x^2 = 0$).

Finally, we should remark that treatment of the strangeness-conserving inelastic neutrino processes which we have presented can be paralleled in an obvious way for the $|\Delta Y| = 1$ processes. The magnitude of the Cabibbo angle makes the cross sections for these processes a factor of 25 smaller than those we have considered, and they have not yet been subjected to a detailed experimental study. We shall note the results for these processes in 4.3.5 .

4.2.12 ELECTROMAGNETIC CORRECTIONS

In any process where the theory makes precise predictions it is important to compute electromagnetic corrections, so that one can compare theory and experiment. Such processes, for which there is also accurate data, include muon decay, $\pi_{\ell 2}$ and superallowed beta decays.

The corrections to muon decay present no problem in principle, since the particles involved all have electromagnetic properties which are believed to be well understood, as indicated by the success of quantum electrodynamics. All of the other processes, however, involve hadrons which are known not to have point-like electromagnetic structure (even though they may have point-like constituents in <u>some</u> kinematic regions). Nevertheless, it has been the practice to compute the electromagnetic corrections assuming that the nucleon and pion are point particles like the electron and muon. In calculating these corrections one encounters divergent integrals, just as in QED, and these are controlled by introducing a cut-off Λ. However, unlike QED, the cut-off Λ appears in the expressions for observable quantities; for example, the ratio of the lifetimes of the muon and the neutron depends logarthimically upon Λ when one includes electromagnetic corrections[70, 71]. Let us see how this arises. Consider the decay of a point particle A with charge Q-1 into another, B, with charge Q together with an electron and a neutrino:

$$A \to B + e^- \bar{\nu}_e \ . \tag{4.440}$$

We assume that A and B have spin ½ and that the interaction Hamiltonian has the same form as that for muon decay:

$$\mathcal{H}_I = \frac{G}{\sqrt{2}} \bar{B} \gamma^\alpha (1-\gamma_5) A \bar{\psi}_{(e)} \gamma_\alpha (1-\gamma_5) \psi_{(\nu_e)} + \text{h. c.} \ . \tag{4.441}$$

In the absence of radiative corrections the invariant matrix element has the same form:

$$M_o = \frac{G}{\sqrt{2}} \left[\bar{B} \gamma^\alpha (1-\gamma_5) A \right] \left[\bar{e} \gamma_\alpha (1-\gamma_5) \nu \right] \tag{4.442}$$

and is represented by the diagram (0) in Fig. 4. 6.

The remaining diagrams (1) (2) (3) show the radiative corrections which arise from the exchange of a virtual photon between two of the charged particles, while (4) (5) (6) show the corrections arising from the charged particles' self energy parts. It is convenient to work in the Landau gauge, in which the photon propagator has the form

$$D_{\mu\nu}(k) = (g_{\mu\nu} - k_\mu k_\nu / k^2)(k^2 - i\epsilon)^{-1} \ . \tag{4.443}$$

In this gauge the wave function renormalization constant is finite as we shall now show. The self energy of any of the charged particles is proportional

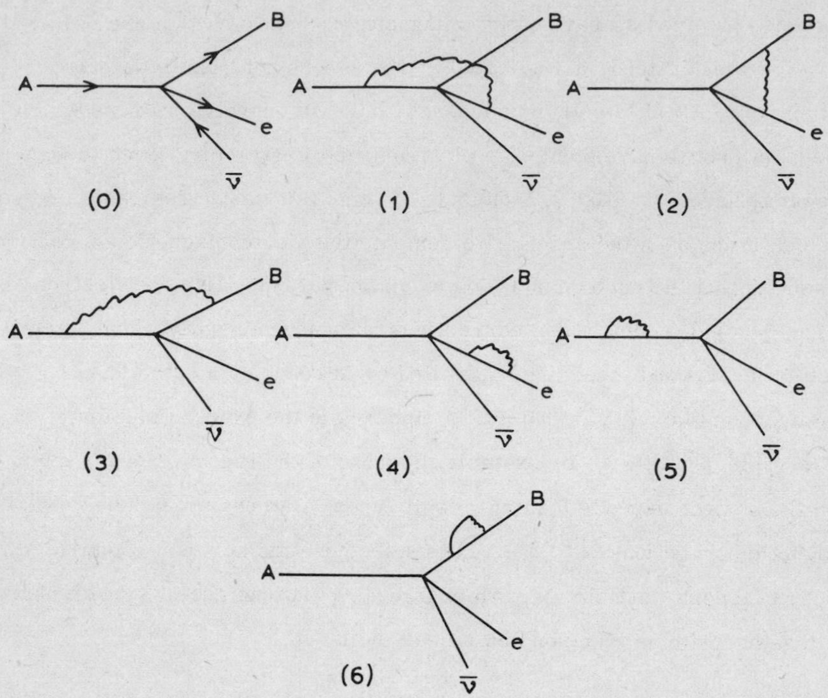

Fig. 4.6 Radiative corrections to $A \to B e \bar{\nu}$

to

$$\Sigma(p) = \int d^4k \; \gamma_\mu \frac{1}{\not{p} - \not{k} - m} \gamma_\nu \, D^{\mu\nu}(k) \,. \tag{4.444}$$

Since $D \sim k^{-2}$ for <u>large</u> k, and the spinor propagator $\sim k^{-1}$, we see that the above integral is linearly divergent. Thus we may write

$$\Sigma(p) = A + B(\not{p} - m) + \Sigma_f(p) \,,$$

where B is at most logarithmically divergent and Σ_f is finite. We may drop the term A by redefining (renormalizing) m, so we are concerned only to evaluate B. This is most easily done by considering

$$\gamma_\rho \frac{\partial \Sigma}{\partial p^\rho} = \gamma_\rho B \gamma^\rho + \gamma_\rho \frac{\partial \Sigma_f}{\partial p_\rho}$$

$$= 4B + \text{finite} \,. \tag{4.445}$$

Now it is easy to verify the following identity, which is left as an exercise for the reader :

$$\frac{\partial}{\partial p_\rho} \frac{1}{\not{p} - m} \equiv \frac{-1}{\not{p} - m} \gamma^\rho \frac{1}{\not{p} - m} \,. \tag{4.446}$$

Thus from (4.444) we deduce that

$$\gamma_\rho \frac{\partial \Sigma}{\partial p_\rho} = - \int d^4k\, \gamma_\rho \gamma_\mu \frac{1}{\not p + \not k - m} \gamma^\rho \frac{1}{\not p + \not k - m} \gamma_\nu D^{\mu\nu}(k) ,$$

and using (4.445) we see that the <u>divergent part</u> $\tilde{B}$ of B is given by

$$\begin{aligned} 4\tilde{B} &= - \int d^4k\, \gamma_\rho \gamma_\mu \frac{1}{\not k} \gamma^\rho \frac{1}{\not k} \gamma_\nu D^{\mu\nu}(k) \\ &= - \int d^4k \frac{1}{(k^2)^2} \gamma_\rho \gamma_\mu \not k \gamma^\rho \not k \gamma_\nu D^{\mu\nu}(k) \\ &= \int d^4k\, (k^2)^{-2} [\gamma_\rho \gamma_\mu \gamma^\rho \gamma_\nu k^2 - 2 \not k \gamma_\mu \not k \gamma_\nu]\, D^{\mu\nu}(k) \\ &= \int d^4k\, (k^2)^{-2} [-2\gamma_\mu \gamma_\nu k^2 + 2\gamma_\mu \gamma_\nu k^2 - 4k_\mu \not k \gamma_\nu]\, D^{\mu\nu}(k) \\ &= 0 , \end{aligned} \tag{4.447}$$

since $k_\mu D^{\mu\nu} = 0$, as is apparent from the definition (4.443). Thus B, the wave function renormalization constant, is finite, and so, consequently, are the diagrams (4) (5) (6). The contribution from diagram (3) is proportional to

$$M_3 = \int d^4k\, [\bar{B} \gamma_\mu \frac{1}{\not B - \not k - m_B} \gamma_\alpha (1-\gamma_5) \frac{1}{\not A - \not k - m_A} \gamma_\nu A]\, D^{\mu\nu}(k) .$$

Thus to find the divergent part we need only the integral

$$I_\alpha = \int d^4k\, \gamma_\mu \frac{1}{\not k} \gamma_\alpha (1-\gamma_5) \frac{1}{\not k} \gamma_\nu D^{\mu\nu}(k) ,$$

which satisfies

$$\gamma^\alpha I_\alpha = - 4\tilde{B}(1-\gamma_5) = 0$$

using (4.447). Hence I_α is zero and M_3 is <u>finite</u>.

The expression for diagram (2) is

$$M_2 = \frac{iQe^2}{(2\pi)^4} \frac{G}{\sqrt{2}} \int d^4k \left[\bar{B} \gamma_\mu \frac{1}{\not B + \not k - m_B} \gamma_\alpha (1-\gamma_5) A\right] D^{\mu\nu}(k)$$
$$\left[\bar{e} \gamma_\nu \frac{1}{\not e - \not k - m_e} \gamma^\alpha (1-\gamma_5) \nu\right] , \tag{4.448}$$

so to calculate the divergent part we need the integral

$$I_2 = -\int d^4k\,(k^2)^{-2}\gamma_\mu \not{k}\gamma_\alpha(1-\gamma_5) \otimes \gamma_\nu \not{k}\gamma^\alpha(1-\gamma_5)\,D^{\mu\nu}(k)$$

$$= -\int d^4k\,(k^2)^{-2}\left[k_\mu\gamma_\alpha + k_\alpha\gamma_\mu - g_{\mu\alpha}\not{k} + i(\mu k\alpha\rho)\gamma_5\gamma^\rho\right](1-\gamma_5) \otimes$$

$$\otimes \left[k_\nu\gamma^\alpha + k^\alpha\gamma_\nu - g^\alpha_\nu\not{k} + i(\nu k\alpha\sigma)\gamma_5\gamma^\sigma\right](1-\gamma_5)\,D^{\mu\nu}(k), \tag{4.449}$$

using the identity (2.167). Now $k_\mu D^{\mu\nu} = 0 = k_\nu D^{\mu\nu}$, as before, and

$$(k_\alpha\gamma_\mu - g_{\mu\alpha}\not{k})\,k^\mu = 0 = (k^\alpha\gamma_\nu - g^\alpha_\nu\not{k})\,k^\nu ,$$

so we need keep only the $g_{\mu\nu}$ part of $D_{\mu\nu}$. Thus

$$I_2 = -\int d^4k\,(k^2)^{-3}\left\{(k_\alpha\gamma_\mu - g_{\mu\alpha}\not{k})(1-\gamma_5) \otimes (k^\alpha\gamma_\mu - g^\alpha_\mu\not{k})(1-\gamma_5) - \right.$$

$$\left. - (\mu k\alpha\rho)(\mu k\alpha\sigma)\,\gamma^\rho(1-\gamma_5) \otimes \gamma^\sigma(1-\gamma_5)\right\} ,$$

since $\gamma_5(1-\gamma_5) = -(1-\gamma_5)$. Further, the symmetry of the k-integration requires

$$\int d^4k\, f(k^2)\, k_\alpha k_\beta = \tfrac{1}{4} g_{\alpha\beta}\int d^4k\, k^2 f(k^2) ,$$

and, using the identity

$$(\mu k\alpha\rho)(\mu k\alpha\sigma) = -2\,(k^2 g^{\rho\sigma} - k^\rho k^\sigma) ,$$

we see that

$$I_2 = -\int d^4k\,(k^2)^{-2}\left\{\tfrac{3}{2}\gamma_\alpha(1-\gamma_5) \otimes \gamma^\alpha(1-\gamma_5) + \tfrac{3}{2}\gamma_\alpha(1-\gamma_5) \otimes \gamma^\alpha(1-\gamma_5)\right\} . \tag{4.450}$$

The divergent k-integration is rendered finite by regulating it. That is to say, we replace it by the integral

$$D \equiv \int d^4k\,\frac{1}{(k^2)^2}\,\frac{-\Lambda^2}{k^2-\Lambda^2} \to \int d^4k\,\frac{1}{(k^2)^2} \quad \text{as } \Lambda \to \infty ,$$

where Λ is an ultra-violet cut-off having the effect of making the integration convergent as $k \to \infty$. To carry out this integration we rotate the k_0-integration (from $-\infty$ to $+\infty$) contour to lie along the imaginary axis $k_0 = i\,k_4$ (from $-i\infty$ to $+i\infty$). Then changing to the Euclidean metric,

$$x \equiv k_1^2 + k_2^2 + k_3^2 + k_4^2$$

$$d^4k \equiv i\,dk_1\,dk_2\,dk_3\,dk_4$$

$$= i\pi^2\,x\,dx\,.$$

Thus

$$D = i\pi^2 \int_0^\infty \frac{dx}{x} \frac{\Lambda^2}{x+\Lambda^2} = 2\,i\pi^2 \ln \frac{\Lambda}{0}\,.$$

As anticipated D depends logarithmically upon Λ and diverges as $\Lambda \to \infty$. It also has a divergence from the small k region, but this is removed by the ultra-violet finite pieces we have dropped. Putting all this together we find the divergent part $\tilde{M}_2$ of M_2 is given by

$$\tilde{M}_2 = \frac{3Qe^2}{8\pi^2} \ln \Lambda\, M_o\,. \tag{4.451}$$

We may calculate $\tilde{M}_1$ from

$$M_1 = \frac{i(Q-1)e^2}{(2\pi)^4} \frac{G}{\sqrt{2}} \int d^4k \left[\bar{B}\gamma_\alpha (1-\gamma_5) \frac{1}{\not{A} - \not{k} - m_A} \gamma_\mu A \right] D^{\mu\nu}(k) \times$$
$$\times \left[\bar{e}\gamma_\nu \frac{1}{\not{e} - \not{k} - m_e} \gamma^\alpha (1-\gamma_5)\nu \right] \tag{4.452}$$

in the same way. The γ-matrices in the $[\bar{B} \cdots A]$ term of (4.452) have the opposite order compared with those in (4.448). This has the effect of reversing the sign of the $\gamma^\rho\gamma_5$ term in (4.449), which leads to the cancellation of the $\frac{3}{2}$ factors in (4.450). This cancellation depends crucially upon the fact that we have $1-\gamma_5$ in both currents. If we had $1+\gamma_5$ in the (BA) current M_2 would be finite and M_1 would be divergent <u>and</u> proportional to Q-1. This is a reflection of the fact that the commutators of the space components of the currents are <u>not</u> invariant under the replacement $A_\lambda^i \to -A_\lambda^i$. Thus M_1 is finite, and the divergent part of the radiative corrections (in this gauge) is given by M_2. Consequently the total matrix element

$$M = M_o + \sum_i M_i = \left(1 + \frac{3\alpha}{2\pi} Q \ln \Lambda\right) M_o + \alpha M_f\,, \tag{4.453}$$

where αM_f is the finite part of the radiative correction ΣM_i. Thus the decay rate to order α is given by

$$\Gamma = \Gamma_o \left(1 + \frac{3\alpha}{\pi} Q \ln \Lambda\right) + \Gamma_f \quad ,$$

where Γ_o is the "bare" rate calculated without radiative corrections and $\Gamma_f \propto 2\mathrm{Re}\, M_o^{\dagger} M_f$ is a finite correction.

Now for muon decay $A = \mu^-$ and $B = \nu_\mu$, so $Q = 0$ and we see that the corrected decay rate is finite. In fact for muon decay[70]

$$\Gamma(\mu) = \Gamma_o(\mu) \left[1 - \frac{\alpha}{2\pi}\left(\pi^2 - \frac{25}{4}\right)\right] . \qquad (4.454)$$

On the other hand for neutron decay $A = n$ and $B = p$, so $Q = 1$ and the decay rate $\Gamma(n)$ depends on Λ, as does the ratio $R = \Gamma(n)/\Gamma(\mu)$.

$$\Gamma(n) = \Gamma_o(n) \left[1 + \frac{3\alpha}{\pi} \ln \frac{\Lambda}{m_N}\right] + \Gamma_f \quad , \qquad (4.455)$$

where we have scaled Λ with m_N, as is appropriate for neutron decay. For a time it was generally believed that the appearance of Λ in $\Gamma(n)$ was occasioned by our neglect of the known electromagnetic structure of the nucleons, even though the finiteness of the muon decay corrections is somewhat of a fluke, as the above calculation makes clear. In a more realistic calculation one might expect that Λ would be replaced by some mass characteristic of the degree of non-locality of the electromagnetic interaction with the nucleon, say 0.5f. Since the dependence upon Λ is only logarithmic, the result is insensitive to quite large variations of Λ, so one would expect that the results obtained using $\Lambda \sim m_N$ would be quite realistic.

There are good reasons for supposing that this view is incorrect. The radiative corrections may be recalculated using the non-local form of the <u>weak</u> interaction which results when a vector boson W mediates, but <u>still</u> assuming point electromagnetic structure for the hadrons. That is to say, we use the form (3.104) for the muon decay interaction and the corresponding analogue of (4.1). It turns out that both the muon and the neutron decay rates depend upon Λ, but that their ratio is finite, at least for the simplest assumptions on the electromagnetic properties of the W. Further, as m_W becomes large the ratio has the same form as that derived using the local weak interaction,but the cut-off Λ is replaced by m_W[72] :

$$R = R_o \left(1 + \frac{3\alpha}{\pi} \ln \frac{m_W}{m_N}\right) \text{ for } m_W >> m_N \ . \tag{4.456}$$

In addition, the radiative corrections to the $\pi_{\ell 2}$ branching ratio remains independent of Λ and approaches the value (4.118) (obtained using the point interaction) as m_W becomes large[73]. Thus the inclusion of weak structure can remove the cut-off dependence from observable quantities, but it may be that any structure, not just weak structure, can have the same effect. In any case, it is a priori unreasonable to ignore the electromagnetic structure arising from strong interactions which is known to exist.

However, it now appears that the radiative corrections to the local weak rate are divergent even when all strong interaction effects are included[74]. To include such effects one uses the current algebra discussed in section 4.1.4. The derivation requires a knowledge not only of the "model independent" commutators of the electromagnetic and weak currents ($[j_o, J_\lambda^+]$, $[j_\lambda, J_o^+]$) but also of the "model dependent" commutators ($[j_r, J_s^+]: r, s = 1,2,3$) ; these latter commutators are called model dependent, because, unlike the former, they depend upon the charges of the fundamental triplet of fields q from which the currents are constructed. We shall not repeat the sophisticated derivation using current algebra, since we may derive the result (although not its limitations) from what we have already done.

Firstly we note that our result (4.453) did not utilize the fact that A and B were on-mass-shell (physical) particles. The result therefore states that the effect of the divergent radiative corrections is to renormalize G in (4.441) by the substitution

$$G \to G' = \left(1 + \frac{3\alpha}{2\pi} Q \ln \Lambda\right) G \ , \tag{4.457}$$

where Q is the charge of the field B. Now the interaction (4.441) is of the same form as (4.1), which, we are postulating, describes all semi-leptonic processes, with $\mathcal{J}_\lambda$ defined in (4.50) and (4.65).

Thus

$$\begin{aligned} \mathcal{J}_\lambda^+ &= \tfrac{1}{2} \bar{q} \gamma_\lambda (1-\gamma_5) [(\lambda^1 + i\lambda^2) \cos\theta_c + (\lambda^4 + i\lambda^5) \sin\theta_c] q \\ &= \text{"} \bar{p} \text{"} \gamma_\lambda (1-\gamma_5) (\text{"n"} \cos\theta_c + \text{"}\lambda\text{"} \sin\theta_c) \ , \end{aligned} \tag{4.458}$$

where $q = (\text{"p"}, \text{"n"}, \text{"}\lambda\text{"})$ is the triplet of fundamental fields. In the quark model, the charge of "p" is given in (4.64) and is $\frac{2}{3}$. Thus

$$G' = (1 + \frac{\alpha}{\pi} \ln \frac{\Lambda}{m_N}) \; G \qquad \text{(quark model)} \tag{4.459}$$

if we scale with a typical hadronic mass m_N. In general for $\Delta Y = 0$ processes we are concerned only with the $\cos\theta_c$ part of $\mathcal{J}_\lambda^+$. "p" and "n" must belong to an isodoublet (because of CVC) so

$$Q_p = \overline{Q} + \tfrac{1}{2} , \tag{4.460}$$

where $\overline{Q}$ is the average charge $\frac{1}{2}(Q_p + Q_n)$ of the isodoublet. Thus in general[74]

$$G' = [1 + \frac{3\alpha}{4\pi} (1 + 2\overline{Q}) \ln \frac{\Lambda}{m_N}] \; G \; . \tag{4.461}$$

Comparing this with (4.231) we see that the model dependent constant C is given by

$$C = 3 \left(1 + 2\overline{Q}\right) \ln \frac{\Lambda}{m_N} - 1 \; , \tag{4.462}$$

where we have included a part of the model independent corrections[25]. Thus using (4.233), (4.231) and (3.62) we can finally compute $\cos\theta_c$ under various assumptions about $\overline{Q}$ and Λ. For obvious reasons we take $\overline{Q}$ to have its quark model value, $\overline{Q} = \frac{1}{6}$. Then larger values of Λ are associated with larger values of θ_c

$$\cos\theta_c = 0.985, \quad 0.979, \quad 0.974 \tag{4.463}$$

for

$$\Lambda/m_p = \quad 1 \quad , \quad 10 \quad , \quad 100 \; .$$

The current algebra treatment has been generalized to the case when a W-boson mediates the weak interaction[75]. The divergences in muon and beta decays cancel from the lifetime ratio, thereby generalizing the earlier result. In view of this parallel it seems likely that as m_W becomes large we shall retrieve the result (4.462) with Λ replaced by m_W. Since there is no good reason for supposing that m_W has a value near m_p, plainly we should not restrict our attention to the corresponding large value of $\cos\theta_c$ in (4.463).

In conclusion, it appears that the local weak Hamiltonian (4.1) we have been using leads to insuperable difficulties with the radiative corrections. For this reason we suspect that the theory we have been using is fundamentally wrong and that some form of non-locality will have to be introduced to remedy its defects. The successes of the theory, which we have already encountered, lead us to suppose that at low energies, at least, the (truly) fundamental Hamiltonian cannot be too different from (4.1). We shall return to this problem in Chapter 6.

4.3 HYPERCHARGE NONCONSERVING SEMILEPTONIC PROCESSES

The second type of semileptonic processes predicted by the Cabibbo hypothesis are those which proceed via the part $S^{\pm}_{\lambda}$ of $\mathscr{J}^{\pm}_{\lambda}$. The hadronic selection rules characterizing these processes are given in (4.57b). In addition to these positive predictions, the selection rules also make the negative predictions that hadronic transitions with $|\Delta Y| > 1$ are forbidden, as are those with $\Delta Y = -\Delta Q$. In particular, the following processes are forbidden

$$\Xi^- \rightarrow N \ell^- \bar{\nu}_e \qquad (\text{B.R.} \lesssim 10^{-2}) ,$$

$$K^+ \rightarrow \pi^+ \pi^+ \ell^- \bar{\nu}_e \qquad (\text{B.R.} \lesssim 5.10^{-7}) ,$$

$$\Sigma^+ \rightarrow n \ell^+ \bar{\nu}_e \qquad (\text{B.R.} \lesssim 10^{-5}) ,$$

$$\Xi^o \rightarrow \Sigma^- \ell^+ \bar{\nu}_e \qquad (\text{B.R.} \lesssim 10^{-3}) .$$

The upper limits on the relevant branching ratios (B.R.) are given in parentheses[10].

4.3.1 $K_{\ell 2}$

The simplest examples of the processes with which we are now concerned are

$$K^- \rightarrow \ell^- \bar{\nu}_\ell , \qquad (4.464a)$$

$$K^+ \rightarrow \ell^+ \nu_\ell . \qquad (4.464b)$$

The analysis of these processes precisely parallels that of $\pi_{\ell 2}$. Thus

$$< 0|S_\lambda^+|K^-> = < 0|A_\lambda^4 + iA_\lambda^5|K^-> = if_K K_\lambda ,\qquad (4.465)$$

and in the absence of electromagnetic corrections the decay rate is (cf. (4.113))

$$\Gamma_o(K_{\ell 2}) = \frac{G^2}{8\pi} f_K^2 \sin^2\theta_c \frac{m_\ell^2}{m_K^3}(m_K^2 - m_\ell^2)^2 .\qquad (4.466)$$

As before, this predicts the ratio R_o of the decay rates of the electronic and muonic modes. The electromagnetic corrections yield finite corrections to R_o, as in (4.118), and the final prediction is

$$R \equiv \Gamma(K_{e2})/\Gamma(K_{\mu 2}) = 2.10 \times 10^{-5} .$$

The value measured experimentally is[10]

$$R_{exp} = (2.17 \pm 0.31)\, 10^{-5} ,$$

which again is in complete accord with the above theoretical prediction.

We may also take the ratio of the $K_{\mu 2}$ and $\pi_{\mu 2}$ rates

$$\frac{\Gamma(K_{\mu 2})}{\Gamma(\pi_{\mu 2})} = \frac{f_K^2}{f_\pi^2} \tan^2\theta_c \frac{m_\pi^3}{m_K^3} \frac{(m_K^2 - m_\mu^2)^2}{(m_\pi^2 - m_\mu^2)^2} .$$

Using (4.118) and its analogue to include electromagnetic corrections, the experimental data then determines the quantity

$$f_K f_\pi^{-1} \tan\theta_c \equiv \tan\theta_M^A .\qquad (4.467)$$

(The letters "A" and "M" are to remind us that the value θ_M^A is determined from the experimental data on the axial vector current's contribution to meson decays). The result is that

$$\sin\theta_M^A = 0.2655 \pm 0.0006 .\qquad (4.468)$$

In the limit of SU(3) symmetry it is easy to see that $f_K = f_\pi$. The proof is analogous to that used to derive (4.308). We start from

$$< 0|A_\mu^j|P^\ell> \equiv ig_{j\ell} P_\mu ,\qquad (4.469)$$

where $j, \ell = 1 \dots 8$ and P^j is the octet of pseudoscalar mesons. Then, as before, SU(3) symmetry requires

$$<0|\mathbb{F}^i A^j_\mu|P^\ell> - <0|A^j_\mu \mathbb{F}^i|P^\ell> = if^{ijk} <0|A^k|P^\ell> .$$

Interpreting SU(3) symmetry in the minimal sense discussed before, we find

$$- i f^{i\ell m} g_{jm} = if^{ijk} g_{k\ell} .$$

Since f^{ijk} are totally antisymmetric,

$$g_{j\ell} = f_P \delta_{j\ell} \tag{4.470}$$

is a solution; in fact it is the only solution, since the product $\underline{8} \times \underline{8}$ contains $\underline{1}$ representation just once (see (4.307)). From (4.26) we have

$$|\pi^-> = \frac{1}{\sqrt{2}} (|P^1> - i|P^2>) ,$$

$$|K^-> = \frac{1}{\sqrt{2}} (|P^4> - i|P^5>) .$$

Then comparing (4.105) and (4.465) with (4.469) and using (4.470) we deduce

$$f_\pi = \sqrt{2} f_P = f_K , \quad \text{for SU(3) symmetry.}$$

Then from the definition (4.467) we see that

$$\theta^A_M = \theta_c , \quad \text{for SU(3) symmetry.}$$

Now, in fact, SU(3) symmetry is not exact (and is not even very good as far as the pseudoscalar octet is concerned), so we shall retain the distinction between θ^A_M and θ_c in order that we may later evaluate the status of the Cabibbo hypothesis in the light of the experimental data.

4.3.2 $K_{\ell 3}$

We have already noted that the Cabibbo Hypothesis requires that $S^\pm_\lambda$ has total isospin $I = \frac{1}{2}$. To test this particular prediction we need to consider processes which might a priori be induced by a current with different total isospin. The decay $K_{\ell 2}$, for example, provides no test of the $\Delta I = \frac{1}{2}$ rule, since the initial kaon has $I = \frac{1}{2}$ and the vacuum has $I = 0$. Thus even if $S^\pm_\lambda$ had additional pieces having $I = \frac{3}{2}, \frac{5}{2}, \ldots\ldots$, these could not contribute in $K_{\ell 2}$; (we have to consider half-integral isospin, since if $\Delta Y = \Delta Q = +1$ (say) then $\Delta I^3 = \frac{1}{2}$.) Similar considerations apply to the leptonic decays characterized by the hadronic transitions

$\Lambda \to p$ and $\Xi \to \Lambda$. The $\Sigma \to N$ decays do provide a test in principle; the Σ has $I = 1$ and N has $I = \frac{1}{2}$, so they could be connected by a current with $I = \frac{1}{2}$ and/or $\frac{3}{2}$. The predicted absence of the $I = \frac{3}{2}$ piece relates the $\Sigma^- \to n$ matrix element to that for $\Sigma^o \to p$, but unfortunately the weak decays of the Σ^o are practically unobservable. The inverse transitions $n \to \Sigma^-$ and $p \to \Sigma^o$ can be induced when anti-neutrinos are scattered from the initial nucleons, and such experiments will eventually provide a test of the $\Delta I = \frac{1}{2}$ rule. At the present time, however, the only practical tests of the rule are those provided by the meson decay modes $K_{\ell 3}$ ($K \to \pi$) and $K_{\ell 4}$ ($K \to 2\pi$). As far as isospin is concerned, the $K \to \pi$ transition is the same as the $N \to \Sigma$ transition mentioned above (it could proceed via $I = \frac{3}{2}$). The $K \to 2\pi$ transitions provide tests for the absence of $\frac{3}{2}$ and $\frac{5}{2}$. Let us see how these predictions are derived.

Isospin And TCP-Invariance Relations

Since S^+_λ has $(I, I^3) = (\frac{1}{2}, \frac{1}{2})$, it must commute with the isospin raising operator $v^1 + i v^2$

$$[v^1 + i v^2, S^+_\lambda] = 0 \ . \tag{4.471}$$

Thus

$$<\pi^+|(v^1 + iv^2) S^+_\lambda |K^-> = <\pi^+|S^+_\lambda (v^1 + iv^2)|K^-> \ .$$

So

$$-\sqrt{2} <\pi^o|S^+_\lambda|K^-> = - <\pi^+|S^+_\lambda|\bar{K}^o> \ , \tag{4.472}$$

remembering our phase convention (4.26). In the same way $[v^1 - iv^2, S^-_\lambda] = 0$ gives

$$\sqrt{2} <\pi^o|S^-_\lambda|K^+> = <\pi^-|S^-_\lambda|K^o> \ , \tag{4.473}$$

and these are the basic $\Delta I = \frac{1}{2}$ rule predictions. We have already noted in 4.2.8. that the neutral kaon states K^o and $\bar{K}^o$ are not the relevant states for weak interactions. Instead we must deal with the states having well defined lifetimes which are given in (4.280). Now, $S^\pm_\lambda$ has $I^3 = \pm \frac{1}{2}$, so

$$< \pi^+ | S_\lambda^+ | K^o > \; = \; 0 \; = \; < \pi^- | S_\lambda^- | \bar{K}^o > \quad , \tag{4.474}$$

and we may combine (4.472 , 3, 4) to relate the $K^o_{\substack{S\\L}}$ decay amplitudes. Recall, using only TCP-invariance, that

$$|K^o_{\substack{S\\L}} > \; = \; (1 + |r|^2)^{-\frac{1}{2}} (|K^o > \mp r |\bar{K}^o >) \; . \tag{4.280}$$

Then

$$< \pi^+ | S_\lambda^+ | K^o_{\substack{S\\L}} > \; = \; (1 + |r|^2)^{-\frac{1}{2}} (\mp r) \sqrt{2} \; < \pi^o | S_\lambda^+ | K^- > \tag{4.475a}$$

$$< \pi^- | S_\lambda^- | K^o_{\substack{S\\L}} > \; = \; (1 + |r|^2)^{-\frac{1}{2}} \sqrt{2} \; < \pi^o | S_\lambda^- | K^+ > \; , \tag{4.475b}$$

from which it follows that

$$\Gamma (K_S \to \pi^+ \ell^- \bar{\nu}_\ell) = \Gamma (K_L \to \pi^+ \ell^- \bar{\nu}_\ell) = \frac{2 |r|^2}{1 + |r|^2} \Gamma (K^- \to \pi^o \ell^- \bar{\nu}_\ell) \tag{4.476a}$$

$$\Gamma (K_S \to \pi^- \ell^+ \nu_\ell) = \Gamma (K_L \to \pi^- \ell^+ \nu_\ell) = \frac{2}{1 + |r|^2} \Gamma (K^+ \to \pi^o \ell^+ \nu_\ell) \; . \tag{4.476b}$$

TCP-invariance also tells us that the decay rates of the charged mesons are equal. We denote by $\mathcal{R}$ the operator $\mathcal{TCP}$, which, like $\mathcal{P}$, is anti-unitary. Thus, quite generally, as in (2.231),

$$< \alpha | \mathcal{H}_W | \beta > \; = \; < \alpha^r | \mathcal{H}_W | \beta^r >^* \quad ,$$

where $\mathcal{H}_W$ is the weak Hamiltonian and

$$|\alpha^r > = \mathcal{R} |\alpha > \; , \; |\beta^r > \; = \; \mathcal{R} | \beta > \; .$$

The states $|\alpha^r >$ have all particles in $|\alpha >$ replaced by anti-particles and all spins (but not momenta) reversed. The same is true of $|\beta^r >$ and $|\beta >$. To compute the decay rate $\Gamma(\beta \to \alpha)$ we take $| < \alpha | \mathcal{H}_W | \beta > |^2$ and sum (average) over spins. Then we conclude that

$$\Gamma (\beta \to \alpha) \; = \; \Gamma (\bar{\beta} \to \bar{\alpha}) \quad ,$$

where $|\bar{\alpha} >$, $|\bar{\beta} >$ are the anti-particle states. With this result we may eliminate the (presently) unknown $|r|$ and find

$$\Gamma(K_L \to \pi^+ \ell^- \bar{\nu}_\ell) + \Gamma(K_L \to \pi^- \ell^+ \nu_\ell) = \Gamma(K_S \to \pi^+ \ell^- \bar{\nu}_\ell) + \Gamma(K_S \to \pi^- \ell^+ \nu_\ell)$$

$$= 2\Gamma(K^+ \to \pi^o \ell^+ \nu_\ell) = 2\Gamma(K^- \to \pi^o \ell^- \bar{\nu}_\ell) \ . \tag{4.477}$$

As before, the slight phase space differences and electromagnetic corrections change these a little[76]. With these modifications the predictions tested by experiment are

$$R_1 \equiv \frac{\Gamma(K_{Le3}) + \Gamma(K_{L\mu 3})}{2\Gamma(K^+_{e3}) + 2\Gamma(K^+_{\mu 3})} = 1.012$$

and

$$R_2 \equiv \frac{\Gamma(K_{L\mu 3})}{\Gamma(K^+_{\mu 3})} \left[\frac{\Gamma(K_{Le3})}{\Gamma(K^+_{e3})} \right]^{-1} = 1 \ .$$

Both of these are in reasonable accord with experiment[10]

$$R_1|_{exp} = 0.969 \pm 0.017$$

$$R_2|_{exp} = 1.039 \pm 0.050 \ .$$

TCP-invariance also permits us to relate the <u>amplitudes</u> in (4.475a) and (4.475b); as shown in Chapter 2, since R is anti-unitary,

$$< \mathcal{R}\pi^o | \mathcal{R} S^+_\lambda \mathcal{R}^{-1} \mathcal{R} | K^- > = < \pi^o | S^+_\lambda | K^- >^* \ . \tag{4.478}$$

Now,

$$\mathcal{R} | K^-(\underline{p}) > = - | K^+(\underline{p}) > \tag{4.479a}$$

$$\mathcal{R} | \pi^o(\underline{k}) > = - | \pi^o(\underline{k}) > \ . \tag{4.479b}$$

We showed in 4.2.6 that

$$\mathcal{T} L_\alpha \mathcal{T}^{-1} = L^\alpha \ , \tag{4.480}$$

so, using (3.47), it follows that

$$\mathcal{R} L_\alpha \mathcal{R}^{-1} = - L^\dagger_\alpha \ . \tag{4.481}$$

Thus the TCP invariance of (4.1) requires

$$\mathcal{R} \mathcal{J}^\pm_\alpha \mathcal{R}^{-1} = - \mathcal{J}^\mp_\alpha \ , \tag{4.482}$$

and plainly this is true for $J^\pm_\alpha$ and $S^\pm_\alpha$ separately. Thus, in particular,

$$\mathcal{R}\, S_\lambda^+ \,\mathcal{R}^{-1} = -S_\lambda^- . \qquad (4.483)$$

Combining (4.479),(4.483) and (4.478) we find

$$<\pi^o|S_\lambda^-|K^+> = -<\pi^o|S_\lambda^+|K^->^* . \qquad (4.484)$$

Thus, if we now accept the validity of the $\Delta I = \frac{1}{2}$ rule, all of the amplitudes in (4.475) are related, and we may confine our further analysis to just one process. We consider

$$K_{\ell 3}^- : K^- \to \pi^o \ell^- \bar{\nu}_\ell .$$

Form Factors And SU(3) Symmetry

As for $\pi_{\ell 3}$, this can proceed only by the vector part of S_λ^+, and the hadronic matrix element is determined by two form factors

$$<\pi^o|S_\lambda^+|K^-> = <\pi^o|V_\lambda^4 + iV_\lambda^5|K^-> = f_+(q^2)(p+p')_\lambda + f_-(q^2)\, q_\lambda , \qquad (4.485)$$

where p and p' are the momenta of K and π and $q = p - p'$. The mass difference between the kaon and the pion is about 360 MeV, so that it is possible to detect the q^2-variation of the form factors. The f_+ form factor is directly measurable in the electronic decay mode, since the q_λ gives a factor m_e when we use the Dirac equation, and this is negligible. Further, the $(p+p')_\lambda$ coupling of f_+ is that of a pure spin one object (like the photon); so we might expect $f_+(q^2)$ to have its shape determined by the $K^*(893)$ pole. At any rate, the variation of f_+ is observed to be small, and is parametrized by the linear form

$$f_+(t) = f_+(0)\left(1 + \frac{\lambda_+ t}{m_\pi^2}\right) . \qquad (4.486)$$

Information on the f_- form factor can be extracted from $K_{\mu 3}$ decay to which both form factors contribute. Historically a linear fit to $f_-(t)$ was assumed, but from a physical point of view it seems more sensible to parametrize the form factor corresponding to pure spin zero. This is given by the divergence of the current. Now,

$$<\pi^o|i\partial^\lambda(V^4_\lambda + iV^5_\lambda)|K^-> = q_\lambda <\pi^o|V^4_\lambda + iV^5_\lambda|K^->$$

$$= f_+(q^2)(m_K^2 - m_\pi^2) + q^2 f_-(q^2) \ , \qquad (4.487)$$

and we give this combination a linear parametrization

$$f_+(t) + t(m_K^2 - m_\pi^2)^{-1} f_-(t) \equiv f_o(t) = f_+(0)\left(1 + \frac{\lambda_o t}{m_\pi^2}\right) . \qquad (4.488)$$

Note that $f_o(0) = f_+(0)$ and that such a fit requires $f_-(t)$ to be a constant, since we have taken $f_+(t)$ to be linear. Thus the hadronic amplitude, and hence the transition probability, is determined by three quantities which it is the theorist's task to predict.

In the limit of SU(3) symmetry V^i_λ $(i = 1 \ldots 8)$ are conserved and $m_K = m_\pi$, so from (4.487) we see that

$$f_-(q^2) = 0 \quad \text{for SU(3) symmetry,} \qquad (4.489)$$

and in this limit

$$<P^j|V^i_\lambda(x)|P^k> = g_{ijk}\, G(q^2)\,(p + p')_\lambda\, e^{-iqx} . \qquad (4.490)$$

Thus, since

$$\mathbb{F}^i = \int d^3x\; V^i_o(x) \ ,$$

we have

$$<P^j|\mathbb{F}^i|P^k> = g_{ijk}\, G(q^2)\,(E_j + E_k)\, e^{-iq_o x_o}\,(2\pi)^3\,\delta(\underline{q})$$

$$\text{(no summation)} \ .$$

Hence

$$if^{ik\ell} <P^j|P^\ell> = if^{ikj}\,(2\pi)^3\, 2E_j\,\delta(\underline{q}) = g_{ijk}\, G(0)\, 2E_j\,(2\pi)^3\,\delta(\underline{q})$$

$$\text{(no summation)} \ ,$$

since $\underline{q} = 0$ implies $E_j = E_k$ and $q^2 = 0$. Thus we may take

$$g_{ijk} = -if^{ijk} \quad \text{and} \quad G(0) = 1 \ . \qquad (4.491)$$

In this way we may relate $f_+(t)$ to the form factor $g_+(t)$ defined in (4.128). Specifically we find

$$f_+(q^2) = \frac{1}{\sqrt{2}}\, G(q^2) = \tfrac{1}{2}\, g_+(q^2) \qquad \text{for SU(3) symmetry.} \tag{4.492}$$

Thus in this symmetry limit $f_+(0) = \frac{1}{\sqrt{2}}$, and the shape is probably fixed by the 1^- vector meson pole. However SU(3) symmetry is badly broken by the pseudoscalar mesons, so we must try and do better.

CTMOP Relation

What we can do is to use PCAC to derive a "soft pion theorem". First we use the LSZ reduction formula to define the matrix element (4.485) when the pion is off-mass-shell

$$\begin{aligned} F_\lambda(q,p') &\equiv i \int dx\, e^{ip'x}\, (m_\pi^2 - p'^2) < 0|T\, \{\varphi^3(x),\, V_\lambda^4 + i V_\lambda^5\}|K^- > \\ &\equiv f_+(q^2, p'^2)\,(p+p')_\lambda + f_-(q^2, p'^2)\, q_\lambda \ , \end{aligned} \tag{4.493}$$

where $\varphi^3(x)$ is the field operator of π^o. When $p'^2 = m_\pi^2$, we have the physical matrix element, so

$$f_\pm(q^2,\, m_\pi^2) = f_\pm(q^2) \ . \tag{4.494}$$

Next we use PCAC, given in (4.157), so

$$\begin{aligned} F_\lambda(q, p') &= (m_\pi^2 - p'^2)\, \frac{\sqrt{2}}{f_\pi m_\pi^2} \int dx\, e^{ip'x} <0|T\, \{\partial^\mu A_\mu^3(x),\, V_\lambda^4 + i V_\lambda^5\}|K^- > \\ &= \frac{\sqrt{2}\,(m_\pi^2 - p'^2)}{f_\pi m_\pi^2} \left\{ - i p'_\mu \int dx\, e^{ip'x} < 0|T\, \{A_\mu^3(x),\, V_\lambda^4 + i V_\lambda^5\}|K^- > - \right. \\ &\qquad \left. - \int dx\, e^{ip'x} <0|\delta(x^o)\,[\, A_o^3(x),\, V_\lambda^4 + i V_\lambda^5\,]\, |K^- > \right\} \ , \end{aligned} \tag{4.495}$$

using Gauss' theorem, as before. We use the SU(3) x SU(3) algebra (4.69) to evaluate the commutator, and take the limit $p' \to 0$ $(q \to p)$. Then

$$\begin{aligned} - i F_\lambda(p, 0) &= - \sqrt{2}\, f_\pi^{-1} < 0|\, \tfrac{1}{2}\,(A_\lambda^4 + i A_\lambda^5)\,|K^- > \\ &= \frac{-i}{\sqrt{2}\, f_\pi}\, f_K\, p_\lambda \ , \end{aligned} \tag{4.496}$$

using (4.465). (Note that there is no contribution from the term proportional to p'_μ when we take $p' \to 0$, since the single pion intermediate state has momentum p' and $(p'^2 - m_\pi^2)^{-1}$ does not diverge.) Combining (4.493) and (4.496) we obtain the CTMOP relation[77]

$$f_+(m_K^2, 0) + f_-(m_K^2, 0) = \frac{f_K}{\sqrt{2}\, f_\pi} .$$

Thus the combined use of PCAC and the SU(3) x SU(3) current algebra enables us to derive a "soft pion theorem" which relates the $K_{\ell 3}$ form factors for zero mass pions to the $K_{\ell 2}$ decay amplitude. Apart from the fact that zero mass pions are not available in the laboratory, the test of the CTMOP relation is complicated by the fact that $q^2 = m_K^2$ is outside the physical region for $K_{\ell 3}$ which satisfies

$$m_\ell^2 < q^2 < (m_K - m_\pi)^2 .$$

The best we can do is essentially to ignore these difficulties. If $f_\pm(q^2, p'^2)$ are regular functions of p'^2, we see from (4.494) that

$$f_\pm(q^2, 0) = f_\pm(q^2) + O(m_\pi^2) ,$$

and our experience of PCAC and the Goldberger-Treiman formula suggests that $O(m_\pi^2)$ is about 10%. Thus the CTMOP formula gives

$$f_+(m_K^2) + f_-(m_K^2) = \frac{f_K}{\sqrt{2}\, f_\pi} + O(m_\pi^2) . \qquad (4.497)$$

We "ignore" the second problem, alluded to above, by assuming that the linear forms for $f_+(t)$ and $f_o(t)$ are valid outside the physical region and up to $t = m_K^2$. Substituting in (4.488) gives

$$f_o(m_K^2) = f_+(m_K^2) + f_-(m_K^2) + \frac{m_\pi^2}{m_K^2 - m_\pi^2} f_-(m_K^2) . \qquad (4.498)$$

Now, since $m_\pi^2/m_K^2 \simeq 0.07$, the last term is small and in any case $O(m_\pi^2)$. Further, if there is any relic of SU(3) symmetry in the form factor $f_-(q^2)$, it too is expected to be small, as we have shown in (4.489). Neglecting the last term in (4.498), we see that the CTMOP relation is essentially a statement about $f_o(m_K^2)$. Thus, with our linear fit, it gives one relation between the two parameters

$$f_+(0)\left(1+\lambda_o \frac{m_K^2}{m_\pi^2}\right) = f_o(m_K^2) \simeq f_+(m_K^2) + f_-(m_K^2) \simeq \frac{f_K}{\sqrt{2}\, f_\pi} \ . \tag{4.499}$$

Digression

Before seeing what else we can say about the form factors, and before evaluating the consistency of these predictions with experiment, let us examine (4.499) in more detail. Since the CTMOP relation is essentially a statement about $f_o(t)$, which is <u>defined</u> by the matrix element of the current divergence $\partial^\lambda(V_\lambda^4 + iV_\lambda^5)$ in (4.487, 8), could we not also have derived the relation from a soft pion theorem for <u>this</u> quantity rather than for the matrix element of $V_\lambda^4 + iV_\lambda^5$? Consider the off-shell continuation of (4.487)

$$F(q^2, p'^2) \equiv i\int dx\, e^{ip'x}(m_\pi^2 - p'^2) < 0|T\{\varphi^3(x), \partial^\lambda(V_\lambda^4 + iV_\lambda^5)\}|K^- > . \tag{4.500}$$

Proceeding as before, we derive the soft pion theorem

$$F(m_K^2, 0) = -\sqrt{2}\, f_\pi^{-1}\, i\int dx\, \delta(x^o) < 0|\left[A_o^3(x), \partial^\lambda(V_\lambda^4 + iV_\lambda^5)\right]|K^- > . \tag{4.501}$$

The current divergence in (4.501) is in general not known. To explore this further then, we may consider a particular model in which the strong interaction Hamiltonian $\mathcal{H}$ is known. Plainly such a model must ensure that the current algebra we have already used remains true. A popular model of this type has

$$\mathcal{H} = \mathcal{H}_o + \epsilon_o S^o + \epsilon_8 S^8 \ , \tag{4.502}$$

where $\mathcal{H}_o$ is SU(3) x SU(3) invariant. The quantities S^i are scalar densities which transform in the quark model notation of (4.65) as

$$S^i(x) = \bar{q}(x)\, \tfrac{1}{2}\lambda^i q(x) \qquad (i = 0 \cdots 8) \ . \tag{4.503}$$

Plainly the Hamiltonian (4.502) leaves the canonical anti-commutation relations (4.66) unaltered, so that the SU(3) x SU(3) algebra defined by v^i and a^i is also unaltered, although the symmetry is broken, as we shall see.

The densities S^i, together with the analogous pseudoscalar densities

$$P^i(x) = \bar{q}(x)\,\tfrac{1}{2}\lambda^i\gamma_5\,q(x) \quad , \tag{4.504}$$

define $(\underline{3}, \overline{\underline{3}})$ and $(\overline{\underline{3}}, \underline{3})$ representations of the group SU(3) x SU(3). Under ordinary classification SU(3) (with generators v^i) (4.68) enables us to prove that

$$[\,v^i(x^o),\, S^j(x)\,] = i f^{ijk} S^k(x) \ . \tag{4.505}$$

Thus S^o is an SU(3) invariant, while S^8 belongs to an octet representation of SU(3). Then the S^8 term of (4.502) breaks the SU(3) symmetry and gives rise to mass differences between the pseudoscalar mesons. Further, as Gell-Mann and Okubo[(78)] showed, such a breaking leads to experimentally well satisfied relations between the hadron masses. Let us now calculate the current divergence we need in this model

$$\begin{aligned}\int d^3x\, \partial^\lambda V^i_\lambda(x) &= \int d^3x\, \partial^o V^i_o(x) = \partial^o v^i(x^o) \\ &= i\left[H\,,\, v^i(x^o)\right] \\ &= -i\int d^3x \left[v^i(x^o)\ ,\ \mathcal{H}\right] \\ &= \epsilon_8 f^{i8k}\int d^3x\ S^k(x) \qquad (i = 1 \cdots 8)\ ,\end{aligned}$$

where we have discarded the integral of the spatial divergence $\partial_r V^i_r$ and used (2.202), (4.502) and (4.505). Thus

$$\partial^\lambda V^i_\lambda = \epsilon_8 f^{i8k} S^k \ , \tag{4.506a}$$

so

$$\partial^\lambda (V^4_\lambda + iV^5_\lambda) = i\epsilon_8 \frac{\sqrt{3}}{2}\ (S^4 + i S^5)\ . \tag{4.506b}$$

To evaluate (4.501), therefore, we need also the commutator $[a^i, S^j]$. Using (4.68) again, gives

$$[a^i(x^o), S^j(x)] = -\ d^{ijk} P^k(x) - \sqrt{\tfrac{2}{3}}\ \delta^{ij} P^o(x) \quad , \tag{4.507a}$$

so

$$[a^3, S^4 + iS^5] = -\ \tfrac{1}{2}\ (P^4 + iP^5)\ . \tag{4.507b}$$

In the same way as (4.506) was derived we find also

$$\partial^\lambda A^i_\lambda = -i\left[a^i, \mathcal{H}\right] = -i\left[a^i, \epsilon_o S^o + \epsilon_8 S^8\right] \quad , \tag{4.508a}$$

so

$$\partial^\lambda \left(A^4_\lambda + i A^5_\lambda\right) = i \frac{1}{2\sqrt{3}} \left(2\sqrt{2}\,\epsilon_o - \epsilon_8\right) \left(P^4 + i P^5\right) . \qquad (4.508b)$$

With the aid of these formulae we may evaluate (4.501)

$$F(m_K^2, 0) = \frac{3\epsilon_8\, i}{\sqrt{2}\, f_\pi (2\sqrt{2}\,\epsilon_o - \epsilon_8)} <0|\, \partial^\lambda \left(A^4_\lambda + i A^5_\lambda\right) |K^-> $$

$$= \frac{f_K}{\sqrt{2}\, f_\pi} \frac{3\,\epsilon_8}{(2\sqrt{2}\,\epsilon_o - \epsilon_8)} \; i\, m_K^2 . \qquad (4.509)$$

But comparing (4.500) with (4.487) and (4.493) we see that

$$F(q^2, m_\pi^2) = -i \left\{ f_+(q^2)\,(m_K^2 - m_\pi^2) + q^2 f_-(q^2) \right\}$$

$$= -i\, q^\lambda F_\lambda(q, p') \quad \text{when } p'^2 = m_\pi^2 . \qquad (4.510)$$

Now, from (4.496) we see that

$$-i q^\lambda F_\lambda(q, p') \to -i \frac{f_K}{\sqrt{2}\, f_\pi} m_K^2 \quad \text{when } p' \to 0 , \qquad (4.511)$$

so that the mass-shell relationship (4.510) can only be maintained off-mass-shell at $p' = 0$ provided

$$\epsilon_8 + \sqrt{2}\,\epsilon_o = 0 . \qquad (4.512)$$

This corresponds to the limit of <u>exact</u> $SU(2) \times SU(2)$ symmetry in which V^i_λ, A^i_λ $(i = 1,2,3)$ are conserved. This follows from (4.506) and (4.508); the V^i_λ $(i = 1,2,3)$ are conserved anyway since (4.502) is an isoscalar. For the same reason $\partial^\lambda A^i_\lambda$ $(i = 1,2,3)$ is plainly proportional to P^i, and using (4.508) we find

$$\partial^\lambda A^3_\lambda = i \left(\epsilon_o \sqrt{\tfrac{2}{3}} + \epsilon_8 \frac{1}{\sqrt{3}}\right) P^3 .$$

So when (4.512) is true the A^i_λ are conserved and $SU(2) \times SU(2)$ is exact. Further, by PCAC, if $\partial^\lambda A^i_\lambda = 0$ (for $i = 1,2,3$) the pion has zero mass. So what we have shown is that the relation $F = -i q^\lambda F_\lambda$, which is true on-shell, is maintained off-shell only if the pion has zero mass. Actually this is quite reassuring, since the pion is by far and away the lightest hadron, and it lends further support to the validity of the CTMOP relation. But it is interesting to explore what happens in the (hypothetical) case that classification $SU(3)$ is a good symmetry. In this case $\epsilon_8 = 0$, the V^i_λ are conserved, so $F = 0$ on-shell, and by (4.509), off-shell. On the other hand, when SU(3) is exact

$$F_\lambda(q, p') = \frac{1}{\sqrt{2}} G(q^2)(p + p')_\lambda \quad , \tag{4.513}$$

using (4.489) and (4.492), and, as shown in the previous section,

$$\frac{f_K}{\sqrt{2}\, f_\pi} = \frac{1}{\sqrt{2}} \ . \tag{4.514}$$

Thus in <u>this</u> case (4.496) yields

$$G(m_K^2) = 1 = G(0) \ .$$

This badly broken relation is presumably true only if $m_K^2 \simeq 0$, which is also required by the SU(3) symmetry if we continue the pion mass to zero. It is also required to maintain the relation $-i q^\lambda F_\lambda = F = 0$ in the limit of SU(3) symmetry. On-shell it is satisfied since $m_K^2 = m_\pi^2$. Off-shell (4.513) gives

$$F_\lambda(p, 0) = \frac{1}{\sqrt{2}} G(m_K^2)\, p_\lambda \quad ,$$

so as $p' \to 0$

$$q^\lambda F_\lambda(p, 0) \to \frac{1}{\sqrt{2}} G(m_K^2)\, m_K^2 = 0 \quad ,$$

since $m_K^2 = 0$ in this limit.

Form Factor Shapes (Continued)

The only other statements which we can make about the form factor shapes depend upon additional assumptions. If we <u>assume</u> an unsubtracted dispersion relation, we might expect the shape of $f_+(t)$ to be well-approximated by the nearest 1^- pole. There is only one such vector meson having the right hypercharge, namely the $K^*(893)$. In this picture the f_+ form factor arises essentially from diagram (a) in Fig. 4.7, and its shape is

$$f_+(t) = f_+(0) \left(\frac{m_{K^*}^2}{m_{K^*}^2 - t} \right) . \tag{4.515}$$

The denominator $m_{K^*}^2 - t$ arises from the propagator of the exchanged K^*. Similarly we might assume that the shape of $f_0(t)$ was determined by the exchange of a $0^+ \varkappa$ meson, as shown in diagram (b) of Fig. 4.7. Although there is no definite $\varkappa$ resonance, there is an enhancement in the 1.2 - 1.4

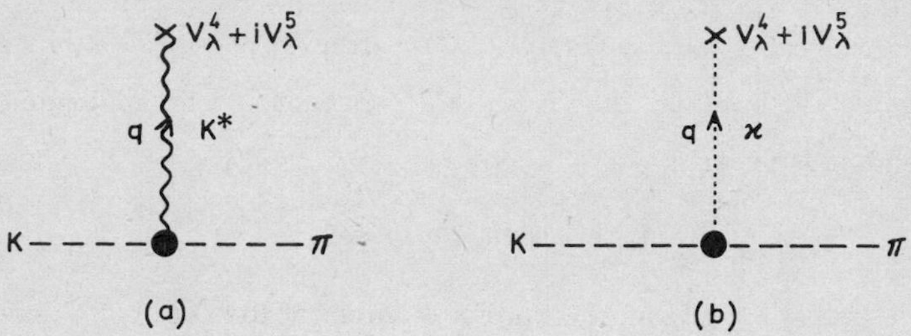

Fig. 4.7 Diagrams contributing to $<\pi|V_\lambda^4 + iV_\lambda^5|K>$

GeV range. With such an exchange

$$f_o(t) = f_+(0)\left(\frac{m_\varkappa^2}{m_\varkappa^2 - t}\right). \tag{4.516}$$

Neither of these expressions is linear in t, so the value of $\lambda_{+,0}$ obtained by making the "best fit" to such a shape depends upon how this fit is done. If (4.515) is fitted by the linear form (4.486) in the physical region, we find

$$\lambda_+ = 0.029 \quad , \tag{4.517}$$

and this is the prediction for experiments with uniform sensitivity over the physical region. An experiment which was most sensitive at low values of t, for example, would see

$$f_+(t) = f_+(0)\left(1 + \frac{t}{m_{K^*}^2} + \cdots\right) \quad ,$$

so that a value closer to

$$\lambda_+ = \frac{m_\pi^2}{m_{K^*}^2} = 0.024 \tag{4.518}$$

would be observed. Since we have no definite candidate for $\varkappa$, we shall reverse the above procedure and deduce $m_\varkappa$ from the data. Notice, however, that this picture will certainly predict λ_o to be positive.

Finally, we might try and parallel our derivation of the Adler-Weisberger sum rule (4.211) to obtain an expression for $|f_+(0)|^2$. To do this we would need an analogue of PCAC for the vector current $V_\lambda^4 + i V_\lambda^5$; we might try PCVC, which would relate the current divergence to the $\varkappa$ field. But the resulting sum rule would involve $\varkappa$-K scattering cross sections. However we can get a little information from this approach. The analogue of (4.184) is evidently of the form

$$|f_+(0)|^2 = \tilde{d}(0, 0) + \tfrac{1}{2} .$$

The term $\frac{1}{2}$, arising from the matrix element of the equal time commutator $[V_o^4 - i V_o^5, V_\nu^4 + i V_\nu^5] = - V_\nu^3$ taken between K^- states, gives the value of $|f_+(0)|^2$ in the limit that $V_\lambda^4 + i V_\lambda^5$ is conserved. The remainder $\tilde{d}$ arises since the conservation is not exact and is defined by the analogue of D^{ij}. It is therefore proportional to the square of the current divergence. We see from (4.506b) that this divergence is proprtional to ϵ_8, so

$$|f_+(0)|^2 = \tfrac{1}{2} + O(\epsilon_8^2) . \tag{4.519}$$

This is an example of the Ademollo-Gatto theorem[79]. It lends the hope that the SU(3) prediction

$$f_+(0) = \frac{1}{\sqrt{2}} \quad \text{for SU(3) symmetry,} \tag{4.520}$$

following from (4.491, 2), may not be too badly violated; even if SU(3) is violated by 30% this relation could be true to 10%. No such theorem applies to $f_-(0)$. Actually, the Ademollo-Gatto theorem is modified somewhat by corrections arising from the SU(3) x SU(3) symmetry-breaking Hamiltonian (4.502). Langacker and Pagels[79] have shown that the coefficient of ϵ_8^2 in (4.519) is proportional to ϵ_o^{-1}. Thus if SU(2) x SU(2) is a good symmetry, so that (4.512) is valid, the corrections to (4.520) are $O(\epsilon_8)$. However, a model independent calculation shows that even so the deviations are only 2%.

Experimental Data

Let us now turn to the data. The experimental information is derived from measurements of decay rates and branching ratios, Dalitz plot densities, and, in the case of muonic decay modes, the polarization of the final state muon. At the present time the data derived from $K_{\mu 3}$ polarization measurements is not consistent with that obtained by the other methods. For the moment, therefore, we ignore this data.

With this exclusion the values of λ_+ as determined from five different types of experiment are all consistent. We quote the result of a recent very high statistics experiment[80]

$$\lambda_+ = 0.030 \pm 0.003 \; . \tag{4.521}$$

Plainly this is in excellent agreement with the value (4.517) obtained assuming K^* dominance of $f_+(t)$. Knowing λ_+, the total K_{e3} lifetime is entirely determined by $f_+(0)$, since $f_-(t)$ contributes only negligibly in the electronic decays, as we have already noted. The result is that

$$\sin\theta_M^V \equiv \sqrt{2}\, f_+(0)\, \sin\theta_c = 0.216 \pm 0.002 \quad , \tag{4.522}$$

where, as before, 'M' and 'V' indicate that the value is derived from meson decays induced by the vector current. From (4.520) we see that

$$\theta_M^V = \theta_c \quad \text{for SU(3) symmetry} \; . \tag{4.523}$$

Combining (4.522) and (4.467) we find

$$\begin{aligned} \frac{f_K}{\sqrt{2}\, f_\pi f_+(0)} &= \frac{\tan\theta_M^A}{\sin\theta_M^V} \cos\theta_c \\ &\simeq \frac{\tan\theta_M^A}{\tan\theta_M^V} \quad (\text{taking } \sqrt{2}\, f_+(0) \simeq 1) \\ &= 1.25 \pm 0.02 \; . \end{aligned}$$

Thus the prediction from the CTMOP relation (4.499) is that

$$1 + \lambda_o \frac{m_K^2}{m_\pi^2} = 1.25 \pm 0.02 \quad ,$$

which implies

$$\lambda_o = 0.019 \; . \tag{4.524}$$

The data on λ_o is not in such good shape as that for λ_+, but this value is certainly consistent with the data other than the polarization measurements. We quote the high statistics Dalitz plot measurement[80]

$$\lambda_o = 0.019 \pm 0.004 \; . \tag{4.525}$$

Thus the CTMOP relation seems to be well confirmed by experiment. The same experiment also fits the form factors $f_{+,0}$ using the pole shapes (4.515, 6), with the result

$$m_{K^*} = (867 \pm 18) \text{ MeV} \tag{4.526a}$$

$$m_{\varkappa} = (1109 \pm 42) \text{ MeV} , \tag{4.526b}$$

and these two are in excellent agreement with the masses of the known $K^*(890)$ and the 0^+ enhancement.

Let us turn briefly to the muon polarization experiments. The quantity measured directly in these experiments is

$$\xi(q^2) \equiv \frac{f_-(q^2)}{f_+(q^2)} ,$$

and the polarization experiments have consistently given a value of

$$\xi(0) \simeq -1 .$$

Now, with the assumed linear fits

$$\xi(0) = m_\pi^{-2} (m_K^2 - m_\pi^2)(\lambda_o - \lambda_+) , \tag{4.527}$$

so using the predicted values of $\lambda_{+,0,}$ or those observed in the other types of experiment, we would expect

$$\xi(0) = -0.11 \pm 0.02 . \tag{4.528}$$

To obtain the values actually observed requires λ_o to be negative, in contrast to the values (4.525) as well as those predicted by any pole dominated form for $f_o(t)$. At the present time this discrepancy has not been understood, and until it is we cannot be certain that the theory of $K_{\ell 3}$ is well understood.

Finally we remark that the kaon has other leptonic decay modes which have been observed, $K_{\ell 4}$ and $K_{\ell 5}$, in which the final hadrons contain respectively two and three pions. The techniques we have used may be applied to these processes which are complicated by the fact that there are more form factors to deal with. We shall not pursue the analysis of these processes, and the interested reader is referred to any of the excellent treatments in the literature(81).

4.3.3 HYPERON DECAYS

The current $S_\lambda^{\pm}$ also induces semileptonic decays of the hyperons

$$C \to D \ell^- \bar{\nu}_\ell$$

$$C \to D \ell^+ \nu_\ell \quad ,$$

where C and D are baryon states differing by one unit of hypercharge. In particular, the following hadronic transitions are allowed by the selection rules and also kinematically: $\Lambda \to N$, $\Sigma \to N$, $\Xi \to \Lambda$, $\Xi \to \Sigma$.

SU(3) symmetry is violated in the baryon masses only by about 10%, so we may expect use of the symmetry in hyperon decays to provide a powerful test of the Cabibbo hypothesis. Consider first the matrix element of the vector current V_λ^i between two baryon states B^j, B^k $(i,j,k = 1 \cdots 8)$. On general grounds

$$\langle B^j | V_\lambda^i | B^k \rangle = \bar{B}^j \left\{ g_{1\,ijk}^V (q^2)\, \gamma_\lambda + g_{2\,ijk}^V (q^2)\, i\sigma_{\lambda\mu} q^\mu + g_{3\,ijk}^V (q^2)\, q_\lambda \right\} B^k \quad \text{(no summation)} \; , \tag{4.529}$$

just as in neutron decay. As before, we apply SU(3) symmetry in the minimal sense (cf. section 4.2.8), so the argument leading to (4.308) gives

$$g_{m\,ijk}^V(t) = -\,i f^{ijk} F_m^V(t) + d^{ijk} D_m^V(t) \qquad (m = 1,2,3). \tag{4.530}$$

Since the electromagnetic current j_λ may be written $V_\lambda^3 + \frac{1}{\sqrt{3}} V_\lambda^8$, we may determine F_m^V, D_m^V from knowledge of two of its matrix elements. Taking $B^j = B^k = p$, and comparing with (4.145), we find

$$F_1^p(t) = F_1^V(t) + \tfrac{1}{3} D_1^V(t)$$

$$-F_2^p(t) = F_2^V(t) + \tfrac{1}{3} D_2^V(t)$$

$$0 = F_3^V(t) + \tfrac{1}{3} D_3^V(t) \quad .$$

Likewise, for $B^j = B^k = n$, we obtain

$$F_1^n(t) = -\tfrac{2}{3} D_1^V(t)$$

$$-F_2^n(t) = -\tfrac{2}{3} D_2^V(t)$$

$$0 = -\tfrac{2}{3} D_3^V(t) \quad .$$

(Recall the absence of $F_3^N(t)$ follows because j_λ is exactly conserved.) Thus $F_m^V(t)$ and $D_m^V(t)$ are completely determined by the nucleons' electromagnetic form factors. If we take the charge form factor of the neutron $F_1^n(t)$ to be identically zero, as indicated by experiment, we may invert these to give

$$F_1^V(t) = F_1^p(t) \;, \; D_1^V(t) = 0 \qquad (4.531a)$$

$$-F_2^V(t) = F_2^p(t) + \tfrac{1}{2} F_2^n(t) \;, \; D_2^V(t) = \tfrac{3}{2} F_2^n(t) \qquad (4.532b)$$

$$F_3^V(t) = 0 \;, \; D_3^V(t) = 0 \;. \qquad (4.532c)$$

Similarly we may write the matrix elements of the axial vector current

$$< B^j | A_\lambda^i | B^k > = \bar{B}^j \left\{ g_{1\,ijk}^A (q^2) \gamma_\lambda \gamma_5 + \right.$$
$$\left. + g_{2\,ijk}^A (q^2) q_\lambda \gamma_5 + g_{3\,ijk}^A (q^2) i\sigma_{\lambda\mu} q^\mu \gamma_5 \right\} B^k \;, \qquad (4.533a)$$

and SU(3) symmetry requires

$$g_{m\,ijk}^A(t) = -i F_m^A(t) f^{ijk} + D_m^A(t) d^{ijk} \quad . \qquad (4.533b)$$

We have already noted that the charge symmetry property (4.247) and T-invariance together imply that $g_3^A = 0$ for neutron decay. Thus

$$F_3^A(t) + D_3^A(t) = 0 \;.$$

By the same argument the analogous form factor for $\Xi^- \to \Xi^0 e^- \bar{\nu}_\ell$ also vanishes, so

$$-F_3^A(t) + D_3^A(t) = 0 \;.$$

Thus we deduce

$$F_3^A(t) = D_3^A(t) = 0 \;. \qquad (4.534)$$

For the electronic decay modes the induced pseudoscalar term contributes only negligibly, as we have noted several times. Thus for these modes the axial vector matrix elements for all baryon decays are determined by two functions $F_1^A(t)$, $D_1^A(t)$. If we ignore the q^2-dependence, as we did in neutron decay, these functions are just constants. We may reasonably claim to have predicted one of these from the success of the Adler-Weisberger sum rule

$$F_1^A(0) + D_1^A(0) = -1.24 \pm 0.03 \quad , \tag{4.535}$$

but recall we did <u>not</u> predict the sign. The technique may be generalized, in principle, to predict another linear combination and thereby predict all baryon matrix elements. As we shall see, however, the required assumptions are more dubious than PCAC, and the resulting sum rule is difficult to evaluate.

To do this we must first extend the PCAC notion. We take

$$\partial^\lambda A_\lambda^i = \frac{f_K}{\sqrt{2}} m_K^2 \varphi_K^i \quad (i = 4, 5, 6, 7) \quad , \tag{4.536}$$

where φ_K^i $(i = 4 \cdots 7)$ are the four field operators used to define the kaon fields. The derivation then parallels that of the Adler-Weisberger formula (4.211), and we shall not reproduce it. A novelty is that there are <u>two</u> single-particle intermediate states which can contribute: Λ and Σ. The result is[82]

$$2 - |g_{1p\Lambda}^A(0)|^2 - |g_{1p\Sigma^o}^A(0)|^2 = \frac{f_K^2}{\pi} \int_{\nu_o}^{\infty} \frac{d\nu}{\nu} \left[\sigma^o(K^- p, \nu) - \sigma^o(K^+ p, \nu) \right] \quad , \tag{4.537a}$$

where

$$\nu_o = (2m_N)^{-1} (m_\Lambda + m_\pi - m_N)(m_\Lambda + m_\pi + m_N) \tag{4.537b}$$

and $\sigma^o(K^\pm p)$ are the total cross sections for zero mass <u>kaons</u> scattering off protons. As before we may simply ignore the effect of this masslessness upon the cross sections, but this is presumably much less justifiable than the neglect of the pion's mass. Another complication is that the threshold ν_o of the integral is below that for physical $K^\pm p$ scattering. Thus we are forced to continue the cross sections below threshold, usually by assuming that the threshold behaviour is dominated by the $Y_o^*(1405)$ and $Y_1^*(1385)$ resonances. We may use (4.536) to obtain generalized Goldberger-Treiman formulae, of the type (4.163), if we neglect the kaon's mass. These then yield

$$g^A_{1p\Lambda}(0)\,(m_N + m_\Lambda)\, g^{-1}_{Kp\Lambda} = g^A_{1p\Sigma^o}(0)\,(m_N + m_\Sigma)\, g^{-1}_{Kp\Sigma^o} \quad . \tag{4.538}$$

If we know the strong coupling constants, we may solve (4.537) and derive the axial vector constants $g^A_{1p\Lambda}$ and $g^A_{1p\Sigma^o}$. Unfortunately, the strong coupling constants are not well known, so results of such an analysis differ according to the assumptions made. The published results are

$$\alpha \equiv D^A_1/(D^A_1 + F^A_1) = 0.73 \quad \text{(Amati et al.}^{(82)}\text{)} \tag{4.539a}$$

$$= 0.63 \quad \text{(Levinson et al.}^{(82)}\text{)} \quad . \tag{4.539b}$$

For this reason it is customary to fit <u>all</u> available baryon decay data (including the hypercharge conserving decays) and to compare the values D^A_1, F^A_1 thus obtained to the somewhat dubious theoretical predictions (4.539). The Cabibbo hypothesis predicts the existence of the following hadronic transitions, which are also kinematically allowed :

$$\langle p|\mathcal{J}^+_\lambda|n\rangle \sim \cos\theta_c \;(F + D) \tag{4.540a}$$

$$\langle \Lambda|\mathcal{J}^+_\lambda|\Sigma^-\rangle = \langle \Lambda|\mathcal{J}^-_\lambda|\Sigma^+\rangle \sim \cos\theta_c \sqrt{\tfrac{2}{3}}\, D \tag{4.540b}$$

$$\langle \Sigma^o|\mathcal{J}^+_\lambda|\Sigma^-\rangle = -\langle \Sigma^o|\mathcal{J}^-_\lambda|\Sigma^+\rangle \sim \cos\theta_c \sqrt{2}\, F \tag{4.540c}$$

$$\langle \Xi^o|\mathcal{J}^+_\lambda|\Xi^-\rangle \sim \cos\theta_c \;(D - F) \tag{4.540d}$$

$$\langle p|\mathcal{J}^+_\lambda|\Lambda\rangle \sim \sin\theta_c \frac{-1}{\sqrt{6}} \;(3F + D) \tag{4.540e}$$

$$\langle n|\mathcal{J}^+_\lambda|\Sigma^-\rangle \sim \sin\theta_c \;(D - F) \tag{4.540f}$$

$$\langle p|\mathcal{J}^+_\lambda|\Sigma^o\rangle \sim \sin\theta_c \;(D - F) \tag{4.540g}$$

$$\langle \Lambda|\mathcal{J}^+_\lambda|\Xi^-\rangle \sim \sin\theta_c \frac{1}{\sqrt{6}} \;(3F - D) \tag{4.540h}$$

$$\langle \Sigma^+|\mathcal{J}^+_\lambda|\Xi^o\rangle \sim \sin\theta_c \;(F + D) \tag{4.540i}$$

$$\langle \Sigma^o|\mathcal{J}^+_\lambda|\Xi^-\rangle \sim \sin\theta_c \frac{1}{\sqrt{2}} \;(F + D) \quad . \tag{4.540j}$$

We have indicated the combination of F and D corresponding to each decay. Notice that the hypercharge nonconserving decays all proceed via S_λ^+ rather than S_λ^-. This is because the heavier baryons all have lower hypercharge than the lighter ones, so that the $\Delta Y = \Delta Q$ rule, following from the Cabibbo hypothesis, ensures that only ℓ^- decays are allowed. (The anti-particles have ℓ^+ decays, of course.) The vector current matrix elements are then given by (4.529) and (4.532), while the axial vector matrix elements are given in terms of F_1^A and D_1^A by (4.533) and (4.534). Unfortunately data is not available on all of the above transitions, in particular the $\Sigma^- \to \Sigma^o$, $\Xi^- \to \Xi^o$ transitions are hard to detect for the reasons given earlier. The $\Sigma^o \to p$ transition is masked by the fast electromagnetic decay, as is the $\Xi^o \to \Sigma^+$ by the nonleptonic decay $\Xi^o \to \Lambda\pi^o$.

The calculation of the experimentally observable distributions is straightforward but tedious[(83)]. The magnitude of the vector and axial vector constants are determined, as in neutron decay (4.221), by measurement of the total decay rate and the lepton-neutrino correlation coefficient of $\underline{v}_\ell \cdot \underline{v}_\nu$. Their relative sign can be determined by the shape of the lepton spectrum from an unpolarized hyperon, the asymmetry of the decay products from a polarized hyperon, or from the polarization of the final state baryon from an unpolarized hyperon.

With (4.532) as input, a fit to all the data available was recently made by Kleinknecht[(84)] with the results

$$\sin\theta_B = 0.230 \pm 0.003 \qquad (4.541a)$$

$$F_1^A = -0.437 \pm 0.010 \qquad (4.451b)$$

$$D_1^A = -0.809 \pm 0.010 \qquad (4.451c)$$

$$\alpha = 0.658 \pm 0.007 \; . \qquad (4.451d)$$

A two angle fit, in which we allow θ_B^V to differ from θ_B^A because of SU(3) symmetry-breaking effects, does not alter the conclusions significantly.

With these values we may check the consistency of the sum rule (4.537). Using (4.540e, g) and substituting (4.541b, c) we find

$$2 - |g^A_{1p\Lambda}|^2 - |g^A_{1p\Sigma^o}|^2 = 2(1 - |F^A_1|^2 - \tfrac{1}{3}|D^A_1|^2) \simeq 1.18 \quad . \qquad (4.542a)$$

The integral on the right of (4.537) has been estimated as[85]

$$\frac{f_K^2}{\pi}\int_{\nu_o}^{\infty}\frac{d\nu}{\nu}\left[\sigma(K^- p, \nu) - \sigma(K^+ p, \nu)\right] \simeq 1.15 \qquad (4.542b)$$

using physical cross sections above threshold and assuming pole dominance below. In view of the inherent approximations the agreement between the two is remarkably good. Finally, turning to the generalized Goldberger-Treiman formula (4.538), we may use (4.541) to predict the strong coupling constant ratio. This gives

$$(g_{Kp\Lambda}/g_{Kp\Sigma^o})^2 \simeq 10.1 \quad ,$$

which is consistent with most of the available data[86].

Thus we conclude that the baryon decays are well described by three parameters θ_B, F^A_1, D^A_1, and that two of these F^A_1, D^A_1 might reasonably be predicted by the theory, if and when definitive values of the strong coupling constants are available. θ_B, the effective Cabibbo angle for baryon decays, remains unpredicted and is regarded at present as a constant of nature just like G.

4.3.4 STATUS OF CABIBBO THEORY

By allowing for SU(3) symmetry-breaking effects we have derived the following "effective" Cabibbo angles

$$\sin\theta^A_M = 0.2655 \pm 0.0006 \quad , \qquad (4.543a)$$

where

$$\tan\theta^A_M \equiv f_K f_\pi^{-1} \tan\theta_c \quad . \qquad (4.543b)$$

$$\sin\theta^V_M \equiv \sqrt{2}\, f_+(0)\sin\theta_c = 0.216 \pm 0.002 \quad . \qquad (4.543c)$$

$$\sin\theta_B = 0.230 \pm 0.003 \quad , \qquad (4.543d)$$

where

$$\sin\theta_B = \beta_B \sin\theta_c \qquad (4.543e)$$

and β_B is an overall "renormalization" resulting from the symmetry-breaking. The Ademollo-Gatto theorem[79] indicates that these symmetry-breaking effects are of second order for vector current matrix elements, as we have seen. Thus we should certainly expect θ_M^V and $\theta_B^V = \theta_B$ to be closer to the true θ_c than θ_M^A. On the other hand, SU(3) is rather badly broken for the mesons but quite good for the baryons. On this basis, therefore, we expect θ_B to be closer to θ_c than θ_M^V, which itself should be closer than θ_M^A. So if we assume for the moment that θ_B is very close to θ_c, we should expect

$$|\theta_M^A - \theta_B| > |\theta_M^V - \theta_B| \ .$$

It is reassuring that the data (4.543) do indeed have this property.

However, we can do better than this by including some symmetry-breaking effects in a purely phenomenological way. We stick to the baryon decays for the reasons given above. Roos[87] has proposed that we simulate the symmetry-breaking effects by replacing the vector coupling F_1^V and the axial vector couplings F_1^A, D_1^A in the hadronic transition $C \to D$ by

$$|F_1^V|^2 \to \left\{ 1 + k \left[1 - \left(\frac{m_D}{m_C}\right)^r \right] \right\}^2 |F_1^V|^2 \qquad (4.544a)$$

$$\left(|F_1^A|^2, |D_1^A|^2\right) \to \left(\frac{m_D}{m_C}\right)^r \left(|F_1^A|^2, |D_1^A|^2\right) \ , \qquad (4.544b)$$

where r and k are free parameters. In the SU(3) symmetry limit $r \to 0$. When $r, k = 1$ we see that the vector symmetry-breaking is of second order in the baryon mass difference while the axial symmetry breaking is first order. For r and k large the symmetry is badly broken, so that all conceivable corrections can be included. When the data are analysed with θ_c, α, r, k free parameters ((4.535) is imposed), the range of acceptable values of θ_c is extended. At the 95% confidence level the results are[88]

$$0.224 \le \sin\theta_c \le 0.252 \qquad (95\% \text{ C.L.}) \ , \qquad (4.545a)$$

$$0.226 \le \sin\theta_c \qquad (95\% \text{ C.L.}) \ . \qquad (4.545b)$$

However even these rather limited conclusions on the actual value of θ_c lead to interesting results, when compared with the information on θ_c which can be derived from the superallowed beta decays. Using (4.233), (4.231), (4.462) and (3.62), the lower bound on θ_c leads to a lower bound on Λ ; we conclude

$$\Lambda \geq 105 \text{ GeV } (!) \; . \tag{4.546}$$

Plainly a cut-off of this magnitude is <u>not</u> a strong interaction cut-off. In view of the discussion in 4.2.12 it seems more plausible to interpret this as a measure of the <u>weak</u> structure, such as is supplied if a W-boson mediates the weak interactions: as noted earlier, there are no a priori grounds for supposing that m_W has a value close to m_N.

Finally we note that if the naive model used in 4.2.12 is used to estimate the divergent part of the electromagnetic corrections to hypercharge nonconserving processes, then the same result (4.459) is obtained; the reason for this is that the $|\Delta Y| = 1$ processes occur in this model via the "λ" → "p" transition, rather than the "n" → "p". But it is the charge of the final particle "p" which determines the radiative corrections, so these are unaltered. We note that even for the value of Λ given above these radiative corrections are still of the order of one percent or so, and are easily accommodated by the all embracing correction factors (4.544).

4.3.5 HYPERCHARGE NONCONSERVING NEUTRINO PROCESSES

It is a straightforward exercise to generalize the treatment given in sections 4.2.9, 10, 11 for hypercharge conserving processes to those induced by $S_\lambda^\pm$, which do not conserve hypercharge. The analogues of the quasi-elastic processes (4.323) are those in which the final hadron is a hyperon rather than a nucleon. According to the Cabibbo hypothesis this hyperon must have hypercharge differing by at most one unit from the target nucleon. Thus $|\Delta Y| = 2$ processes are forbidden, so that

$$\bar{\nu}_\ell n \not\to \ell^+ \; \Xi^- \; . \tag{4.547}$$

The remaining hyperons have zero hypercharge, so, since the nucleons have $Y = 1$, the $\Delta Y = \Delta Q$ rule requires the final hyperons to have one unit of charge less than the initial nucleon. Thus the only quasi-elastic $|\Delta Y| = 1$ processes allowed by the Cabibbo hypothesis have an ℓ^+ in the final state

$$\bar{\nu}_\ell p \rightarrow \ell^+ \Lambda \qquad (4.548a)$$

$$\bar{\nu}_\ell p \rightarrow \ell^+ \Sigma^o \qquad (4.548b)$$

$$\bar{\nu}_\ell n \rightarrow \ell^+ \Sigma^- \ . \qquad (4.548c)$$

In particular, the hypothesis predicts

$$\nu_\ell n \not\rightarrow \ell^- \Sigma^+ \ . \qquad (4.549)$$

Since N has $I = \frac{1}{2}$ and Σ has $I = \frac{3}{2}$, we may connect them by an operator having $I = \frac{1}{2}$ and/or $\frac{3}{2}$. The Cabibbo hypothesis predicts that S_λ^- has $I = \frac{1}{2}$, and this restriction enables us to relate the two Σ-processes (4.548b, c)

$$<\Sigma^-|[v^1 - iv^2, S_\lambda^-]|p> = 0 \ .$$

Thus,

$$\sqrt{2} <\Sigma^o|S_\lambda^-|p> = <\Sigma^-|S_\lambda^-|n>$$

and

$$d\sigma(\bar{\nu}_\ell n \rightarrow \ell^+ \Sigma^-) = 2\, d\sigma(\bar{\nu}_\ell p \rightarrow \ell^+ \Sigma^o) \ . \qquad (4.550)$$

The derivation of the differential cross section differs from the $\Delta Y = 0$ case only in that the masses of the initial and final hadrons are no longer equal. The resulting expression is then slightly more complicated than (4.338) and (4.339). It will be a long time before accurate measurements of these cross sections can be made, so we shall not quote the result[89]. The form factors occurring in this expression have been written down already in (4.529), (4.533) and (4.540e, f, g). (Note that the matrix elements $<Y|S_\lambda^-|N>$, required for hyperon production, are simply the complex conjugates of those $<N|S_\lambda^+|Y>$ occurring in hyperon decay.) Thus, eventually we shall be able to test the consistency of the fit (4.541) to the hyperon decays, by comparing its predictions for hyperon production with experiment.

As before, the exclusive inelastic processes also present opportunities for us to test the selection rules following from the assumed quantum numbers of $S_\lambda^\pm$. Analogously to (4.550), the $I = \frac{1}{2}$ property of S_λ^-

implies[90]

$$d\sigma\,(\bar{\nu}_\ell n \to \ell^+ Y^{*-}) = 2\, d\sigma\,(\bar{\nu}_\ell p \to \ell^+ Y^{*o}) \qquad (4.551a)$$

$$d\sigma\,(\bar{\nu}_\ell n \to \ell^+ \Lambda\pi^-) = 2\, d\sigma\,(\bar{\nu}_\ell p \to \ell^+ \Lambda\pi^o) \qquad (4.551b)$$

$$d\sigma\,(\bar{\nu}_\ell n \to \ell^+ \Sigma^o \pi^-) = d\sigma\,(\bar{\nu}_\ell p \to \ell^+ \Sigma^- \pi^o)\ . \qquad (4.551c)$$

We may also derive triangular relationships between the hadronic matrix elements, which lead to triangular inequalities between the square roots of the corresponding cross sections, as in (4.358). For example

$$\sqrt{\sigma\,(\nu_\ell p \to \ell^- p K^+)} + \sqrt{\sigma\,(\nu_\ell n \to \ell^- p K^o)} \le \sqrt{\sigma\,(\nu_\ell n \to \ell^- n K^+)}, \text{ etc.} \qquad (4.552)$$

Other examples are given by Lee and Yang[90].

Finally we may consider the inclusive inelastic processes. There is no analogue of the PCAC test (4.366), unless we assume SU(3) symmetry, since the vector current $V_\lambda^4 \pm i V_\lambda^5$ is not conserved. Even so we should only be testing the generalized PCAC relation (4.536). However these processes do provide a test of the current algebra, analogous to (4.367, 8). As before, this is a special case $q^2 = 0$ of a general q^2 result proved by Adler[58]. The only difference in the derivation is the value of the commutator

$$\delta(x^o)\,[\,S_o^-(x),\, S_o^+\,] = -2\delta(x)\left(V_o^3 + A_o^3 + \sqrt{3}\,V_o^8 + \sqrt{3}\,A_o^8\right). \qquad (4.553)$$

This then leads to the sum rule (recall $\sqrt{3}\,Y = 2\,v^8$)

$$\int_o^\infty d\nu\,[\,w_2^{(\nu)}(q^2,\nu) - w_2^{(\bar{\nu})}(q^2,\nu)\,] = -2 < I^3 > - 3 < Y > , \qquad (4.554)$$

where $w_2^{(\nu,\bar{\nu})}$ are structure functions defined like $W_2^{(\nu,\bar{\nu})}$ but referring to the processes

$$\nu_\ell N \to \ell^- X \;(Y = 2) \qquad (4.555a)$$

$$\bar{\nu}_\ell N \to \ell^+ X \;(Y = 0) \qquad (4.555b)$$

and $X(Y)$ are any hadronic states having hypercharge Y. As before this may be written explicitly in terms of the cross section

$$\lim_{E_\nu\to\infty}\left[\frac{d\sigma}{d|q^2|}(\bar{\nu}_\ell N \to \ell^+ X(Y=0)) - \frac{d\sigma}{d|q^2|}(\nu_\ell N \to \ell^- X(Y=2))\right]$$

$$= \frac{G^2}{\pi}\sin^2\theta_c\left[<I^3> + \tfrac{3}{2}<Y>\right] . \qquad (4.556)$$

Analogously to (4.395), we define

$$m_N w_1(q^2,\nu) = g_1\left(x, \frac{q^2}{m_N^2}\right) \qquad (4.557a)$$

$$\nu w_2(q^2,\nu) = g_2\left(x, \frac{q^2}{m_N^2}\right) \qquad (4.557b)$$

$$\nu w_3(q^2,\nu) = g_3\left(x, \frac{q^2}{m_N^2}\right) . \qquad (4.557c)$$

Then Bjorken's scaling hypothesis states that

$$\lim_{\substack{-q^2\to\infty \\ x \text{ fixed}}} g_i\left(x, \frac{q^2}{m_N^2}\right) = f_i(x) \neq 0 . \qquad (4.558)$$

We can then derive analogues of (4.398, 9) and (4.403) with $\cos^2\theta_c$ replaced by $\sin^2\theta_c$ and F_i by f_i.

The application of the quark-parton model to these processes is also a trivial generalization of what we did before. The only way for a $|\Delta Y| = 1$ process to proceed in this model is by the creation or destruction of a "λ". Taking a nucleon as target, we see from (4.407) that there is no "λ" present to be destroyed, so <u>all</u> of the $|\Delta Y| = 1$ processes are induced by the scattering

$$\bar{\nu}_\ell \text{ "p"} \to \ell^+ \text{ "}\lambda\text{"} .$$

It then follows, as before, that the positivity inequalities are all saturated, and that

$$f_2^{(\nu p)}(x) = f_2^{(\nu n)} = 0 , \qquad (4.559a)$$

$$f_2^{(\bar{\nu}p)}(x) = 2xu_p(x) , \quad f_2^{(\bar{\nu}n)}(x) = 2xu_n(x) . \qquad (4.559b)$$

The left-hand side of the Adler sum rule (4.554) is, for a proton target,

$$\int_0^1 \frac{dx}{x} \left[f_2^{(\nu p)}(x) - f_2^{(\bar{\nu} p)}(x) \right] = -2 \int_0^1 dx\, u_p(x) = -4 \ ,$$

using (4.429). The right-hand side is

$$-2 < I_3 >_p - 3 < Y >_p = -4 \qquad \text{(Q. E. D.)}$$

It will be most interesting, eventually, to test the predictions of this simple model, but it will plainly be some time before these contributions can be separated from the dominant $\Delta Y = 0$ states.

Finally, we note the predictions for the isospin averaged total cross sections (4.430)

$$\sigma(\nu N)_{Y=2} = 0$$

$$\sigma(\bar{\nu} N)_{Y=0} = \tan^2 \theta_c \, \sigma(\bar{\nu} N)_{Y=1} \ ,$$

where $\sigma(\bar{\nu} N)_{Y=1}$ is the previously calculated $\Delta Y = 0$ cross section given in (4.431).

Chapter 5

PURELY HADRONIC PROCESSES

The purely hadronic weak interactions are defined as those weak processes which do not involve the leptons. In practice they can only be distinguished from the other interactions which also involve only hadrons, if they violate some selection rules or invariances which are believed to be conserved by the other known interactions. So far as is known, the only other interactions are strong or electromagnetic (we ignore gravity), and both of these conserve hypercharge and parity. Both of these invariances are violated by the weak interactions already considered. Thus it is perhaps not entirely surprising that the purely hadronic processes, which also violate hypercharge and/or parity conservation, have rates characteristic of weak processes. Of course, for all we know there might be weak processes which violate neither of these invariances, but, since they are indistinguishable from strong or electromagnetic processes, they are beyond experimental analysis. Most of the experimental data on purely hadronic weak interactions have been derived from processes in which hypercharge is not conserved. There is also a certain amount of data on hypercharge conserving processes, which has been obtained by the observation of parity violation effects in nuclear decays. In this chapter we shall consider both of these processes.

If we accept (tentatively perhaps) that the purely leptonic and semileptonic processes are described by the Hamiltonians in (3.1) and (4.1), we see that they are both contained in the "universal current-current Hamiltonian"

$$\mathcal{H}_W = \frac{G}{\sqrt{2}} \left(\mathcal{J}^+_\lambda + L_\lambda\right)\left(\mathcal{J}^{-\lambda} + L^\lambda\right) . \qquad (5.1)$$

It is therefore tempting to conjecture that $\mathcal{H}_W$ is the Hamiltonian responsible for <u>all</u> weak processes. The untested part of this hypothesis is contained in the purely hadronic piece $\mathcal{H}_W^H$

$$\mathcal{H}_W^H = \frac{G}{\sqrt{2}} \mathcal{J}_\lambda^+ \mathcal{J}^{-\lambda} \quad . \tag{5.2}$$

There is no particular reason for writing $\mathcal{J}_\lambda^+$ first, so to make the symmetry apparent we may write

$$\mathcal{H}_W^H = \frac{G}{2\sqrt{2}} \left\{ \mathcal{J}_\lambda^+ , \mathcal{J}^{-\lambda} \right\} . \tag{5.3}$$

For the majority of this chapter we shall explore the hypothesis that $\mathcal{H}_W^H$ is the Hamiltonian responsible for all purely hadronic weak processes. However, it is wise from the outset to be somewhat sceptical that this is indeed so ; the property (4.253) of $\mathcal{J}_\lambda^\pm$ under the time reversal operation $\mathcal{T}$ plainly implies that $\mathcal{H}_W^H$ is time reversal invariant. Then by the TCP theorem, or directly, it is also CP-invariant. But, as noted in the Introduction, there is by now no doubt that CP-violation has been observed in some purely hadronic processes. Thus it might appear unwise to start from an hypothesis which assuredly cannot explain this effect. It turns out, as we shall see in Chapter 7, that the T-violating effects are of order 10^{-3}, so in this chapter we ignore them. In any case, it is possible that the observed T-violation does not originate in the weak Hamiltonian, even though, for special reasons, it has only been observed in weak processes. Further, we now know that the Hamiltonian (5.1) does not predict the recently observed neutral current phenomena. These <u>might</u> be described by another term in $\mathcal{H}_W$ having the form of a "universal neutral current-current" interaction - the universal neutral current being the sum of leptonic and hadronic pieces. If so, the Hamiltonian for the purely hadronic processes would have, in addition to that given in (5.2), a contribution from the neutral hadronic current coupled to its hermitian conjugate. However, we saw in 2.5.2 that this neutral hadronic current appears to conserve hypercharge, in which case it cannot contribute to the hypercharge nonconserving hadronic processes discussed in the next section. It <u>can</u> contribute to the hypercharge conserving processes, which we deal with in section 5.2 , so we must admit from the outset that we really have no very strong reasons for thinking that (5.2) <u>might</u> describe these processes. Unfortunately nothing else is known about this current and for the remainder of this chapter we shall ignore it.

5.1 HYPERCHARGE NONCONSERVING PURELY HADRONIC DECAYS

We recall first the Cabibbo hypothesis that $\mathcal{J}_\lambda^\pm$ has the form

$$\mathcal{J}_\lambda^\pm = J_\lambda^\pm \cos\theta_c + S_\lambda^\pm \sin\theta_c \quad , \tag{5.4}$$

where $J_\lambda^\pm$ has $Y = 0$ and $S_\lambda^\pm$ creates hypercharge $Y = \pm 1$. Thus $\mathcal{H}_W^H$ may be split into two pieces, one of which conserves hypercharge, while the other creates plus or minus one unit

$$\mathcal{H}_W^H = \mathcal{H}_W^H(\Delta Y = 0) + \mathcal{H}_W^H(|\Delta Y| = 1) \ , \tag{5.5a}$$

where

$$\mathcal{H}_W^H(\Delta Y = 0) = \frac{G}{2\sqrt{2}}\left[\cos^2\theta_c \left\{J_\lambda^+, J^{-\lambda}\right\} + \sin^2\theta_c \left\{S_\lambda^+, S^{-\lambda}\right\}\right] \tag{5.5b}$$

$$\mathcal{H}_W^H(|\Delta Y| = 1) = \frac{G}{2\sqrt{2}} \sin\theta_c \cos\theta_c \left[\left\{J_\lambda^+, S^{-\lambda}\right\} + \left\{S_\lambda^+, J^{-\lambda}\right\}\right] \ . \tag{5.5c}$$

In this section we shall be concerned only with processes in which $\Delta Y \neq 0$, and we see that the only such processes predicted by our hypothesis have $|\Delta Y| = 1$. Thus

$$\Xi \not\to N\pi \qquad (\text{B.R.} \ \tilde{<}\ 10^{-3}) \ .$$

Other tests of this selection rule will become available when Ω^- events are more plentiful. In particular the following processes are forbidden if $|\Delta Y| \leq 1$:

$$\Omega \to N\pi, \ \Lambda\pi \ , \ \Sigma\pi \ .$$

Now if $\Delta Y = \pm 1$, then it follows from the Gell-Mann Nishijima relation that $\Delta I^3 = \mp \frac{1}{2}$, since charge is conserved ($\Delta Q = 0$). Thus the Hamiltonian responsible for purely hadronic weak interactions could, a priori, have components with isospin $I = \frac{1}{2}, \frac{3}{2}, \frac{5}{2}, \ldots$. Since $J_\lambda^\pm$ has $I = 1$ and $S_\lambda^\pm$ has $I = \frac{1}{2}$, it follows from (5.5c) that our hypothesis predicts that $\mathcal{H}_W^H(|\Delta Y| = 1)$ is a mixture only of $I = \frac{1}{2}$ and $I = \frac{3}{2}$

$$\mathcal{H}_W^H(|\Delta Y| = 1) = \mathcal{H}_W^H(1, \tfrac{1}{2}, -\tfrac{1}{2}) + \mathcal{H}_W^H(1, \tfrac{3}{2}, -\tfrac{1}{2}) + \mathcal{H}_W^H(-1, \tfrac{1}{2}, \tfrac{1}{2}) + \mathcal{H}_W^H(-1, \tfrac{3}{2}, \tfrac{1}{2}) \ , \tag{5.6}$$

where $\mathcal{H}_W^H(Y, I, I^3)$ has hypercharge Y, isospin I, and third component of of isospin I^3. The predicted absence of $I = \frac{5}{2}, \frac{7}{2}, \ldots$ transitions leads to relationships between decay amplitudes which would otherwise have been arbitrary, as we shall see.

Finally, we may analyse the structure of $\mathcal{H}_W^H$ according to its SU(3) transformation properties. The Cabibbo currents $\mathcal{J}_\lambda^\pm$ belong to an $\underline{8}$ representation of SU(3), in the sense described in section (4.1). We denote by $\mathcal{J}^i$ $(i = 1 \ldots\ldots 8)$ the general element of this octet - we have suppressed the Lorentz suffix λ, which is irrelevant to SU(3) considerations. The product of two octets will in general contain 64 elements, which may be decomposed into irreducible representations of SU(3) in the way given in (4.307). However, if we consider the product of two identical octets $\mathcal{J}^i \mathcal{J}^j$ $(i, j = 1 \ldots\ldots 8)$, this will only contain 36 independent elements, since

$$\mathcal{J}^i \mathcal{J}^j = \mathcal{J}^j \mathcal{J}^i \quad , \tag{5.7}$$

and these may be decomposed into $\underline{1}$, $\underline{8}$ and $\underline{27}$ representations of SU(3). We can see how this comes about from what we have done already. Plainly the 36 surviving elements of the general 64 correspond to the symmetrical combinations. One of these is obviously the SU(3) singlet $\mathcal{J}^j \mathcal{J}^j$; that this is indeed an SU(3) representation can be seen as follows

$$\left[v^i, \mathcal{J}^j \mathcal{J}^j\right] = i f^{ijk} \left\{\mathcal{J}^j, \mathcal{J}^k\right\} = 0 \quad , \tag{5.8}$$

since the f^{ijk} are totally antisymmetric. Since $\mathcal{J}^j \mathcal{J}^j$ commutes with all of the generators v^i of SU(3), it is plainly SU(3) invariant. It therefore corresponds to the trivial one-dimensional representation of the group, in which the infinitesimal generators are all zero. We have already encountered the symmetrical octet combination $d^{ijk} T^j U^k$ of two general octets T^j and U^j, and this survives if $T^j = U^j$; the other octet combination $f^{ijk} T^j U^k$ vanishes if $T^j = U^j$, because of the antisymmetry of f^{ijk}. The remaining 27 of the 36 elements do in fact constitute an irreducible representation of SU(3), although we shall not prove it. We can make all of this explicit by decomposing $\left\{\mathcal{J}^i, \mathcal{J}^j\right\}$ as follows :

$$\left\{\mathcal{J}^i, \mathcal{J}^j\right\} \equiv T^{ij} + \frac{3}{5} d^{ijk} D^k + \frac{1}{8} \delta^{ij} S , \tag{5.9a}$$

where

$$S = \left\{\mathcal{J}^k, \mathcal{J}^k\right\} \tag{5.9b}$$

$$D^k = d^{k\ell m} \left\{\mathcal{J}^\ell, \mathcal{J}^m\right\} \tag{5.9c}$$

$$T^{ij} = \left\{\mathcal{J}^i, \mathcal{J}^j\right\} - \frac{3}{5} d^{ijk} D^k - \frac{1}{8} \delta^{ij} S . \tag{5.9d}$$

The reason for writing the identity this way (that is with the particular coefficients $\frac{3}{5}, \frac{1}{8}$) is that the identities

$$\delta^{ij} \delta^{ij} = 8 \tag{5.10a}$$

$$d^{ijk} d^{ij\ell} = \frac{5}{3} \delta^{k\ell} \tag{5.10b}$$

and

$$\delta^{ij} d^{ijk} = 0 \tag{5.10c}$$

guarantee that the decomposition is "orthonormal". Referring back to (5.5c) we see that we are concerned only with the combination

$$\left\{\mathcal{J}^1, \mathcal{J}^4\right\} + \left\{\mathcal{J}^2, \mathcal{J}^5\right\} , \tag{5.11}$$

since $J^{\pm} \sim \mathcal{J}^1 \pm i \mathcal{J}^2$ and $S^{\pm} \sim \mathcal{J}^4 \pm i \mathcal{J}^5$. Thus the singlet S in (5.9a) cannot contribute; this is obvious in any case, since the singlet must have zero quantum numbers (Y, I, I_3), and we are concerned with $Y \neq 0$. Thus with the d^{ijk} given in (4.13) we see

$$\mathcal{H}_W^H(|\Delta Y| = 1) = T^{14} + T^{25} + \frac{3}{5} D^6 . \tag{5.12}$$

From (4.26) we see that the sixth component of an octet is a linear combination of states with $(Y, I, I_3) = (-1, \frac{1}{2}, \frac{1}{2})$ and $(1, \frac{1}{2}, -\frac{1}{2})$. So we may write

$$\frac{3}{5} D^6 = \mathcal{H}(\underline{8}, 1, \tfrac{1}{2}, -\tfrac{1}{2}) + \mathcal{H}(\underline{8}, -1, \tfrac{1}{2}, \tfrac{1}{2}) . \tag{5.13}$$

This too was fairly clear from the outset. Given that the $\underline{8}$ actually contributes, the only possibility is that its contribution is pure $I = \frac{1}{2}$, since the octet has no component with $I = \frac{3}{2}, \frac{5}{2}, \ldots$.

Let us consider the $Y = -1$ part of the $\underline{27}$ representation, i.e. the contribution from $\left\{J^+, S^-\right\}$. From (5.9d) we see that

$$T^{1+i2,4-i5} = \left\{ \mathcal{J}^1 + i\mathcal{J}^2, \mathcal{J}^4 - i\mathcal{J}^5 \right\} - \frac{3}{5}(D^6 - iD^7)$$

$$= \frac{2}{5}\left\{\mathcal{J}^1 + i\mathcal{J}^2, \mathcal{J}^4 - i\mathcal{J}^5\right\} + \frac{3}{5}\left\{\mathcal{J}^3, \mathcal{J}^6 - i\mathcal{J}^7\right\} + \frac{\sqrt{3}}{5}\left\{\mathcal{J}^6 - i\mathcal{J}^7, \mathcal{J}^8\right\}. \tag{5.14}$$

It is apparent from (4.26) that $\mathcal{J}^1 \pm i\mathcal{J}^2, \mathcal{J}^3$ define an isovector, while $\mathcal{J}^4 - i\mathcal{J}^5$ and $\mathcal{J}^6 - i\mathcal{J}^7$ define an isospinor. Thus we may decompose the first two terms of (5.14) into a linear combination of objects with isospin $\frac{1}{2}$ and $\frac{3}{2}$. Defining

$$(Y, I, I_3) = (-1, \tfrac{3}{2}, \tfrac{3}{2}) = (0,1,1)(-1, \tfrac{1}{2}, \tfrac{1}{2}) = -\tfrac{1}{2}\left\{\mathcal{J}^1 + i\mathcal{J}^2, \mathcal{J}^6 - i\mathcal{J}^7\right\},$$

we leave it as an exercise in isospin addition to show that

$$\frac{2}{5}\left\{\mathcal{J}^1 + i\mathcal{J}^2, \mathcal{J}^4 - i\mathcal{J}^5\right\} + \frac{3}{5}\left\{\mathcal{J}^3, \mathcal{J}^6 - i\mathcal{J}^7\right\} = \frac{2}{\sqrt{3}}(-1, \tfrac{3}{2}, \tfrac{1}{2}) - \frac{\sqrt{2}}{5\sqrt{3}}(-1, \tfrac{1}{2}, \tfrac{1}{2})_1 . \tag{5.15}$$

The third term of (5.14) is plainly a different (orthogonal) state with isospin $\frac{1}{2}$, since $\mathcal{J}^8$ is an isoscalar :

$$\frac{\sqrt{3}}{5}\left\{\mathcal{J}^6 - i\mathcal{J}^7, \mathcal{J}^8\right\} = \frac{\sqrt{6}}{5}(-1, \tfrac{1}{2}, \tfrac{1}{2})_2 ,$$

and combining the two $I = \frac{1}{2}$ states gives

$$\frac{-\sqrt{2}}{5\sqrt{3}}(-1, \tfrac{1}{2}, \tfrac{1}{2})_1 + \frac{\sqrt{6}}{5}(-1, \tfrac{1}{2}, \tfrac{1}{2})_2 = \frac{2}{\sqrt{15}}(-1, \tfrac{1}{2}, \tfrac{1}{2}) . \tag{5.16}$$

Thus finally we have

$$T^{1+i2,4-i5} = \frac{2}{\sqrt{15}}\left[\sqrt{5}(-1, \tfrac{3}{2}, \tfrac{1}{2}) + (-1, \tfrac{1}{2}, \tfrac{1}{2})\right] . \tag{5.17}$$

We have to worry about the <u>relative</u> normalization of the $I = \frac{1}{2}$ and $\frac{3}{2}$ pieces of the $\underline{27}$, since (by the Wigner Eckart theorem) they both contribute to the same reduced amplitude. The normalization relative to the $\underline{8}$ component is immaterial, if we are assuming SU(3) invariance, since they can never mix. This rather tedious derivation of the factor $\sqrt{5}$ in (5.17) is just the SU(3) analogue of decomposing a product of two isospin states into a linear combination of states with definite isospin. In practice, of course, one uses the tables of SU(2) Clebsch-Gordan coefficients. Such tables also exist(1) for the group SU(3). $T^{1+i2,4-i5}$ is the $\underline{27}$ component of the product of two octet states

$$(\underline{8}, 0, 1, 1) \otimes (\underline{8}, -1, \tfrac{1}{2}, -\tfrac{1}{2}) .$$

The reader is invited to verify directly from the tables that the coefficients of $(\underline{27}, -1, \frac{3}{2}, \frac{1}{2})$ and $(\underline{27}, -1, \frac{1}{2}, \frac{1}{2})$ do indeed differ by a factor $\sqrt{5}$.

Finally we exhibit the SU(3) properties of $\mathcal{H}_W^H(|\Delta Y| = 1)$:

$$\mathcal{H}_W^H(|\Delta Y| = 1) = \sin\theta_c \cos\theta_c \Big\{ \mathcal{H}(\underline{8}, 1, \tfrac{1}{2}, -\tfrac{1}{2}) + \mathcal{H}(\underline{8}, -1, \tfrac{1}{2}, \tfrac{1}{2}) + \\ + \mathcal{H}(\underline{27}, 1, \tfrac{1}{2}, -\tfrac{1}{2}) + \mathcal{H}(\underline{27}, -1, \tfrac{1}{2}, \tfrac{1}{2}) + \\ + \sqrt{5}\, \mathcal{H}(\underline{27}, 1, \tfrac{3}{2}, -\tfrac{1}{2}) + \sqrt{5}\, \mathcal{H}(\underline{27}, -1, \tfrac{3}{2}, \tfrac{1}{2}) \Big\} . \tag{5.18}$$

We have absorbed the factor $\frac{G}{2\sqrt{2}}$ into the various $\mathcal{H}$'s but have displayed the dependence on θ_c so that we may use the same normalizations when we come to consider $\mathcal{H}_W^H(\Delta Y = 0)$.

5.1.1. $K \to 2\pi$

Charge conservation permits three decay modes of the kaon into two pions

$$K^{\pm} \to \pi^{\pm} \pi^{o} \tag{5.19a}$$

$$K^{o}, \overline{K}^{o} \to \pi^{+} \pi^{-} \tag{5.19b}$$

$$K^{o}, \overline{K}^{o} \to \pi^{o} \pi^{o} . \tag{5.19c}$$

Since the kaon has spin zero, the final two-pion states must also have total angular momentum zero. The pions have spin zero, so the final state must have zero orbital angular momentum ($\ell = 0$). The pions are bosons so the final state must be totally symmetric with respect to interchanging the space and charge labels of the two pions. Since the orbital state is an s-wave, it is symmetric under interchange of the spatial labels, so we deduce that the final state must also be symmetric under interchange of the charge labels of the pions. The pions have isospin $I = 1$ so in principle the final state could contain $I = 0, 1, 2$. The symmetry property under interchange of charge labels excludes the $I = 1$ state leaving only the possibility of the symmetric $I = 0, 2$.

Let us make all of this more explicit by considering the decay of a kaon K into two pions π_1, π_2. In the centre-of-mass frame of the kaon π_1, π_2 have equal and opposite momenta $\underline{p}$, $-\underline{p}$, whose magnitude is determined by energy-momentum conservation, as in any two-body decay mode. Lorentz invariance shows that the invariant matrix element is just a constant; it must be independent of the direction of $\underline{p}$, since there is no other vector in the problem. Thus

$$M\left(K \rightarrow \pi_1(\underline{p})\ \pi_2(-\underline{p})\right) = f \ , \tag{5.20}$$

where f depends only on the masses etc.. Bose symmetry requires

$$M\left(K \rightarrow \pi_2(-\underline{p})\ \pi_1(\underline{p})\right) = f \ . \tag{5.21}$$

Then, since f is independent of the direction of $\underline{p}$ we may replace $\underline{p}$ by $-\underline{p}$, so

$$M\left(K \rightarrow \pi_2(\underline{p})\ \pi_1(-\underline{p})\right) = f \ . \tag{5.22}$$

(5.21) and (5.22) exhibit the symmetry of f under interchange of spatial labels ($\underline{p}$ and $-\underline{p}$), while (5.20) and (5.22) exhibit the symmetry of f under interchange of charge labels (1 and 2).

CP-Invariance and Isospin Restrictions

The CP-invariance of $\mathcal{H}(|\Delta Y| = 1)$ requires that only one linear combination of K^o and $\overline{K}^o$ may participate in the two-pion decay modes. We can see this as follows. Consider for example the $\pi^+\pi^-$ final state

$$\begin{aligned} \mathcal{C}\mathcal{P}\,|\pi^+(\underline{p})\ \pi^-(-\underline{p})> &= \mathcal{C}\,|\pi^+(-\underline{p})\ \pi^-(\underline{p})> \\ &= |\pi^-(-\underline{p})\ \pi^+(\underline{p})> \\ &= |\pi^+(\underline{p})\ \pi^-(-\underline{p})> \ , \end{aligned} \tag{5.23}$$

where we have used the facts that both pions have intrinsic parity -1, and that the two-pion state is totally symmetric. The same is true of the $\pi^o\pi^o$ state. On the other hand

$$\mathcal{C}\mathcal{P}|K^o> = -\,\mathcal{C}|K^o> = -\,|\overline{K}^o> \tag{5.24a}$$

$$\mathcal{C}\mathcal{P}|\overline{K}^o> = -\,|K^o> \ . \tag{5.24b}$$

Thus defining

$$|K^0_{\substack{1\\2}}> \equiv \frac{1}{\sqrt{2}}\left[|K^o> \pm |\overline{K}^o>\right] \ , \tag{5.25a}$$

$$\mathcal{C}\mathcal{P}|K^0_{\substack{1\\2}}> = \mp\,|K^0_{\substack{1\\2}}> \ . \tag{5.25b}$$

Since the final two-pion state has parity +1 under $\mathcal{CP}$, CP-invariance requires

$$K_1^o \not\to 2\pi \ . \tag{5.26}$$

Experimentally, both of the states $K^o_{\substack{L\\S}}$ with definite lifetimes are observed to decay into two pions. This proves that $\mathcal{CP}$ is not an exact invariance of nature and also that the states $K^o_{\substack{1\\2}}$ are linear combinations of $K^o_{\substack{L\\S}}$. However, the decay of K_L^o into two pions has an amplitude only 0.1% of that for K_S^o. So to good approximation

$$K_1^o \simeq K_L^o \tag{5.27a}$$

$$K_2^o \simeq K_S^o \ . \tag{5.27b}$$

CP-invariance also relates the K^+ amplitude to that for K^-

$$< \pi^+ \pi^o | \mathcal{H}_W^H | K^+ > \ = \ - < \pi^- \pi^o | \mathcal{H}_W^H | K^- > \ ,$$

so we are left with three independent processes to consider

$$K^+ \to \pi^+ \pi^o \tag{5.28a}$$

$$K_2^o \to \pi^+ \pi^- \tag{5.28b}$$

$$K_2^o \to \pi^o \pi^o \ . \tag{5.28c}$$

The two pions in the final state have I = 0, 2, while the decaying kaon of course has $I = \frac{1}{2}$. In general, therefore, the decay could go via a Hamiltonian containing a mixture of $I = \frac{1}{2}, \frac{3}{2}, \frac{5}{2}$. Thus the already noted fact (5.6) that $\mathcal{H}_W^H$ contains only $I = \frac{1}{2}, \frac{3}{2}$ imposes some restriction upon amplitudes which would otherwise be arbitrary. The $|\Delta \underline{I}| \leq \frac{3}{2}$ character of $\mathcal{H}_W^H (|\Delta Y| = 1)$ implies that

$$< \pi^+ \pi^+ | \left[v^1 + i v^2, [v^1 + i v^2, \mathcal{H}_W^H] \right] | K^o > \ = \ 0 \ , \tag{5.29}$$

since only the Y = - 1 part of $\mathcal{H}_W^H$ can connect K^o and $\pi^+\pi^+$. Then, proceeding as before, we find (remembering the $\frac{1}{\sqrt{2}}$ in the normalization of $|\pi^o \pi^o>$)

$$< \pi^+ \pi^- | \mathcal{H}_W^H | K^o > - \sqrt{2} < \pi^o \pi^o | \mathcal{H}_W^H | K^o > = \sqrt{2} < \pi^+ \pi^o | \mathcal{H}_W^H | K^+ >$$

$$= - < \pi^+ \pi^- | \mathcal{H}_W^H | \bar{K}^o > + \sqrt{2} < \pi^o \pi^o | \mathcal{H}_W^H | \bar{K}^o > ,$$

using CP-invariance. Thus the invariant matrix elements satisfy the triangular relationship

$$M(K_2^o \to \pi^+ \pi^-) - \sqrt{2}\, M(K_2^o \to \pi^o \pi^o) = 2 M(K^+ \to \pi^+ \pi^o) . \tag{5.30}$$

The amplitudes in (5.30) are not expected to have the same phase, because of differences in the final state interactions between the pions. The strong interactions, which are responsible for these phase differences, conserve isospin, so we consider the (s-wave) amplitudes with definite isospin $I = 0,2$

$$M(K \to 2\pi, I, \text{out}) = | M(K \to 2\pi, I) | \exp(i \delta_o(I)) .$$

We are concerned with "outgoing" two-pion states in (5.30). T-invariance, which is implied by TCP + CP invariance, implies that

$$< 2\pi, I, \text{out} | \mathcal{H}_W^H | K >^* = < 2\pi, I, \text{in} | \mathcal{H}_W^H | K > .$$

So

$$M(K \to 2\pi, I, \text{in}) = | M(K \to 2\pi, I) | \exp(-i \delta_o(I))$$

$$= \exp(-2 i \delta_o(I))\, M(K \to 2\pi, I, \text{out}) .$$

Thus

$$< 2\pi, I, \text{out} | \exp(-2 i \delta_o(I)) = < 2\pi, I, \text{in} | ,$$

and we identify $\delta_o(I)$ as the s-wave, I-channel phase shifts for $\pi\pi$ scattering at centre-of-mass energy of one kaon mass. If we assume $\delta_o(I)$ are known, we may translate (5.30) into a statement about the decay rates. First, writing $| \pi^+ \pi^- >$ and $| \pi^o \pi^o >$ in terms of $I = 0,2$ states we have

$$M(K_S^o \to \pi^+ \pi^-) = - a_2 \frac{1}{\sqrt{3}} e^{i \delta(2)} - a_o \sqrt{\frac{2}{3}} e^{i \delta(0)}$$

$$M(K_S^o \to \pi^o \pi^o) = a_2 \sqrt{\frac{2}{3}} e^{i \delta(2)} - a_o \frac{1}{\sqrt{3}} e^{i \delta(0)} ,$$

where

$$a_I = |M(K_S^o \to 2\pi, I)| \qquad I = 0,2 .$$

Thus

$$B(K_S^o) \equiv \Gamma(K_S^o \to \pi^+\pi^-)/\Gamma(K_S^o \to \pi^o\pi^o) \tag{5.31a}$$

$$= \left\{2 + \omega^2 + 2\sqrt{2}\,\omega\cos\Delta\right\}\left\{2\omega^2 + 1 - 2\sqrt{2}\,\omega\cos\Delta\right\}^{-1} , \tag{5.31b}$$

with

$$\omega \equiv a_2/a_o \qquad \text{and} \qquad \Delta \equiv \delta(2) - \delta(0) . \tag{5.31c}$$

Also,

$$\Gamma(K_S^o \to \pi^+\pi^-) + \Gamma(K_S^o \to \pi^o\pi^o) = a_2^2 + a_o^2 .$$

In the same way, since the $\pi^+\pi^o$ state must have $I = 2$ (recall $I = 1$ is excluded),

$$M(K^+ \to \pi^+\pi^o) = - A_2\, e^{i\delta(2)} ,$$

where A_2 is related to a_2 by the sum rule (5.30) :

$$- \sqrt{3}\, a_2\, e^{i\delta(2)} = - 2\, A_2\, e^{i\delta(2)} .$$

Thus

$$b^+ \equiv \Gamma(K^+ \to \pi^+\pi^o)\left\{\Gamma(K_S^o \to \pi^+\pi^-) + \Gamma(K_S^o \to \pi^o\pi^o)\right\}^{-1}$$

$$= A_2^2\,(a_2^2 + a_o^2)^{-1}$$

$$= \tfrac{3}{4}\,\omega^2\,(1 + \omega^2)^{-1} . \tag{5.32}$$

Thus, eliminating ω, we see that the sum rule (5.30) enables us to relate the branching ratios (5.31) and (5.32), provided the phase shift difference Δ is known. Let us check the consistency of this prediction with the data.

In fact the branching ratio b^+ is very small (of order 10^{-3}). This small value is of great significance and we shall return to it shortly. For the moment, however, the magnitude of b^+ means that ω is small, so we may work to lowest order. Then

$$B(K_S^o) \simeq 2\,(0.986)\,[1 + 3\sqrt{2}\;\omega\cos\Delta]$$

$$\simeq 2\,(0.986)\,[1 + 2\sqrt{6b^+}\cos\Delta] , \tag{5.33}$$

where we have included a factor 0.986 to take into account the fact that the $\pi^+\pi^-$ phase space integral is not quite equal to that for $\pi^o\pi^o$.

Now experimentally

$$b^+ = (1.486 \pm 0.05)\,10^{-3} \quad . \tag{5.34}$$

The phase shifts are not known accurately enough for us to insert Δ , but since $-1 < \cos\Delta < 1$, (5.33) and (5.34) enable us to bound $B(K^o_S)$:

$$1.8 < B(K^o_S) < 2.34 \quad .$$

This inequality is satisfied by the experimentally measured value(2) :

$$B(K^o_S) = 2.207 \pm 0.029 \quad .$$

Thus there is no doubt that the data is consistent with the prediction following from (5.6). The trouble is that the dominant feature of the data, namely the smallness of b^+, is not predicted. As already noted, the $\pi^+\pi^o$ state is pure $I = 2$, so the decay can only proceed by the part of $\mathcal{H}^H_W$ having $I = \frac{3}{2}$. So the smallness of b^+ suggests that there is some damping (unpredicted so far) of the $I = \frac{3}{2}$ amplitudes relative to the $I = \frac{1}{2}$ amplitudes. Of course, we have not used all of the information in (5.5c) - only that following from the predicted absence of $|\Delta I| = \frac{5}{2}$ transitions.

The question of how to use all of the information contained in (5.5c) is a continuing problem in the purely hadronic decays. It is tempting to approach the problem from the other end and try and find a phenomenological Hamiltonian which incorporates the dominant features of the data. Thus one might assume simply that the smallness of b^+ was explained by having $\mathcal{H}^H_W$ purely $I = \frac{1}{2}$ with no admixture of $I = \frac{3}{2}$, ... $K^+ \to \pi^+\pi^o$ could then proceed only via electromagnetic effects, so we would expect b^+ to be of order α^2 . But this is considerably smaller than the observed value (5.34); so one then has to explain why b^+ is so large ! Rather than do this we shall stick with (5.5c).

SU(3) Restrictions

One obvious way to use more (but not all) of the information in (5.5c) is to use its SU(3) symmetry properties as contained in (5.18). This would require us to utilize the fact that the kaons and pions are in the same pseudoscalar meson octet. From past experience we know that SU(3) is rather badly broken by the meson octet, so we cannot be optimistic that the resulting predictions will be very realistic. In fact SU(3) symmetry requires all of the amplitudes to be zero[(3)]. We are not referring to the fact that they decay is obviously forbidden kinematically if the kaon and pions all have the same mass. Rather, it is the vanishing of the Clebsch-Gordan coefficient which is causing the prohibition. Let us see how this comes about.

Firstly we note that the decaying kaon is a pseudoscalar and has parity -1. The final two pion state is s-wave, as already noted, so has parity +1 (c.f. (5.23)). Thus the decay proceeds via the parity violating part of $\mathcal{H}_W^H$, which arises from the vector-axial vector interference. Since $\mathcal{H}_W^H$ is CP-invariant, plainly its parity violating piece must have C-parity -1.

We can see this explicitly from (5.11). The parity violating part of $\{\mathcal{J}^1, \mathcal{J}^4\}$, for example, is $\{V^1, A^4\} + \{A^1, V^4\}$. Now under $\mathcal{C}$ the currents V^i and A^i transform as follows

$$\mathcal{C}\, V^i\, \mathcal{C}^{-1} = -\epsilon_i V^i \quad \text{(no summation)} \tag{5.35a}$$

$$\mathcal{C}\, A^i\, \mathcal{C}^{-1} = +\epsilon_i A^i \quad \text{(no summation)}\ , \tag{5.35b}$$

where

$$\begin{aligned} \epsilon_i &= +1 \quad i = 1,3,4,6,8 \\ &= -1 \quad i = 2,5,7\ . \end{aligned} \tag{5.35c}$$

The ϵ_i arise so that the charged components of the octets are transformed into their charge conjugates as required :

$$\mathcal{C}\,(A^1 + i A^2)\, \mathcal{C}^{-1} = A^1 - i A^2\ .$$

In the quark model, for example, in which the currents are given by (4.65) the minus signs arise because $\lambda^2, \lambda^5, \lambda^7$ are antisymmetric, see (4.9); so when we take the transpose involved in charge conjugation, recall (2.224b, c), these provide factors of -1. We say that the octet V^i has intrinsic C-parity -1 while A^i has intrinsic C-parity +1. Thus it follows that

$$\mathcal{C}\,\{V^1, A^4\}\,\mathcal{C}^{-1} = -\,\{V^1, A^4\}\ ,$$

as required. The same is true of the octet and $\underline{27}$ parts separately, of course. For example, the parity violating part of D^k defined in (5.9c) is

$$D^k_{pv} = d^{k\ell m} \{ V^\ell, A^m \} .$$

So

$$\mathcal{C} D^k_{pv} \mathcal{C}^{-1} = - \sum_{\ell, m} \epsilon_\ell \epsilon_m \; d^{k\ell m} \{ V^\ell, A^m \} .$$

Now we see from (4.13) that $d^{k\ell m}$ is non-zero only when $\epsilon_k \epsilon_\ell \epsilon_m = +1$. Thus the intrinsic C-parity of D^k_{pv} is -1 and the C-parity of D^6_{pv} is also -1 (since $\epsilon_6 = +1$). The same is true of the parity violating part of the $\underline{27}$; it has intrinsic C-parity -1 and (14), (25) components with C-parity -1.

Returning to the problem in hand, we note that Bose symmetry requires us to couple the three identical SU(3) meson octets totally symmetrically. This then implies that they are coupled only to representations with intrinsic C-parity +1, since the meson octet has intrinsic C-parity +1; this is a simple generalization of what we have just observed; if we couple two identical octets (symmetrically) using d^{ijk} then the resultant octet has intrinsic C-parity +1. Hence SU(3) symmetry predicts $K \not\to 2\pi$ under our hypothesis. The only way SU(3) would allow the decay would be if the parity violating part of $\mathcal{H}^H_W$ contained the <u>seventh</u> component of an octet with intrinsic C-parity +1, for example; this would ensure that it still had actual C-parity -1.

Soft Pion Programme

In view of the success of the CTMOP relation in the $K_{\ell 3}$ decays, another approach is to use PCAC to derive soft pion theorems. As in $K_{\ell 3}$ there are no pole terms to worry about, and taking π^+ to be soft in $K^+ \to \pi^+ \pi^0$ we find

$$\langle \pi^+ \pi^0 | \mathcal{H}^H_W | K^+ \rangle = - i f_\pi^{-1} \langle \pi^0 | [a^1 - i a^2 , \mathcal{H}^H_W] | K^+ \rangle \quad (5.36a)$$

$$= - i f_\pi^{-1} \langle \pi^0 | [v^1 - i v^2 , \mathcal{H}^H_W] | K^+ \rangle \quad (5.36b)$$

$$= i f_\pi^{-1} \{ \sqrt{2} \langle \pi^+ | \mathcal{H}^H_W | K^+ \rangle + \langle \pi^0 | \mathcal{H}^H_W | K^0 \rangle \} . \quad (5.36c)$$

(5.36b) follows simply from the fact that $\mathcal{H}_W^H$ is constructed from the currents $V_\lambda^i + A_\lambda^i$ and (5.36c) utilizes the fact that $v^i (i = 1,2,3)$ are isospin generators. Thus in terms of the invariant matrix elements we have, as $\pi^+ \to 0$,

$$M(K^+ \to \pi^+\pi^o) = i f_\pi^{-1} [\sqrt{2}\, M(K^+ \to \pi^+) + M(K^o \to \pi^o)] . \qquad (5.37)$$

On the other hand, taking $\pi^o \to 0$ gives, in the same way,

$$M(K^+ \to \pi^+\pi^o) = - i f_\pi^{-1} \frac{1}{\sqrt{2}} M(K^+ \to \pi^+) . \qquad (5.38)$$

Thus, if we assume that $M(K^+ \to \pi^+\pi^o)$ has a value well approximated by both soft pion extrapolations, we obtain a consistency condition

$$-\frac{3}{\sqrt{2}} M(K^+ \to \pi^+) = M(K^o \to \pi^o) = \frac{1}{\sqrt{2}} M(K_1^o \to \pi^o) , \qquad (5.39)$$

where the last equality follows from CP-invariance, which requires

$$M(K^o \to \pi^o) = M(\bar{K}^o \to \pi^o) \qquad (5.40a)$$

$$M(K^+ \to \pi^+) = M(K^- \to \pi^-) . \qquad (5.40b)$$

Suppose now we consider the process $K_1^o \to \pi^+\pi^-$, which is <u>forbidden</u> by CP-invariance, see (5.26). Soft pion theory then gives as $\pi^+ \to 0$

$$M(K_1^o \to \pi^+\pi^-) = - i f_\pi^{-1} [\sqrt{2}\, M(K_1^o \to \pi^o) + \frac{1}{\sqrt{2}} M(K^- \to \pi^-)] , \qquad (5.41a)$$

and as $\pi^- \to 0$

$$M(K_1^o \to \pi^+\pi^-) = + i f_\pi^{-1} [\sqrt{2}\, M(K_1^o \to \pi^o) + \frac{1}{\sqrt{2}} M(K^+ \to \pi^+)] . \qquad (5.41b)$$

Thus using (5.40b) we see that the simultaneous validity of these two does indeed require $M(K_1^o \to \pi^+\pi^-) = 0$, as required by CP-invariance. But the consistency condition

$$-\frac{1}{2\sqrt{2}} M(K^+ \to \pi^+) = \frac{1}{\sqrt{2}} M(K_1^o \to \pi^o) \qquad (5.42)$$

is inconsistent with (5.39) which was obtained from $K^+ \to \pi^+\pi^o$. They can only be satisfied if all $M(K \to \pi)$ are zero, which implies all $M(K \to 2\pi)$ are zero. Evidently the trouble stems from the fact that the soft pion treatment destroys the Bose symmetry of the two pions. If we explicitly symmetrize the final state, so as to guarantee the Bose symmetry, we obtain for $M(K^+ \to \pi^+\pi^o)$ the average of the two values (5.37) and (5.38) :

$$M(K^+ \to \pi^+\pi^o) = i f_\pi^{-1} \frac{1}{2\sqrt{2}} [M(K^+ \to \pi^+) + M(K_1^o \to \pi^o)] \tag{5.43}$$

and $M(K_1^o \to \pi^+\pi^-) = 0$, as required. Proceeding as before we find that the $K_2^o \to 2\pi$ amplitudes are explicitly Bose symmetric, and

$$M(K_2^o \to \pi^+\pi^-) = i f_\pi^{-1} \frac{1}{\sqrt{2}} M(K^+ \to \pi^+) \tag{5.44a}$$

$$M(K_2^o \to \pi^o\pi^o) = -\, i f_\pi^{-1} \tfrac{1}{2} M(K_1^o \to \pi^o) \ . \tag{5.44b}$$

Then eliminating the unknown $M(K \to \pi)$ from (5.43) and (5.44) we find

$$M(K_2^o \to \pi^+\pi^-) - \sqrt{2}\, M(K_2^o \to \pi^o\pi^o) = 2 M(K^+ \to \pi^+\pi^o) \ . \tag{5.45}$$

But this is just (5.30) which, we have seen, expresses the fact that $\mathcal{H}_W^H$ has no $I = \frac{5}{2}$ component. However, we have not used that fact in the derivation of (5.45). So this last derivation suggests that in the soft pion limits any $|\Delta I| = \frac{5}{2}$ transitions are suppressed if $\mathcal{H}_W^H$ has a $V + A$ structure. This immediately suggests that if we take both pions to be soft we may explain the inhibition of the $|\Delta I| = \frac{3}{2}$ transitions which was noted earlier. Maintaining the Bose symmetry, we find in the limits $\pi^i, \pi^j \to 0$

$$M(K \to \pi^i\pi^j) \propto \langle 0|[v^i,[v^j, \mathcal{H}_W^H]]|K\rangle + \langle 0|[v^j,[v^i, \mathcal{H}_W^H]]|K\rangle . \tag{5.46}$$

Since the $v^i (i = 1,2,3)$ are just the isospin generators, which just rotate the various elements within any isomultiplet, plainly only the $I = \frac{1}{2}$ part of $\mathcal{H}_W^H$ can contribute, since K has $I = \frac{1}{2}$. (This argument also explains why we obtained the $|\Delta I| \leq \frac{3}{2}$ relation in (5.45), since only $I = \frac{1}{2}, \frac{3}{2}$ can connect a kaon ($I = \frac{1}{2}$) to a single pion ($I = 1$).) Thus in this limit both sides of (5.45) are zero, and the fact that $K^+ \to \pi^+\pi^o$ does occur physically is attributed to the non-zero mass of the pion(4). So the use of soft pion theory suggests that any $V + A$ Hamiltonian will yield an approximate $|\Delta I| = \frac{1}{2}$ rule for $K \to 2\pi$, which leads us to believe that these decays are not the best area in which to test our hypothesis (5.5c).

5.1.2 $K \to 3$

Consider first the decays of the charged kaon

$$(\tau^+) : K^+ \to \pi^+\pi^+\pi^- \tag{5.47a}$$

$$(\tau^{+'}) : K^+ \to \pi^+ \pi^o \pi^o \quad . \tag{5.47b}$$

The invariant amplitude f for a general three-body decay is a function of the energies ω_1, ω_2 of two of the particles, the energy ω_3 of the remaining particle being determined by energy conservation. In the decays under consideration we choose ω_1, ω_2 to be the energies of the two π^+s in τ^+ or the two π^os in $\tau^{+'}$. Then the energy ω_3 of the remaining pion (π^- in τ^+, π^+ in $\tau^{+'}$) is

$$\omega_3 = m_K - \omega_1 - \omega_2 \quad . \tag{5.48}$$

Writing f as a function of ω_1 and ω_2, Bose symmetry requires

$$f(\omega_1, \omega_2) = f(\omega_2, \omega_1) \; . \tag{5.49}$$

Without loss of generality we may split f into a totally symmetric part f^S and a non-symmetric part f^N

$$f(\omega_1, \omega_2) = f^S(\omega_1, \omega_2) + f^N(\omega_1, \omega_2) \quad , \tag{5.50a}$$

where

$$3f^S(\omega_1, \omega_2) = f(\omega_1, \omega_2) + f(\omega_2, \omega_3) + f(\omega_3, \omega_1) \quad , \tag{5.50b}$$

$$3f^N(\omega_1, \omega_2) = 2f(\omega_1, \omega_2) - f(\omega_2, \omega_3) - f(\omega_3, \omega_1) \; . \tag{5.50c}$$

As for $K \to 2\pi$, the $|\Delta \underline{I}| \le \frac{3}{2}$ property of $\mathcal{H}_W^H$ implies some restrictions upon the $K \to 3\pi$ amplitudes. In particular it implies that the three pions in the final state cannot have isospin $I = 3$. But the $I = 3$ states are totally symmetric, so their predicted absence imposes a condition upon the amplitudes f^S for the two modes. As before, we start from

$$<\pi_1^+ \pi_2^+ \pi_3^+ | \left[v^1 + i v^2, [v^1 + i v^2, \mathcal{H}_W^H (|\Delta Y| = 1)] \right] | K^+ > = 0,$$

where π_i^+ has energy ω_i ($i = 1,2,3$). Then using the properties of the isospin operator $v^1 + i v^2$ we find[(5)]

$$f^S_{\tau^+}(\omega_1, \omega_2) = 2 f^S_{\tau^{+'}}(\omega_1, \omega_2) \; . \tag{5.51}$$

Experimentally the amplitudes f are well fitted by a linear energy dependence, which, by virtue of (5.49), has two arbitrary parameters. Thus we take

$$f(\omega_1, \omega_2) = A \left[1 - \tfrac{1}{3}\, g \,\frac{m_K}{m_\pi^2} \left(2m_K - 3\,\omega_1 - 3\,\omega_2 \right) \right], \tag{5.52}$$

with A and g arbitrary. With this form, A is evidently the value of f at the symmetric point $\omega_1 = \omega_2 = \omega_3 = \frac{1}{3}\, m_K$

$$f(\tfrac{1}{3} m_K,\ \tfrac{1}{3} m_K) = A\ , \tag{5.53}$$

and we may identify the symmetric and non-symmetric contributions to f, defined in (5.50), as

$$f^S(\omega_1, \omega_2) = A \tag{5.54a}$$

$$f^N(\omega_1, \omega_2) = -\tfrac{1}{3}\, A g \,\frac{m_K}{m_\pi^2} \left(2m_K - 3\omega_1 - 3\omega_2 \right) . \tag{5.54b}$$

In principle we may measure A and g directly from the observed energy spectrum. The transition probability in the rest frame of the kaon is given by

$$d\Gamma = (2\pi)^{-5} (16 m_K\, \omega_1 \omega_2 \omega_3)^{-1} |f|^2\, \delta(\omega_1 + \omega_2 + \omega_3 - m_K)\, \delta(\underline{p}_1 + \underline{p}_2 + \underline{p}_3)\, d^3p_1\, d^3p_2\, d^3p_3 ,$$

where $\underline{p}_i$ $(i = 1, 2, 3)$ are the momenta of the three pions. We perform the $\underline{p}_3$ integration using the momentum δ-function, and, since f does not depend on the angular variables, we may write

$$\frac{d^3 p_1}{\omega_1} = 4\pi\, p_1\, d\omega_1 \qquad \left(|\underline{p}_1| \equiv p_1 \right)$$

$$\frac{d^3 p_2}{\omega_2} = p_2\, d\omega_2\, 2\pi dx \qquad \left(|\underline{p}_2| \equiv p_2 \right) ,$$

where $x = \cos\theta$ and θ is the angle between $\underline{p}_1$ and $\underline{p}_2$. But

$$p_3^2 = |\underline{p}_1 + \underline{p}_2|^3 = p_1^2 + p_2^2 + 2\, p_1 p_2\, x\ ,$$

so with p_1 and p_2 fixed

$$2\,\omega_3\, d\omega_3 = 2\, p_3\, dp_3 = 2\, p_1 p_2\, dx\ .$$

Putting all this together we obtain a symmetrical expression for $d\Gamma$

$$\begin{aligned} d\Gamma &= (2\pi)^{-3} (8 m_K)^{-1} |f|^2\, \delta(\omega_1 + \omega_2 + \omega_3 - m_K)\, d\omega_1\, d\omega_2\, d\omega_3 \\ &= (2\pi)^{-3} (8 m_K)^{-1} |f|^2\, d\omega_1\, d\omega_2\ . \end{aligned} \tag{5.55}$$

Thus the spectrum $d\Gamma$ is directly proportional to $|f|^2$. However, using the linear approximation (5.52),

$$|f|^2 \simeq |A|^2 \left[1 - \frac{2}{3} g \frac{m_K}{m_\pi^2} \left(2m_K - 3\omega_1 - 3\omega_2 \right) \right] , \tag{5.56}$$

dropping the already neglected quadratic term. Now, if we consider the total decay rate, which is much easier to measure than the spectrum, the term proportional to g in (5.56) integrates out to a very good approximation. Thus the total decay rate Γ is essentially a measure of $|A|^2$. Let us see how this arises.

Energy momentum conservation requires

$$K = p_1 + p_2 + p_3 ,$$

where K, p_i are the four-momenta of the kaon and pions. Thus

$$4 m_\pi^2 \leq (p_2 + p_3)^2 = (K - p_1)^2 = m_K^2 + m_\pi^2 - 2 m_K \omega_1$$

in the rest frame of the kaon. The maximum energy attainable by any of the pions is thus

$$\omega_{max} \equiv m_\pi + t_{max} ,$$

where

$$t_{max} = (2m_K)^{-1} [(m_K - m_\pi)^2 - 4m_\pi^2] \simeq 50 \text{ MeV}$$

is the maximum kinetic energy of the pion. The maximum momentum is

$$p_{max} = (\omega_{max}^2 - m_\pi^2)^{\frac{1}{2}} = 2m_\pi t_{max} \left(1 + \frac{t_{max}}{2 m_\pi} \right)^{\frac{1}{2}}$$

$$\simeq 2m_\pi t_{max} \left(1 + \frac{t_{max}}{4m_\pi} \right) ,$$

using the binomial expansion of the square root. Since $t_{max}/4m_\pi \sim 10\%$ it is reasonable to make the non-relativistic approximation for all energies and set

$$\omega_i \equiv m_\pi + t_i = m_\pi + \frac{p_i^2}{2m_\pi} \qquad (i = 1,2,3) . \tag{5.58}$$

With this approximation the kinetmatically allowed values of ω_i are particularly easy to describe[6]. Energy-momentum conservation requires

$$m_K = 3m_\pi + t_1 + t_2 + t_3 \tag{5.59a}$$

$$\underline{0} = \underline{p}_1 + \underline{p}_2 + \underline{p}_3 \quad . \tag{5.59b}$$

Thus

$$t_3 = m_K - 3m_\pi - (t_1 + t_2)$$

$$= \frac{\underline{p}_3^2}{2m_\pi} = \frac{(\underline{p}_1 + \underline{p}_2)^2}{2m_\pi} = t_1 + t_2 - 2\sqrt{t_1 t_2}\cos\theta ,$$

where θ is the angle between $\underline{p}_1$ and $\underline{p}_2$. So writing $Q = m_K - 3m_\pi$

$$0 \le \left[2(t_1 + t_2) - Q\right]^2 = 4t_1 t_2 \cos^2\theta \le 4\,t_1 t_2 \quad . \tag{5.60}$$

Thus, if we define new variables x and y (which are zero at the symmetric point)

$$\frac{x}{\sqrt{3}} \equiv \tfrac{2}{3} Q - (t_1 + t_2) = (\tfrac{2}{3} m_K - \omega_1 - \omega_2) \tag{5.61a}$$

$$y \equiv t_1 - t_2 \ , \tag{5.61b}$$

we may describe the physical region (5. 60) by

$$0 \le x^2 + y^2 \le \tfrac{1}{3} Q^2 \quad . \tag{5.62}$$

Thus the physical region is the interior of a circle centred on the symmetric point. The important point is that, when we express $|f|^2$ in terms of the variables (5. 61), we see that it is linear in x

$$|f|^2 \simeq |A|^2 \left[1 - \frac{2}{9}\, g\, \frac{m_K}{m_\pi^2}\, \frac{x}{\sqrt{3}} \right] . \tag{5.63}$$

Thus if we integrate over the whole kinematic region, as we do to obtain the total decay rate Γ, the linear term integrates out, because the x integration is symmetric. (This, of course, is why the particular form (5. 52) was chosen in the first place.) In fact, since

$$dx\,dy = 2\sqrt{3}\ dt_1\,dt_2 = 2\sqrt{3}\ d\omega_1\,d\omega_2 \ ,$$

we find from (5. 55)

$$\Gamma = \Phi\,|A|^2 \ , \tag{5.64a}$$

where the phase space integral Φ is

$$\Phi = \frac{1}{384\sqrt{3}} \frac{(m_K - 3m_\pi)^2}{m_K} \tag{5.64b}$$

Thus to an excellent approximation

$$\gamma \equiv \Gamma \Phi^{-1} \simeq |A|^2 \quad , \tag{5.65}$$

and the physical prediction following form (5.51), (5.54a) and (5.65) is

$$\gamma_{\tau^+} = 4 \gamma_{\tau^{+\prime}} \quad . \tag{5.66}$$

Of course, in reality one can do rather better than (5.64b) for Φ by allowing for relativistic corrections. This means that the terms proportional to g contribute a little. Also, corrections due to the Coulomb scattering of the final charged pions may be included. All of these are included by modifying the value (5.64b) for Φ. With these corrections the experimental data is[2]

$$\tfrac{1}{4}\left(\gamma_{\tau^+} / \gamma_{\tau^{+\prime}}\right) - 1 = 0.076 \pm 0.026 \quad .$$

Even with these corrections the right-hand side could differ from zero by 0.1. Thus it is concluded that (5.66) is consistent with the data.

We may discuss the three-pion decay modes of the neutral kaon similarly. Bose symmetry requires the $3\pi^o$ state to be totally symmetric. Thus the orbital angular momentum ℓ of any two of the pions must be even, and, since the pion is a pseudoscalar, this state is an eigenstate of CP with eigenvalue -1. So this mode can only occur via the decay of $K_1^o \simeq K_L^o$. The $\pi^+ \pi^- \pi^o$ state has CP eigenvalue $(-1)^{\ell+1}$, where ℓ is the relative angular momentum of the π^+ and π^-. Thus the modes allowed by CP-invariance are

$$(\tau_S) \; : \; K_2^o \rightarrow \pi^+ \pi^- \pi^o \qquad (\ell = 1, 3, \ldots) \tag{5.67a}$$

$$(\tau_L) \; : \; K_1^o \rightarrow \pi^+ \pi^- \pi^o \qquad (\ell = 0, 2, \ldots) \tag{5.67b}$$

$$(\tau_L^{\prime}) \; : \; K_1^o \rightarrow \pi^o \pi^o \pi^o \quad . \tag{5.67c}$$

The mode τ_S is doubly inhibited by the angular momentum barrier, since conservation of angular momentum requires the π^o to have angular momentum ℓ relative to the $\pi^+ \pi^-$ system. Since $\ell \geq 1$ for the mode

τ_s, we see that it is doubly inhibited. The invariant amplitudes f for τ_L and τ'_L are again well described by a linear form; in the case of τ'_L, since it is totally symmetric,

$$f_{\tau'_L}(\omega_1, \omega_2) = f^S_{\tau'_L}(\omega_1, \omega_2) \simeq A_{\tau'_L} \quad . \tag{5.68}$$

For the τ_L mode we take ω_1, ω_2 to be the energies of the charged pions, so, since ℓ is even (by CP invariance), the symmetry property (5.49) is satisfied. Thus we may again use the linear form (5.52) for f_{τ_L}. As before, the predicted absence of $I = 3$ final states leads to a prediction namely

$$\sqrt{6}\, f^S_{\tau'_L}(\omega_1, \omega_2) = 3\, f^S_{\tau_L}(\omega_1, \omega_2) \; . \tag{5.69}$$

With the linear approximation this gives

$$2\,\gamma_{\tau'_L} = 3\,\gamma_{\tau_L} \quad ,$$

and this too is consistent with the data[2]

$$\left(2\,\gamma_{\tau'_L}\right)\left(3\,\gamma_{\tau_L}\right)^{-1} - 1 = 0.010 \pm 0.048 \; .$$

Soft Pion Programme

The soft pion programme also makes predictions for the $K \to 3\pi$ modes. In the limit that we take $\pi_k \to 0$,

$$<\pi^i \pi^j \pi^k | \mathcal{H}^H_W | K> = i\sqrt{2}\, f_\pi^{-1} <\pi^i \pi^j | [\, v^k, \mathcal{H}^H_W \,] | K> \quad , \tag{5.70}$$

assuming only the $V + A$ structure of $\mathcal{H}^H_W$. The v^k are isospin generators, so the matrix element on the right is a linear combination of $K \to 2\pi$ matrix elements already discussed. The symmetry condition (5.49), which follows from Bose symmetry or CP-invariance, means that we may obtain information by taking either of two pions to be soft. In the case of $\tau^{+\prime}$, for example, we obtain independent equations by taking π^+ soft and π^0 soft, as in (5.36) and (5.38). Since we are using a two parameter form for the decay amplitude, these equations suffice to determine all of the $K \to 3\pi$ amplitudes in terms of f_π and $M(K \to 2\pi)$. The treatment directly parallels that for $K \to 2\pi$ so we leave it as an exercise to verify the following relations[7]: $(F \equiv f_\pi m_K/m_\pi^2)$

$$3\ f_\pi\ A_{\tau^+} = -\ M(K_S^o \to \pi^+ \pi^-) \tag{5.71a}$$

$$F\ A_{\tau^+}\ g_{\tau^+} = M(K_S^o \to \pi^+ \pi^-) - 6\ M(K^+ \to \pi^+ \pi^o) \tag{5.71b}$$

$$-\ 3\ f_\pi\ A_{\tau^{+\prime}} = M(K^+ \to \pi^+ \pi^o) + \frac{1}{\sqrt{2}}\ M(K_S^o \to \pi^o \pi^o) \tag{5.71c}$$

$$F A_{\tau^{+\prime}} g_{\tau^{+\prime}} = -\ 5\ M(K^+ \to \pi^+ \pi^o) - \sqrt{2}\ M(K_S^o \to \pi^o \pi^o) \tag{5.71d}$$

$$\frac{3}{\sqrt{2}}\ f_\pi\ A_{\tau_L} = -\ M(K^+ \to \pi^+ \pi^o) + \tfrac{1}{2}\ M(K_S^o \to \pi^+ \pi^-) \tag{5.71e}$$

$$\frac{1}{\sqrt{2}}\ F\ A_{\tau_L}\ g_{\tau_L} = M(K_S^o \to \pi^+ \pi^-) + M(K^+ \to \pi^+ \pi^o) \tag{5.71f}$$

$$\sqrt{6}\ f_\pi\ A_{\tau_L'} = M(K_S^o \to \pi^o \pi^o) \tag{5.71g}$$

$$g_{\tau_L'} = 0\ . \tag{5.71h}$$

In principle all of these relations may be tested as they stand. However, as we have already remarked, the phases of the $K \to 2\pi$ amplitudes are not yet known. The results (5.71) used only the $V + A$ structure of the Hamiltonian, so it is interesting to see how these new predictions compare with those already obtained in (5.51) and (5.69) using the $|\Delta \underline{I}| \leq \frac{3}{2}$ character of $\mathcal{H}_W^H$. We find

$$3\ f_\pi\ (A_{\tau^+} - 2\ A_{\tau^{+\prime}}) = -\sqrt{2}\ f_\pi\ (3\ A_{\tau_L} - \sqrt{6}\ A_{\tau_L'}) \tag{5.72a}$$

$$= \sqrt{2}\ M(K_S^o \to \pi^o \pi^o) - M(K_S^o \to \pi^+ \pi^-) + 2\,M(K^+ \to \pi^+ \pi^o)\ . \tag{5.72b}$$

The $|\Delta \underline{I}| \leq \frac{3}{2}$ rule predicts that both sides of (5.72a) are zero, as we have seen, but by itself the soft pion programme does not, unless we also apply it to the $K \to 2\pi$ decays. However we have seen that the data are consistent with all of the combinations in (5.72) being zero. To test further the predictions (5.71) we shall assume the $K \to 2\pi$ amplitudes satisfy the $|\Delta \underline{I}| = \frac{1}{2}$ rule. We know that this is only approximately true; it would predict the absence of $K^+ \to \pi^+ \pi^o$, whereas experimentally its matrix element is almost 3% of those for $K_S^o \to \pi\pi$. However the soft pion extrapolation is almost certainly less accurate than this, so nothing much will be lost. At any rate, with this approximation we find

$$\frac{1}{\sqrt{2}} A_{\tau^+} = \sqrt{2}\, A_{\tau^{+\prime}} = - A_{\tau_L} = \sqrt{\frac{2}{3}}\, A_{\tau_L^{\prime}} = \frac{-1}{f_\pi 3\sqrt{2}} M(K_S^o \to \pi^+ \pi^-) \qquad (5.73a)$$

$$g_{\tau^+} = - \tfrac{1}{2} g_{\tau^{+\prime}} = - \tfrac{1}{2} g_{\tau_L} = - \frac{3 m_\pi^2}{m_K^2} = - 0.24 \ . \qquad (5.73b)$$

Apart from specifying the absolute magnitudes of the quantities concerned, (5.73) also predicts that the quantities have the same ratios as would be the case had we assumed that $\mathcal{H}_W^H$ possessed no $I = \frac{3}{2}$ piece. In other words, if the $K \to 2\pi$ amplitudes satisfy the $|\Delta \underline{I}| = \frac{1}{2}$ rule, then so do those for $K \to 3\pi$ in the soft pion limit. These tests of the $|\Delta \underline{I}| = \frac{1}{2}$ rule provide three new predictions

$$\gamma_{\tau^+} (2 \gamma_{\tau_L})^{-1} = - g_{\tau^{+\prime}} (2 g_{\tau^+})^{-1} = g_{\tau_L} \, g_{\tau^{+\prime}}^{-1} = 1 \ . \qquad (5.74)$$

The consistency of these predictions with the data is only fair. For example

$$\tfrac{1}{2} g_{\tau^{+\prime}} + g_{\tau^+} = 0.048 \pm 0.012 \ ,$$

which indicates that $|\Delta \underline{I}| = \frac{3}{2}$ transitions do actually occur albeit inhibited compared with those with $|\Delta \underline{I}| = \frac{1}{2}$. Finally let us look at the absolute magnitude predictions contained in (5.73). Taking the average of the g-values we find

$$\tfrac{1}{3} (g_{\tau^+} - \tfrac{1}{2} g_{\tau^{+\prime}} - \tfrac{1}{2} g_{\tau_L}) = - 0.26 \pm 0.02 \ .$$

Again the agreement is fair, and one naturally wonders whether the experimentally observed deviations from the approximate predictions (5.73) can be accounted for by relaxing the assumption that the $K \to 2\pi$ amplitudes obey the $|\Delta \underline{I}| = \frac{1}{2}$ rule, which, as we have seen, is only approximately true. In other words, are the original predictions (5.71) more accurate than those contained in (5.73) ? There are some indications that this is indeed the case. Bouchiat and Meyer[(8)] assumed that the $K \to 2\pi$ amplitudes satisfy the $|\Delta \underline{I}| \leq \frac{3}{2}$ rule (5.30) and also neglected phase space differences between the various decays. They concluded that the resulting corrections to the approximate predictions (5.71) are of the right sign and magnitude to reduce the discrepancies between theory and experiment. In addition they conclude that it is unlikely that the observed discrepancies could be accounted for by electromagnetic corrections to an exact $|\Delta \underline{I}| = \frac{1}{2}$ rule. Thus the $K \to 3\pi$ decays are well understood in terms of the $K \to 2\pi$ decays.

Let us turn briefly to the other possible three-pion decay of the neutral kaon τ_S. CP-invariance now requires that f is <u>odd</u> under interchange of the energies ω_1, ω_2 of π^+, π^-. Thus

$$f_{\tau_S}(\omega_1, \omega_2) = - f_{\tau_S}(\omega_2, \omega_1) . \tag{5.75}$$

So, if we assume that f is linear in ω_1, ω_2 as before, it must have the form

$$f_{\tau_S}(\omega_1, \omega_2) = h_{\tau_S}(\omega_1 - \omega_2)\, m_K^{-1} . \tag{5.76}$$

The soft pion programme enables us to calculate h_{τ_S}, and we find

$$\frac{1}{2\sqrt{2}} f_\pi h_{\tau_S} = M(K_S^o \to \pi^+ \pi^-) - \sqrt{2}\, M(K_S^o \to \pi^o \pi^o) - \tfrac{1}{2} M(K^+ \to \pi^+ \pi^o) .$$

This vanishes if the $K \to 2\pi$ amplitudes satisfy an exact $|\Delta \underline{I}| = \frac{1}{2}$ rule, but if we use the more general (5.30) we find

$$f_{\tau_S}(\omega_1, \omega_2) = \frac{3\sqrt{2}}{f_\pi m_K} M(K^+ \to \pi^+ \pi^o)(\omega_1 - \omega_2) . \tag{5.77}$$

Thus it is straightforward to calculate the decay rate for the mode τ_S in terms of f_π and the decay rate for $K^+ \to \pi^+ \pi^o$. Note first that $|f|^2 \propto y^2$ using the variables defined in (5.61); this is the manifestation of the double inhibition of this process by the centrifugal barrier, which we referred to earlier. The phase space integral is only slightly less trivial than for the other decays. Integrating over the allowed region (5.62) we find

$$\int dx\, dy\, y^2 = \frac{1}{72} Q^4 \qquad (Q = m_K - 3m_\pi) , \tag{5.78}$$

so

$$\Gamma(\tau_S) = \frac{1}{512\sqrt{3}\,\pi^3} \frac{Q^4}{f_\pi^2 m_K^3} |M(K^+ \to \pi^+ \pi^o)|^2 . \tag{5.79}$$

$|M|^2$ is trivially related to

$$\Gamma(K^+ \to \pi^+\pi^o) = \frac{1}{8\pi} \frac{p}{m_K^2} |M(K^+ \to \pi^+\pi^o)|^2 , \qquad (5.80a)$$

where

$$p = (\tfrac{1}{4} m_K^2 - m_\pi^2)^{\frac{1}{2}} . \qquad (5.80b)$$

So finally we have

$$\Gamma(\tau_S) = \frac{1}{64\sqrt{3}\,\pi^2} \frac{Q^4}{f_\pi^2 m_K p} \Gamma(K^+ \to \pi^+ \pi^o) . \qquad (5.81)$$

Putting in the experimentally measured quantities this gives a predicted branching ratio to the common τ_L mode of[(8)]

$$\frac{\Gamma(\tau_S)}{\Gamma(\tau_L)} = 1.7 \times 10^{-3} .$$

5.1.3 HYPERON DECAYS

The selection rules (5.6) also predict the existence of the hyperon decays. In particular the following seven decays are all predicted and have been observed

$$(\Lambda_-) : \Lambda \to p\,\pi^-$$

$$(\Lambda_o) : \Lambda \to n\,\pi^o$$

$$(\Sigma_+^+) : \Sigma^+ \to n\,\pi^+$$

$$(\Sigma_-^-) : \Sigma^- \to n\,\pi^-$$

$$(\Sigma_o^+) : \Sigma^+ \to p\,\pi^o$$

$$(\Xi_-^-) : \Xi^- \to \Lambda\,\pi^-$$

$$(\Xi_o^o) : \Xi^o \to \Lambda\,\pi^o .$$

The only other kinematically allowed processes are Σ^o decays

$$(\Sigma^o_o) \;:\; \Sigma^o \to p\,\pi^o$$

$$(\Sigma^o_-) \;:\; \Sigma^o \to p\,\pi^- .$$

But these are swamped by the fast electromagnetic decay $\Sigma^o \to \Lambda\gamma$ and so cannot be observed. Angular momentum conservation tells us that the final state has orbital angular momentum $\ell = 0,1$. The parity of the final state is $(-1)^{\ell+1}$, since the pion is pseudoscalar, so the s-wave ($\ell = 0$) states arise from the parity violating part of $\mathcal{H}^H_W$ while the p-wave states arise from the parity conserving part. The invariant amplitude for each process is therefore characterized by two numbers corresponding to these two transition types. For the process

$$(Y_i) \;:\; Y \to B\,\pi^i \tag{5.82}$$

it is conventional to write the invariant matrix element in the form

$$M(Y_i) = G\,\mu_c^2\,\bar{u}_B(p)\,[\,A(Y_i) - B(Y_i)\,\gamma_5\,]\,u_Y(P) \;, \tag{5.83}$$

where $A(Y_i)$ and $B(Y_i)$ are constants and P, p are the momenta of Y and B. The inclusion of the dimensionless constant $G\,\mu_c^2$, where μ_c is the mass of the charged pion, is to give A and B values that are roughly of order unity. The quantities most easily measured experimentally are the total rate for the decay in question and the asymmetry of the final state baryon with respect to the polarization of the decaying hyperon. It is easy to see that these two quantities are sufficient to determine A and B. Starting from (5.83) we may calculate the probability $d\Gamma$ for the decay of Y at rest with its spin polarised in the direction $\underline{s}$ into a pion π^i and baryon B emitted in the direction $\underline{n}$ at an angle θ with $\underline{s}$. We find

$$d\Gamma = \frac{G^2\mu_c^4\,p}{8\pi\,m_Y^2}\left\{ |A|^2[\,(m_Y+m_B)^2 - m_\pi^2\,] + |B|^2[\,(m_Y-m_B)^2-m_\pi^2\,] + 2\,\mathrm{Re}\,A^*B\,m_Y\,p\;\underline{s}.\underline{n} \right\} \tfrac{1}{2}\sin\theta\,d\theta \;, \tag{5.84a}$$

where p is the momentum of the final baryon

$$p^2 + m_B^2 = \left[\frac{m_Y^2 + m_B^2 - m_\pi^2}{2m_Y}\right]^2 . \tag{5.84b}$$

Thus

$$d\Gamma = \Gamma(1 + \alpha\, \underline{s}.\underline{n})\ \tfrac{1}{2} \sin\theta\, d\theta\ , \tag{5.85}$$

where Γ is the total decay rate and α is the asymmetry parameter. If we assume T-invariance, so that A and B are relatively real, measurement of Γ and α suffices to determine magnitudes of A and B and their relative sign. So the problem facing the theorist is to make predictions about the 14 parameters A and B specifying the seven observed hyperon decays.

Isospin and SU(3) Considerations

We start of course with isospin considerations. A πN state has $I = \frac{1}{2}, \frac{3}{2}$; thus for Λ decay only the parts of $\mathcal{H}_W^H$ having $I = \frac{1}{2}, \frac{3}{2}$ can contribute, since Λ has $I = 0$. Since we are studying the hypothesis (5.6) that $\mathcal{H}_W^H$ has both $I = \frac{1}{2}$ and $I = \frac{3}{2}$, no restrictions are imposed upon the Λ decays by isospin considerations, without making further assumptions. The same applies to the Ξ decays, which again involve the Λ and the isodoublet Ξ. For Σ decays the $I = \frac{1}{2}, \frac{3}{2}$ and $\frac{5}{2}$ parts of $\mathcal{H}_W^H$ could induce the transition from $\Sigma(I = 1)$ to $\pi N\, (I = \frac{1}{2}, \frac{3}{2})$. Since the hypothesis (5.6) states that $\mathcal{H}_W^H$ has no $I = \frac{5}{2}$ component, we should expect restrictions upon the Σ decays. We start from

$$<\pi^- n|[v^1 - iv^2, [v^1 - iv^2, \mathcal{H}_W^H]]\,|\Sigma^+> = 0\ ,$$

since we are concerned with $(Y, I_3) = (1, -\frac{1}{2})$ piece of $\mathcal{H}_W^H$. Applying the isospin lowering operators to the states yields

$$M(\Sigma_+^+) + M(\Sigma_-^-) - \sqrt{2}\, M(\Sigma_0^+) = 2\, M(\Sigma_0^0) + \sqrt{2}\ M(\Sigma_-^0)\ . \tag{5.86}$$

As there is no chance of measuring the Σ^0 decay amplitudes, the relation (5.86) is untestable.

Since isospin considerations yield no restrictions, we are naturally led to consider whether the stronger assumption of SU(3) symmetry is more fruitful. We know that SU(3) is a rather good symmetry for the baryons at least, so any predictions should be reasonably reliable. Of course, in the limit of SU(3) symmetry the decaying Y and the final state B have the same mass, so the decay is kinematically forbidden. As before, we impose the SU(3) symmetry only upon the scalar quantities A and B. This is equivalent to applying the symmetry to the off-shell continuation of the invariant matrix element M. The matrix elements which concern us are

$$\langle B^i P^a | \mathcal{H}_W^H (|\Delta Y| = 1) | B^j \rangle \qquad (i,j,a = 1 \ldots. 8) \ , \tag{5.87}$$

where B^i, P^a are the octets of baryons and pseudoscalar mesons. The SU(3) properties of $\mathcal{H}_W^H (|\Delta Y| = 1)$ are given in (5.12) and (5.18). Since we are concerned only with the SU(3) symmetry properties of the amplitudes (5.87), we may as well consider the "crossed" amplitude

$$\langle B^i \bar{B}^j | \mathcal{H}_W^H (|\Delta Y| = 1) | P^a \rangle \ . \tag{5.88}$$

Let us see first how many reduced amplitudes can contribute to this matrix element. B^i and $\bar{B}^j$ are distinct octets, so they couple to all of the representations contained in the product $\underline{8} \times \underline{8}$,

$$\underline{8}_B \times \underline{8}_{\bar{B}} = (\underline{1}) + \underline{8}_s + \underline{8}_a + \underline{10} + \underline{\overline{10}} + \underline{27} \ . \tag{5.89}$$

In the same way the octet piece D^k_{pv} of the parity violating part of $\mathcal{H}_W^H$ when multiplied by the pseudoscalar octet P^a can also produce all of the above irreducible representations

$$\underline{8}_{\mathcal{H}} \times \underline{8}_P = (\underline{1}) + \underline{8}_s + \underline{8}_a + \underline{10} + \underline{\overline{10}} + 27 \ . \tag{5.90}$$

We have bracketed the singlet part $\underline{1}$ of both of these products, since it cannot contribute to the hypercharge nonconserving decays under consideration; in these decays B and $\bar{B}$ have different hypercharge so cannot couple to a singlet. In addition to the octet piece the parity violating part of $\mathcal{H}_W^H$ also contains the $\underline{27}$ contribution and [1]

$$\underline{27}_{\mathcal{H}} \times \underline{8}_P = \underline{8} + \underline{10} + \underline{\overline{10}} + \underline{27}_s + \underline{27}_a + \cdots \ , \tag{5.91}$$

where $\cdots$ indicates that there are additional representations of higher dimensionality. In the limit of exact SU(3) symmetry each of the representations occurring in (5.89) is coupled only to a representation of the <u>same</u> dimension appearing in the product $\mathcal{H}_W^H P^a$. Thus each of the two octets may couple to each of the three octets in (5.90) and (5.91) yielding a total of six reduced octet amplitudes. Similarly, there are two reduced $\underline{10}$ amplitudes and two $\overline{\underline{10}}$ amplitudes. There are also three reduced $\underline{27}$ amplitudes. Thus, a priori, the parity violating part of (5.89) is expressible as a linear combination of 13 reduced amplitudes with calculable coefficients, which may be obtained from the tables of SU(3) Clebsch-Gordan coefficients. (Precisely similar conclusions apply equally to the parity conserving amplitudes.) Since we only have seven parity violating amplitudes measured by experiment, the situation is not promising. In fact it is not quite as bad as this very general analysis allows it to be. The requirements of C-parity intervene, as they did in $K \to 2\pi$, and prevent some of the reduced amplitudes contributing.

Under the operation of charge conjugation $\mathcal{C}$ a baryon becomes an anti-baryon and conversely :

$$\mathcal{C}\, B^i\, \mathcal{C}^{-1} = \epsilon_i \bar{B}^i \qquad \text{(no summation)}$$

$$\mathcal{C}\, \bar{B}^i\, \mathcal{C}^{-1} = \epsilon_i B^i \qquad \text{(no summation)} \quad .$$

The meson octet P^i is self conjugate and has C-parity +1, thus

$$\mathcal{C}\, P^i\, \mathcal{C}^{-1} = \epsilon_i\, P^i \qquad \text{(no summation)} \quad .$$

The ϵ_i are defined in (5.35c). Now the parity violating part of $\mathcal{H}_W^H$ has C-parity -1, because of CP-invariance, so the parity violating amplitudes must satisfy

$$\langle B^i \bar{B}^j | \mathcal{H}^H_{W,pv} | P^a \rangle = - \epsilon_i \epsilon_j \epsilon_a \langle B^j \bar{B}^i | \mathcal{H}^H_{W,pv} | P^a \rangle \quad . \qquad (5.92)$$

It is this restriction upon the matrix elements, which eliminates some of the potential reduced amplitudes. i and j enter the expressions for the matrix elements only via the various Clebsch-Gordan coefficients coupling i, j to various elements of representations appearing in (5.89). Thus we may write in general

$$\langle B^i \bar{B}^j | \mathcal{H}^H_{W,pv} | P^a \rangle = \sum_{r,\underline{n},\nu} C(i,j;\underline{n}_r,\nu) \langle R^{\nu}_{\underline{n}_r} | \mathcal{H}^H_{W,pv} | P^a \rangle \quad . \qquad (5.93)$$

The $C(i,j;\underline{n}_r,\nu)$ are the Clebsch-Gordan coefficients for coupling the i and j elements of the octet to the ν component of the representation $\underline{n}_r$; the suffix r is necessary since representations of the same dimensionality sometimes occur in the same product, as in (5.89), (5.90) and (5.91).

Then the C-invariance condition (5. 92) implies

$$C(i,j;\underline{n}_r,\nu) = -\epsilon_i\epsilon_j\epsilon_a\, C(j,\, i;\, \underline{\bar{n}}_r,\nu) \quad . \tag{5.94}$$

Consider for instance the contribution of the symmetrical octet $\underline{8}_s$ in (5. 89). At least we know its Clebsch-Gordan coefficients without recourse to tables - they are simply $d^{ij\nu}$ $(\nu = 1\cdots 8)$ and are symmetric under interchange of i and j. Thus (5. 94) yields for this representation

$$\epsilon_i\epsilon_j\epsilon_a = -1 \ .$$

But the $d^{ij\nu}$ are non-zero only when

$$\epsilon_i\epsilon_j\epsilon_\nu = +1 \ . \tag{5.95}$$

Thus the only states of $\underline{8}_s$ which can contribute in (5. 93) are those with

$$\epsilon_\nu = -\epsilon_a \ . \tag{5.96}$$

For the antisymmetric octet $\underline{8}_a$ in (5. 89) we find $\epsilon_i\epsilon_j\epsilon_a = +1$, but, since $\epsilon_i\epsilon_j\epsilon_\nu = -1$ for this coupling, we again find that only states satisfying (5. 96) are allowed in the sum (5. 93). This condition then implies restrictions on the couplings of $\mathcal{H}_W^H P^a$ which can give non-zero matrix elements. For example let us consider the symmetric octet coupling $\underline{8}_s$ in (5. 90), which arises from the octet part D^6_{pv} of $\mathcal{H}^H_{W,pv}$. The Clebsch-Gordan coefficients are $d^{6a\nu}$, which from (5. 95) are non-zero only when

$$\epsilon_\nu = +\epsilon_a \ ,$$

since $\epsilon_6 = +1$. On the other hand the antisymmetric $\underline{8}_a$ in (5. 90) plainly does satisfy the requirement (5.96), but it can be shown that the $\underline{8}$ in (5. 91) does not, basically because only a symmetric coupling of $\underline{8}$ and $\underline{27}$ yields $\underline{8}$. Thus of the 3 octets in (5. 90) and (5. 91) only one can contribute in the parity violating decays. So of the six possible reduced octet amplitudes only two are allowed to be non-zero by C-invariance requirements. In the same way only the antisymmetric $\underline{27}_a$ in (5. 91) can couple, so only one of the three reduced $\underline{27}$ amplitudes is non-zero. For the $\underline{10}$ and $\underline{\overline{10}}$ representations the "bar" on $\underline{\bar{n}}_r$ in (5. 94) is important, since, unlike $\underline{8}$ and $\underline{27}$, the $\underline{10}$ is not self-conjugate; under the operation of charge conjugation the $\underline{10}$ is transformed into a $\underline{\overline{10}}$, which is not an equivalent representation. The symmetry (5.94) has the effect of ensuring that the reduced amplitudes $a(10)$ and $a(\overline{10})$, from (5.89) for example, satisfy $a(10) = -a(\overline{10})$, and so are non-zero only in the linear combination $a(10) - a(\overline{10})$. Thus in all only two of the possible four $\underline{10}$, $\overline{\underline{10}}$ amplitudes survive. So the s-wave amplitudes A are all expressible in terms of 5 unknown reduced amplitudes. As a result two linear relations between the seven observed amplitudes may be derived. They are[9]

$$A(\delta_\Lambda) + A(\delta_\Xi) = 0 \tag{5.97a}$$

$$A(\delta_\Xi) + 2A(\delta_{LS}) = (\tfrac{3}{2})^{\frac{1}{2}} A(\delta_\Sigma) , \tag{5.97b}$$

where

$$A(\delta_\Lambda) \equiv A(\Lambda^o_-) + \sqrt{2}\, A(\Lambda^o_o) \tag{5.97c}$$

$$A(\delta_\Xi) \equiv A(\Xi^-_-) + \sqrt{2}\, A(\Xi^o_o) \tag{5.97d}$$

$$A(\delta_\Sigma) \equiv A(\Sigma^+_+) - A(\Sigma^-_-) - \sqrt{2}\, A(\Sigma^+_o) \tag{5.97e}$$

$$A(\delta_{LS}) \equiv 2A(\Xi^-_-) + A(\Lambda^o_-) - \sqrt{3}\, A(\Sigma^+_o) . \tag{5.97f}$$

We shall not reproduce the derivation of these relations. They follow straightforwardly using the tables of Clebsch-Gordan coefficients[1] and the above noted restriction upon the s-wave amplitudes.

The application of SU(3) symmetry to the p-wave amplitudes (i.e. the parity conserving decays) proceeds analogously. The C-parity requirements lead to a condition like (5.92) with the sign reversed on the right-hand side. Then the p-wave amplitudes are expressible in terms of eight (rather than five) reduced amplitudes, and no restrictions follow for the seven observed processes.

Soft Pion Programme

We shall see later that the rather limited predictions (5.97) are in fact consistent with the data. However, as in the kaon decays, the predictions do not contain the most salient features of the data. We shall see that all of the combinations (5.97c, d, e, f) are small compared with the individual amplitudes from which they are constructed, and the same is true of the corresponding p-wave amplitudes. This suggests that some suppression mechanism is at work, so we naturally turn to the soft pion programme, which provided just such behaviour for the kaon decays. The basic equation is obtained from PCAC as usual. In the limit $q \to 0$

$$\frac{if_\pi}{\sqrt{2}} < B\,\pi^i(q)\,|\,\mathcal{H}_W^H\,|\,Y> \;=\; < B|\,[\,a^i, \mathcal{H}_W^H\,]\,|Y> + \; iq^\lambda\, T_\lambda^i \;, \qquad (5.98a)$$

where

$$T_\lambda^i \;=\; \int dx\; e^{iqx} < B\,|\,T\,\left\{A_\lambda^i(x),\; \mathcal{H}_W^H\right\}\,|\,Y> \qquad (i = 1,2,3) \qquad (5.98b)$$

and q is the momentum of the pion. In the soft pion limit the only terms in T_λ^i which survive are the Born contributions T_λ^{Bi}, which have a pole at $q = 0$. To evaluate these we need to know the matrix elements of $\mathcal{H}_W^H$ between single-baryon states, since the Born terms are those with a single-baryon intermediate state.

We consider first the parity violating decays. The matrix elements of $\mathcal{H}_{W\,pv}^H$ between single-baryon states vanish in the limit of exact SU(3) symmetry for the baryons:

$$< B^j\,|\,\mathcal{H}_{W\;pv}^H\,|\,B^k> \;=\; 0 \qquad\qquad (j,k = 1\cdots 8)\;. \qquad (5.99)$$

As before,this is a statement about the Clebsch-Gordan coefficient. The mechanism causing this is essentially that just discussed which led to the vanishing of eight of the possible s-wave amplitudes. We "cross" B^k, and C-parity considerations yield

$$< B^j\,\overline{B}^k\,|\,\mathcal{H}_{W\;pv}^H\,|\,0> \;=\; -\,\epsilon_j\,\epsilon_k\; < B^k\,\overline{B}^j\,|\,\mathcal{H}_{W\,pv}^H\,|\,0> \;, \qquad (5.100)$$

similarly to (5. 92). Then, as before, the only states ν to which $B^j\,\overline{B}^k$ can be connected have $\epsilon_\nu = -1$. But in this case the "state" is $\mathcal{H}_{W\;pv}^H|0>$, and we have already observed that $\epsilon_{\mathcal{H}} = +1$, so (5. 99) follows. Thus the only contributions from the right of (5. 98a) come from the equal time commutators. The V + A structure of $\mathcal{H}_W^H$ implies that

$$\left[a^i,\, \mathcal{H}_{W\;pv}^H\right] \;=\; \left[v^i,\, \mathcal{H}_{W\;pc}^H\right] \;, \qquad (5.101)$$

where $\mathcal{H}_{W pc}^H$ is the parity conserving piece of $\mathcal{H}_W^H$. So the parity violating Λ decay amplitudes are given by

$$G\mu_c^2 A(\Lambda_i)\bar{u}_N u_\Lambda = -i\frac{\sqrt{2}}{f_\pi} < N|[v^i, \mathcal{H}^H_{W pc}]|\Lambda> . \tag{5.102}$$

The v^i (i = 1,2,3) are isospin generators, so only the $I = \frac{1}{2}$ piece of $\mathcal{H}^H_W$ can contribute, since Λ has $I = 0$ and N $I = \frac{1}{2}$. We find then

$$G\mu_c^2 \ A(\Lambda^o_-) = -if_\pi^{-1} \ a(\Lambda \to n) \tag{5.103a}$$

$$G\mu_c^2 \ A(\Lambda^o_o) = +if_\pi^{-1}\frac{1}{\sqrt{2}} a(\Lambda \to n) , \tag{5.103b}$$

where

$$< n|\mathcal{H}^H_{W pc}|\Lambda> = a(\Lambda \to n)\bar{u}_n u_\Lambda . \tag{5.103c}$$

Thus the soft pion programme yields the sum rule

$$A(\delta_\Lambda) \equiv A(\Lambda^o_-) + \sqrt{2} A(\Lambda^o_o) = 0 , \tag{5.104}$$

which we could have deduced at the outset <u>if</u> $\mathcal{H}^H_{W pv}$ had only $I = \frac{1}{2}$. In the same way the s-wave Ξ decays obey an effective $|\Delta \underline{I}| = \frac{1}{2}$ rule in the soft pion limit. Since these too involve the Λ:

$$A(\delta_\Xi) = 0 , \tag{5.105a}$$

with

$$G\mu_c^2 A(\Xi^-_-) = -if_\pi^{-1} \ a(\Xi^o \to \Lambda) . \tag{5.105b}$$

However we do <u>not</u> predict a $|\Delta \underline{I}| = \frac{1}{2}$ rule for the Σ-decays. Instead, from (5.102) with Λ replaced by Σ, we find

$$G\mu_c^2 A(\Sigma^+_+) = -if_\pi^{-1}[a(\Sigma^+ \to p) + \sqrt{2} a(\Sigma^o \to n)] \tag{5.106a}$$

$$G\mu_c^2 A(\Sigma^-_-) = +if_\pi^{-1}\sqrt{2} \ a(\Sigma^o \to n) \tag{5.106b}$$

$$G\mu_c^2 A(\Sigma^+_o) = if_\pi^{-1}\frac{1}{\sqrt{2}} a(\Sigma^+ \to p) , \tag{5.106c}$$

so

$$A(\delta_\Sigma) = 2 A(\Sigma^+_+) . \tag{5.107}$$

$A(\Sigma^+_+)$ is determined by the matrix element of (5.101) between $< n|$ and $|\Sigma^+>$. Since these states have $|\Delta I^3| = \frac{3}{2}$, we see that in this analysis $A(\Sigma^+_+)$ is non-zero only by virtue of the fact that $\mathcal{H}^H_W$ contains an $I = \frac{3}{2}$ component, which, under our hypothesis, must belong to the <u>27</u> piece.

If $\mathcal{H}_W^H$ were purely octet in character, we could deduce from the outset that that $A(\delta_\Sigma) = 0$, as this simply expresses the $|\Delta \underline{I}| = \frac{1}{2}$ rule. Thus in this picture the value of $A(\Sigma_+^+)$ is a measure of how much $I = \frac{3}{2}$ $\mathcal{H}_W^H$ contains.

It is important to realize that our derivation of (5.103) to (5.107) has utilized the SU(3) properties of $\mathcal{H}_W^H$, even though the equal time commutators use only the V + A property of $\mathcal{H}_W^H$; our neglect of the Born contribution $q^\lambda T_\lambda^B$ was justified by the application of SU(3) symmetry considerations. Thus it is not surprising that (5.104) and (5.105a) are consistent with our earlier prediction (5.97a), which also followed from SU(3) considerations. In fact, using (5.105a) and (5.107), we may deduce from the second prediction (5.97b) that

$$A(\delta_{LS}) = (\tfrac{3}{2})^{\frac{1}{2}} A(\Sigma_+^+) \ . \qquad (5.108)$$

This too could have been deduced directly from what we have already done by using the SU(3) properties of the amplitudes $a(B^j \to B^k)$ defined in (5.103c). If $\mathcal{H}_W^H$ had only an octet component, the vanishing of $A(\Sigma_+^+)$ implies that $A(\delta_{LS}) = 0$, and this too could have been deduced from the outset just using SU(3) symmetry. This is the famous Lee-Sugawara Sum Rule[10].

Thus the soft pion programme leads to four relations among the seven observed s-wave amplitudes[11], two of which are new. All of the relations are consistent with the data[2]:

$$A(\delta_\Lambda) = -0.047 \pm 0.03 \qquad (5.109a)$$

$$A(\delta_\Xi) = +0.12 \pm 0.05 \qquad (5.109b)$$

$$A(\delta_\Sigma) - 2A(\Sigma_+^+) = 0.10 \pm 0.10 \qquad (5.109c)$$

$$A(\delta_{LS}) - (\tfrac{3}{2})^{\frac{1}{2}} A(\Sigma_+^+) = 0.11 \pm 0.12 \ . \qquad (5.109d)$$

In fact the more general predictions (5.97) are slightly more consistent with the data than those above. The experimental value of $A(\Sigma_+^+)$ is[2]

$$A(\Sigma_+^+) = 0.06 \pm 0.02 \ , \qquad (5.110a)$$

whereas

$$A(\Lambda_-^0) = 1.48 \pm 0.01 \ . \qquad (5.110b)$$

The small value of $A(\Sigma_+^+)$ compared with the other s-wave amplitudes (of which $A(\Lambda_-^o)$ is typical) has not been predicted by the theory. As already observed, its existence implies the presence of an $I = \frac{3}{2}$ piece of $\mathcal{H}_W^H$, which under our hypothesis must belong to the 27 representation. The fact that $A(\Sigma_+^+)$ is small thus implies the existence of some dynamical mechanism which is suppressing the 27 contribution or alternatively enhancing the 8 contribution. At any rate, it is certainly not predicted by our analysis. Thus the outstanding problem for the theorist is to predict the three remaining unknowns necessary to determine completely the amplitudes $a(B^j \to B^k)$. In particular, we need to understand why the 27 contribution to these is small compared with the octet contributions.

We turn now to the parity conserving amplitudes. The V + A structure of $\mathcal{H}_W^H$ implies (in addition to (5.101)) that

$$\left[a^i, \mathcal{H}_{Wpc}^H\right] = \left[v^i, \mathcal{H}_{Wpv}^H\right] . \tag{5.111}$$

Thus, since v^i $(i = 1,2,3)$ are isospin generators, the equal time commutator in (5.98a) for the parity conserving Λ_-^o decay is

$$\begin{aligned} \langle p|[a^1 + ia^2, \mathcal{H}_{Wpc}^H]|\Lambda\rangle &= \langle p|[v^1 + iv^2, \mathcal{H}_{Wpv}^H]|\Lambda\rangle \\ &= \langle n|\mathcal{H}_{Wpv}^H|\Lambda\rangle \\ &= 0 \end{aligned}$$

by virtue of (5.99). Plainly the same is true for all p-wave amplitudes, and it follows that the parity conserving decays are completely determined by the Born term $q^\lambda T_\lambda^B$ on the right of (5.98a). T_λ^B is given in general by the sum of the two Feynman diagrams shown in Fig. 5.1. Since $i = 1,2,3$, B' and Y' have the same hypercharge as B and Y respectively.

Consider, for example, the decay Λ_-^o. Then $B' = n$ and $Y' = \Sigma^+$, since $A_\lambda^i = A_\lambda^1 + iA_\lambda^2$. The contribution of diagram (a) is

$$T_\lambda^{B(a)}(\Lambda_-^o) = \bar{u}_p[g_1^A(q^2)\gamma_\lambda\gamma_5 + g_2^A(q^2)q_\lambda\gamma_5]\frac{i}{\not p + \not q - m_n} u_\Lambda a(\Lambda \to n) . \tag{5.112}$$

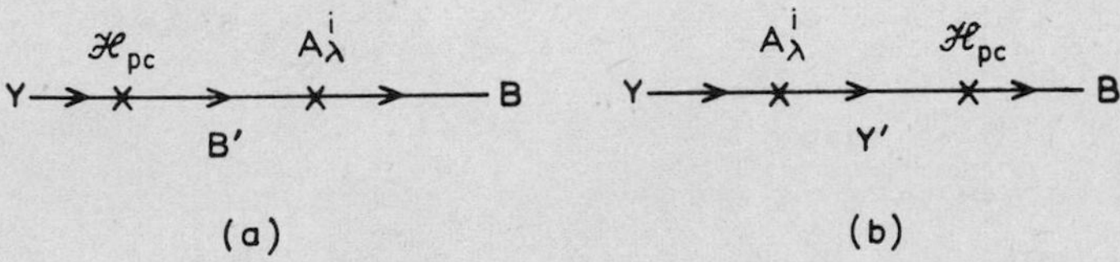

Fig. 5.1 Feynman diagrams determining T_λ^{Bi}

The factor $a(\Lambda \to n)$ is from the $\mathcal{H}_{Wpc}^{H}$ vertex as defined in (5.103c). Thus

$$q^\lambda T_\lambda^{B(a)}(\Lambda_-^o) = \bar{u}_p [\, g_1^A(q^2)\, \not{q} + g_2^A(q^2)\, q^2 \,]\, \gamma_5 \frac{i}{\not{p} + \not{q} - m_n} u_\Lambda a(\Lambda \to n)$$

$$= \bar{u}_p \left\{ g_1^A(q^2)(\not{q} + \not{p} + m_n) + [-2m_n g_1^A(q^2) + q^2 g_2^A(q^2)\,] \right\}$$

$$\gamma_5 \frac{i}{\not{p} + \not{q} - m_n} u_\Lambda a(\Lambda \to n)$$

$$= \bar{u}_p \left[-g_1^A(q^2) i\gamma_5 + \sqrt{2} f_\pi \frac{m_\pi^2}{m_\pi^2 - q^2} g_{\pi NN} K(q^2)\, \gamma_5 \frac{i}{\not{p} + \not{q} - m_n} \right] u_\Lambda a(\Lambda \to n) , \tag{5.113}$$

where we have used the Dirac equation and the general Goldberger-Treiman formula (4.161). We recognise the last term of (5.113) as being proportional to one of the pole contributions to $M(\Lambda_-^o)$. These are shown in Fig. 5.2 for the general process (Y_i).

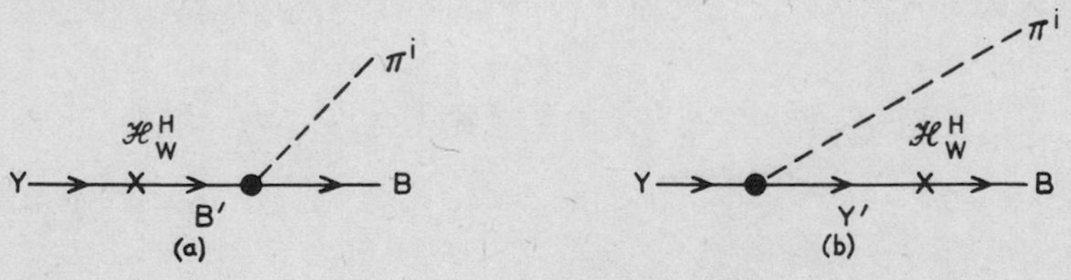

Fig. 5.2 Pole diagrams contributing to $M(Y_i)$

For the process Λ^o_-, Y, B, Y', B' are as before. We do not have to worry about the parity violating part of $\mathcal{H}^H_W$ because of (5.99). Thus the contribution from diagram (a) is

$$M^{B(a)}(\Lambda^o_-) = -i\bar{u}_p\sqrt{2}\,g_{\pi NN}K(q^2)\,i\gamma_5\frac{i}{\not p+\not q-m_n}u_\Lambda\,a(\Lambda\to n)\,, \qquad (5.114)$$

and from (5.113) we deduce

$$iq^\lambda T^{B(a)}_\lambda(\Lambda^o_-) = g^A_1(q^2)\,\bar{u}_p\gamma_5 u_\Lambda a(\Lambda\to n) + i\frac{f_\pi m^2_\pi}{m^2_\pi - q^2}M^{B(a)}(\Lambda^o_-)\,. \qquad (5.115)$$

In the limit $q\to 0$ both $q^\lambda T^B_\lambda$ and M^B are undefined because of the pole, but this equation shows that their difference does have a well defined limit. The same is true of the contributions of the diagrams (b) in Figs. 5.1 and 5.2. Including the contribution from these terms, and referring back to (5.98), we deduce

$$\lim_{q\to 0} if_\pi\left[M_{pc}(\Lambda^o_-) - M^B(\Lambda^o_-)\right] = \lim_{q\to 0}\left[iq^\lambda T^B_\lambda(\Lambda^o_-) - if_\pi M^B(\Lambda^o_-)\right]$$

$$= \left[g^A_1(0)\,a(\Lambda\to n) + f^A_1(0)\,a(\Sigma^+\to p)\right]\bar{u}_p\gamma_5 u_\Lambda$$

$$\simeq -f_\pi\left[\sqrt{2}\,g_{\pi NN}(m_n+m_p)^{-1}a(\Lambda\to n) + g_{\pi\Sigma\Lambda}(m_\Sigma+m_\Lambda)^{-1}a(\Sigma^+\to p)\right]\times$$

$$\times\,(\bar{u}_p\gamma_5 u_\Lambda)\,, \qquad (5.116)$$

where $f^A_1(0)$ is the Σ-Λ axial vector coupling constant and we have used the Goldberger-Treiman formulae (4.161) and (4.296). M_{pc} is the parity conserving part of M, proportional to B in (5.83). So, if we assume that $M_{pc} - M^B$ is a slowly varying function of q, we find

$$M_{pc}(\Lambda^o_-) \simeq M^B(\Lambda^o_-) + \Big[\sqrt{2}\,g_{\pi NN}(m_n+m_p)^{-1}a(\Lambda\to n) +$$

$$+\,g_{\pi\Sigma\Lambda}(m_\Sigma+m_\Lambda)^{-1}a(\Sigma^+\to p)\Big]\bar{u}_p i\gamma_5 u_\Lambda\,. \qquad (5.117)$$

In fact, from (5.114), we see, using the Dirac equation and $p+q=P$, that

$$M^{B(a)}(\Lambda^o_-) = \sqrt{2}\,g_{\pi NN}(m_\Lambda - m_n)^{-1}a(\Lambda\to n)\,\bar{u}_p i\gamma_5 u_\Lambda\,,$$

and a similar expression obtains for $M^{B(b)}(\Lambda^o_-)$. Thus we may write[12]

$$M_{pc}(\Lambda^o_-) \simeq M^B(\Lambda^o_-)\left[1 + \frac{\Delta m_B}{m_B}\right] ,$$

where Δm_B is a typical baryon mass difference, $m_\Lambda - m_n$ or $m_\Sigma - m_p$ in the case under discussion. So the upshot is that to a very good approximation $M_{pc}(\Lambda^o_-)$ is equal to its Born term $M^B(\Lambda^o_-)$. Plainly a similar analysis holds for the other p-wave amplitudes. We see from (5.114) that these Born terms are determined by the amplitudes $a(B^j \to B^k)$, the strong interaction coupling constants $g_{\pi B^j B^k}$, and the observed baryon masses. If we assume, for the moment, that the amplitudes $a(B^j \to B^k)$ and $g_{\pi B^j B^k}$ are unknown, we could use SU(3) symmetry to express the $a(B^j \to B^k)$ terms of three reduced amplitudes, and the $g_{\pi BB}$ in terms of the two parameters f and d. We should thus obtain two relations between the seven p-wave amplitudes. In the approximation $m_\Lambda = m_\Sigma$ and $m_\Lambda - m_N = m_\Xi - m_\Lambda$ these take the simple form[13]

$$B(\delta_\Lambda) + B(\delta_\Xi) = 0 \tag{5.119a}$$

$$B(\delta_\Xi) + 2B(\delta_{LS}) = (\tfrac{3}{2})^{\frac{1}{2}} B(\delta_\Sigma) , \tag{5.119b}$$

where the amplitudes are defined as in (5.97) with A replaced by B. These are the p-wave analogues of (5.97), which were derived using SU(3) symmetry alone. Like these, the relations (5.119) are consistent with the data but do not contain the most salient feature, which is that each of the combinations appearing above is very small compared with the amplitudes from which they are constructed. The latest data give[2]

$$B(\delta_\Lambda) = -0.13 \pm 0.85 \tag{5.120a}$$

$$B(\delta_\Xi) = -1.61 \pm 1.62 \tag{5.120b}$$

$$B(\delta_\Sigma) = 2.7 \pm 1.1 \tag{5.120c}$$

$$B(\delta_{LS}) = 2.8 \pm 2.1 , \tag{5.120d}$$

whereas a typical p-wave amplitude is

$$B(\Lambda^o_-) = 10.17 \pm 0.24 . \tag{5.121}$$

But the amplitudes $a(B^j \to B^k)$ are <u>not</u> unknown, since they arose in the analysis of the s-wave decays. From the <u>experimental</u> value of $A(\Sigma^+_+)$

we know that the $\underline{27}$ contribution to a $(B^j \to B^k)$ must (for unknown theoretical reasons) be small. Hence, in the approximation that we neglect $A(\Sigma_+^+)$, the p-wave Born terms are determined by the octet part of $\mathcal{H}_W^H$ alone, and must satisfy the $|\Delta \underline{I}| = \frac{1}{2}$ rule, since the strong coupling constants $g_{\pi BB}$ are SU(3) symmetric. (Remember $\underline{8}$ has no component with $I = \frac{3}{2}$). Thus the smallness of $A(\Sigma_+^+)$ requires

$$B(\delta_\Lambda) = B(\delta_\Xi) = B(\delta_\Sigma) \simeq 0 \ , \tag{5.122}$$

and from (5.119b) these imply

$$B(\delta_{LS}) \simeq 0 \ . \tag{5.123}$$

We have already noted that these are consistent with the data. However the logic of this requires us to recognize that all of the $a(B^j \to B^k)$ are determined by the s-wave data, and the strong coupling constants are determined by the SU(3) fit to the strong interaction data[14]. Thus the p-wave amplitudes are completely determined. Unfortunately the actual values of the p-wave amplitudes predicted in this way are wrong. They have the correct signs but are smaller, by a factor of about two, than those actually measured[15]. The situation is improved somewhat by including the $\frac{\Delta m_B}{m_B}$ term in (5.118), which is simply the last term in (5.117). And, of course, we can use the actual value of $A(\Sigma_+^+)$ rather than zero. But when all this is done, there is still substantial disagreement between theory and experiment.

In summary then, if we take the s- and p-wave amplitudes separately, the theory makes accurate predictions, which do not include all of the salient features of the data. Taking all 14 amplitudes together theory and experiment are inconsistent. Actually, there is another feature of the data which suggests that things cannot be quite as our hypothesis would suggest. This relates to the absolute magnitudes of the decay amplitudes.

Consider for example Λ_-^0. We might estimate the invariant matrix element as follows

$$\langle p\pi^- | \mathcal{H}_W^H | \Lambda \rangle = \frac{G}{\sqrt{2}} \sin\theta_c \cos\theta_c \langle p\pi^- | S_\lambda^+ J^{-\lambda} | \Lambda \rangle$$

$$\sim \frac{G}{\sqrt{2}} \sin\theta_c \cos\theta_c \langle p | S_\lambda^+ | \Lambda \rangle \langle \pi^- | J^{-\lambda} | 0 \rangle \ . \tag{5.124}$$

Then using what we know from semileptonic processes (4.540) and (4.105)

$$M(\Lambda^o_-) \sim \frac{G}{\sqrt{2}} \sin\theta_c \cos\theta_c \bar{u}_p \left[g^V_{p\Lambda}\gamma_\lambda - g^A_{p\Lambda}\gamma_\lambda\gamma_5\right] u_\Lambda \, i f_\pi q^\lambda$$

$$= \frac{G}{\sqrt{2}} \sin\theta_c \cos\theta_c \, i f_\pi \frac{3}{\sqrt{6}} \bar{u}_p \left[m_\Lambda - m_p + 0.7\,(m_\Lambda + m_p)\gamma_5\right] u_\Lambda$$

$$= i\,\frac{3G}{\sqrt{12}} \sin\theta_c \,(0.93\,\mu_c^2)\,\bar{u}_p \left[1.27 + 10.30\,\gamma_5\right] u_\Lambda$$

$$= G\,\mu_c^2 \sin\theta_c \,\bar{u}_p\,(1.0 + 8.3\,\gamma_5)\,u_\Lambda\,. \tag{5.125}$$

Experimentally it is found that

$$M(\Lambda^o_-) = G\mu_c^2\,\bar{u}_p\,(1.48 - 10.17\,\gamma_5)\,u_\Lambda\,. \tag{5.126}$$

Of course, this is a crude estimate - note that the predicted sign of B is wrong. But we are concerned with the magnitude and we see that the experimental values approximate those we estimated if we omit the factor $\sin\theta_c \simeq 0.23$. The case we have chosen is not atypical. It is a general feature of the nonleptonic decays that their rates are larger (by a factor of about $\mathrm{cosec}^2\theta_c$) than we might reasonably anticipate.

5.1.4 ALTERNATIVE APPROACHES TO HADRONIC DECAYS

While our current-current hypothesis has met with some success, it also has some limitations as well as some wrong predictions, as we have seen. It is natural, therefore, to wonder whether alternative hypotheses might remedy some of the defects. In this section we shall describe some of the alternative hypotheses which have been advanced. We shall not go into very much detail. The interested reader is referred to the original papers.

The simplest way to improve the predictions for the s-wave amplitudes is to demand that the current-current Hamiltonian contains no 27 piece. This can only be done by introducing neutral currents to remove the unwanted part. We are left then with

$$\mathcal{H}^H_W(|\Delta Y| = 1) \propto D^6\,, \tag{5.127}$$

using the notation of (5.9c). This guarantees the $|\Delta \underline{I}| = \frac{1}{2}$ rule for both s- and p-waves a priori, as well as the Lee-Sugawara sum rule for the s-waves using SU(3) symmetry. The soft pion programme then yields

$A(\Sigma^+_+) = 0$, since in this picture the decay proceeds only via the 27 piece of $\mathcal{H}^H_W$ which we have removed. These predictions are certainly consistent with the data, but we have paid the price of introducing neutral currents, which appear to play no role in the hypercharge nonconserving semileptonic decays. In addition, the $K \to 2\pi$ amplitudes still vanish in the SU(3) limit, and the p-wave problem is not alleviated.

If we abandon the current-current picture, we note that $\mathcal{H}^H_W$ is a mixture of scalar and pseudoscalar pices, which respectively induce the p- and s-wave decays. One simple form for a Hamiltonian is to express it in terms of the quark model scalar and pseudoscalar densities[16] S^i, P^i $(i = 1 \cdots 8)$ already introduced in (4.503, 4). CP-invariance requires

$$\mathcal{H}^H_W (|\Delta Y| = 1) = G_s \, i P^7 + G_p S^6 . \qquad (5.128)$$

Unlike V^i_λ, A^i_λ the octets S^i, P^i have the same intrinsic C-parity (+1)

$$\mathcal{C} S^i \mathcal{C}^{-1} = \epsilon_i S^i \qquad (5.129a)$$

$$\mathcal{C} P^i \mathcal{C}^{-1} = \epsilon_i P^i \quad . \qquad (5.129b)$$

So the negative C-parity of $\mathcal{H}^H_{W pv}$ requires us to choose P^7, since $\epsilon_7 = -1$. (The factor i is introduced because as defined the densities P^i are antihermitian .) This immediately solves the $K \to 2\pi$ problem, which was caused by the intrinsic parity of $\mathcal{H}^H_{W pv}$ being -1. The $|\Delta I| = \frac{1}{2}$ rule is also guaranteed for all hadronic decays. The soft pion programme then yields the additional s-wave predictions

$$A(\delta_{LS}) = A(\Sigma^+_+) = 0 . \qquad (5.130)$$

In addition, if one further assumes that S^6 is in the same octet as the SU(3) symmetry breaking S^8 appearing in (4.502), then the s-wave amplitudes are entirely given in terms of one unknown parameter, since the matrix elements of S^i are determined by the baryon octet mass differences

$$-A(\Lambda^o_-) = \sqrt{2}\, A(\Lambda^o_o) = C\left(\tfrac{3}{2}\right)^{\frac{1}{2}}(m_\Lambda - m_N) \qquad (5.131a)$$

$$A(\Xi^-_-) = -\sqrt{2}\, A(\Xi^o_o) = C\left(\tfrac{3}{2}\right)^{\frac{1}{2}}(m_\Xi - m_\Lambda) \qquad (5.131b)$$

$$- A(\Sigma^-_-) = \sqrt{2}\, A(\Sigma^+_o) = C\,(m_\Sigma - m_N) \; . \qquad (5.131c)$$

These predictions agree well with the data. In contrast the p-wave predictions vanish in the SU(3) limit, so only arise in this picture because the symmetry is not exact.

Some further insight into what is going on may be obtained by taking the quark structure of the weak currents and of the hadrons more seriously. The trouble with regarding the baryons as bound states of three quarks is that the quarks cannot have Fermi statistics even though they have spin $\frac{1}{2}$. For example, if we consider the Δ (1236) having $J = I = \frac{3}{2}$, it must be an s-wave state of three "p" quarks all with spins parallel. Plainly this is forbidden by the exclusion principle. One way to avoid this problem is to say that there are three types[17] of each quark: p_i, n_i, λ_i ($i = 1,2,3$), which transform as the $\underline{3}$ representation of <u>another</u> SU(3) group called "colour" and denoted SU(3)$'$. Then by taking known baryons as colour singlets the Fermi statistics may be restored. This is because the only way to construct a singlet representation of SU(3) from three triplets is via the totally antisymmetric coupling ϵ_{ijk}. Thus in this picture the Δ (1236) is $\epsilon_{ijk}\, p_i\, p_j\, p_k$ (with spins parallel). Plainly we now require Fermi statistics. Actually, the possibility of nine rather than three quarks is welcome for other reasons, but we shall not dwell on these here. In this picture the Cabibbo current has the form

$$\mathcal{J}^+_\mu = \bar{p}_i \gamma_\mu (1-\gamma_5) n_i \cos\theta_c + \bar{p}_i \gamma_\mu (1-\gamma_5) \lambda_i \sin\theta_c \qquad (5.132a)$$

$$\mathcal{J}^-_\mu = \bar{n}_i \gamma_\mu (1-\gamma_5) p_i \cos\theta_c + \bar{\lambda}_i \gamma_\mu (1-\gamma_5) p_i \sin\theta_c \quad . \qquad (5.132b)$$

The summation over the colour index i ensures that the currents are colour singlets, so that (unobserved) coloured states are not produced by the weak interactions. The purely hadronic Hamiltonian in such a model is evidently

$$\mathcal{H}^H_W(|\Delta Y| = 1) = \frac{G}{\sqrt{2}} \sin\theta_c \cos\theta_c \left\{ \bar{n}_i \gamma_\mu (1-\gamma_5) p_i \bar{p}_j \gamma^\mu (1-\gamma_5) \lambda_j + \text{h.c.} \right\} \qquad (5.133a)$$

$$= \frac{G}{\sqrt{2}} \sin\theta_c \cos\theta_c \left\{ \bar{p}_j \gamma_\mu (1-\gamma_5) p_i\, \bar{n}_i \gamma^\mu (1-\gamma_5) \lambda_j + \text{h. c.} \right\} \quad , \qquad (5.133b)$$

using the Fierz Identity which demonstrates the symmetry under the exchange of $\bar{n}_i$ and $\bar{p}_j$. Now suppose the quark fields $\bar{n}_i$, $\bar{p}_j$ create quarks in the same baryon. This baryon is, as we have explained, a colour singlet, so only the antisymmetric colour combination $\bar{n}_i \bar{p}_j - \bar{n}_j \bar{p}_i$ can contribute. But the above noted symmetry under $\bar{n}_i \leftrightarrow \bar{p}_j$ then ensures that only the antisymmetric classification SU(3) combination

$$(\bar{n}_i \bar{p}_j - \bar{n}_j \bar{p}_i) + (\bar{p}_j \bar{n}_i - \bar{p}_i \bar{n}_j)$$

$$= (\bar{n}_i \bar{p}_j - \bar{p}_i \bar{n}_j) + (\bar{p}_j \bar{n}_i - \bar{n}_j \bar{p}_i)$$

can contribute. Plainly such a combination has $I = 0$ and, since λ has $I = 0$, this ensures that only the $I = \frac{1}{2}$ piece of $\mathcal{H}_W^H$ can contribute when acting in this way. In fact the antisymmetric combinations above belong to the $\underline{3}$ representation of the classification SU(3). A similar argument shows that if p_i and λ_j destroy quarks in the same baryon, then only the $\underline{\bar{3}}$ combination can contribute. As a result we conclude that $\mathcal{H}_W^H$ behaves like an $\underline{8}$ $(\underline{3} \otimes \underline{\bar{3}} = \underline{1} + \underline{8})$ between single baryon states[18]. This is just what was needed to supply the predictive power missing from the soft pion treatment of the s-wave amplitudes. Of course, it does not solve the problem of the predicted magnitudes of the p-wave amplitudes, and for the argument to work one has to discount the possibility of quark-antiquark excitations. Nevertheless it is an attractive idea, particularly in view of the success of the quark model in the deep inelastic neutrino scattering processes discussed in Chapter 4.

5.2 HYPERCHARGE CONSERVING WEAK PROCESSES

The Hamiltonian $\mathcal{H}_W^H$ responsible for all purely hadronic weak processes contains a piece which conserves hypercharge, $\mathcal{H}_W^H(\Delta Y = 0)$ given in (5.5b), in addition to the $|\Delta Y| = 1$ piece considered in the previous section. The interactions generated by this part of $\mathcal{H}_W^H$ are distinguishable from the strong interactions only to the extent that they give rise to parity violation. Of course, the semileptonic processes already considered would certainly generate parity violating nonleptonic interactions by exchange of an electron neutrino pair; but these interactions are of order

G^2, and we are presently concerned with the possible existence of such processes in order G.

The isospin properties of the currents $J_\lambda^\pm$, $S_\lambda^\pm$ enable us to decompose $\mathcal{H}_W^H(\Delta Y = 0)$ into pieces with definite isospin

$$\mathcal{H}_W^H(\Delta Y = 0) = \cos^2\theta_c\,[\mathcal{H}_W^H(2) + \mathcal{H}_W^H(0)] + \sin^2\theta_c[\mathcal{H}_W^H(1) + \tilde{\mathcal{H}}_W^H(0)] \, . \quad (5.134)$$

The important thing to notice is that the $I = 1$ piece resides entirely in the term proportional to $\sin^2\theta_c$; this is because J_λ^+ and J_λ^- belong to the same isovector. We may also characterize the Hamiltonian in terms of its SU(3) properties.

As in (5.18), $\mathcal{H}_W^H(\Delta Y = 0)$ contains elements belonging to $\underline{8}$ and $\underline{27}$ representations. It also contains a singlet component. Proceeding as before we find

$$\begin{aligned}\mathcal{H}_W^H(\Delta Y=0) = & \sqrt{5}\cos^2\theta_c\,\mathcal{H}_W^H(\underline{27},0,2,0) + \sqrt{3}\sin^2\theta_c\,\mathcal{H}_W^H(\underline{27},0,1,0) + \\ & + \tfrac{1}{2}(3\sin^2\theta_c - \cos^2\theta_c)\mathcal{H}_W^H(\underline{27},0,0,0) - \frac{1}{\sqrt{2}}\sin^2\theta_c\,\mathcal{H}_W^H(\underline{8}_s,0,1,0) + \\ & + \frac{1}{\sqrt{6}}(\sin^2\theta_c - 2\cos^2\theta_c)\mathcal{H}_W^H(\underline{8}_s,0,0,0) + \mathcal{H}_W^H(\underline{1},0,0,0) \, , \end{aligned} \quad (5.135)$$

using the notation $\mathcal{H}_W^H(\underline{n}, Y, I, I_3)$ of (5.18).

As noted previously, the only realistic way of observing $\Delta Y = 0$ weak hadronic processes is by the measurement of parity violating effects in nuclear forces. Let us see how this arises. Consider two nucleons bound inside a nucleus. The existence of $\mathcal{H}_W^H(\Delta Y = 0)$ implies that these nucleons can exchange a boson in the way shown in Fig. 5.3.

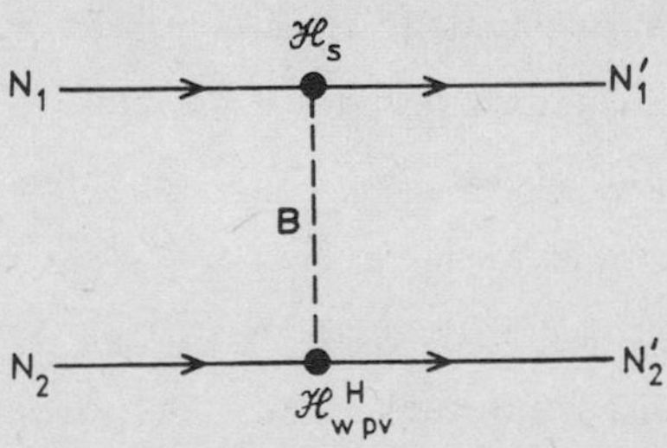

Fig. 5.3 Contribution to parity violation in a nucleus.

Note that one interaction is weak and parity violating, while the other is strong and (therefore) conserves parity. The largest contributions to the parity violating internucleon potential V_{pv} will arise from the exchange of the lowest mass bosons.

We consider first the contribution to V_{pv} from single pion exchange. To determine the parity violating vertex in Fig. 5.3 we need to know the matrix element

$$< N_2' \, \pi^a \, | \, \mathcal{H}^H_{W\,pv} \, | \, N_2 > \qquad (a = 1,2,3) \quad ,$$

where the nucleons N_1, N_2 are bound inside the nucleus. To discuss the isospin properties of this amplitude we may cross the nucleon N_2 and pion π^a.

The $N\bar{N}$ system may be decomposed into $I = 0, 1$ combinations

$$I = 0 \;:\; (\bar{N}N) \tag{5.136a}$$

$$I = 1 \;:\; (\bar{N}\tau^i N) \qquad i = 1,2,3, \tag{5.136b}$$

where N is the nucleons' isospinor, and τ^i the Pauli matrices. The $I = 0$ combination can only couple to π^o, because of charge conservation, but the C-invariance properties of $\mathcal{H}^H_{W\,pv}$ $(C = -1)$ forbid this

$$< (\bar{N}N) \, | \, \mathcal{H}^H_{W\,pv} \, | \, \pi^o > \; = \; - \; < (\bar{N}N) \, | \, \mathcal{H}^H_{W\,pv} \, | \, \pi^o > \; . \tag{5.137}$$

Similarly C-invariance requires

$$< (\bar{N}\tau^i N) \, | \, \mathcal{H}^H_{W\,pv} \, | \pi^a > \; = - \, \epsilon_i \, \epsilon_a < (\bar{N}\tau^i N) \, | \, \mathcal{H}^H_{W\,pv} \, | \pi^a > \; , \tag{5.138}$$

where $\epsilon_1 = - \epsilon_2 = \epsilon_3 = 1$. Thus $\epsilon_i = - \epsilon_a$, and this requires an anti-symmetric coupling of $\mathcal{H}^H_{W\,pv}$ and π^a. But only the $I = 1$ part of $\mathcal{H}^H_{W\,pv}$ can be coupled antisymmetrically to an isovector π^a to give another vector. So we conclude that <u>only</u> the $I = 1$ part $\mathcal{H}^H_W(1)$ of $\mathcal{H}^H_{W\,pv}$ can contribute to single pion exchange[(19)]. So the effective interaction of the pion and nucleons is proportional to $\bar{N}(\underline{\tau} \times \underline{\pi})N$. It is easy to verify that the $I^3 = 0$ component of this isovector has C-parity -1, while the other possible couplings all have C-parity +1. But we recall from (5.134) that $\mathcal{H}^H_W(1)$ occurs only in the piece of $\mathcal{H}^H_W$ that is multiplied by $\sin^2 \theta_c$. Thus under our hypothesis single pion exchange is damped by a factor $\sin^2\theta_c$ ($\sim$ 0.05), and is therefore much less important than it would otherwise have been.

This is an important feature of our hypothesis, so let us explore it a little further. There are four possible (non-derivative parity violating) couplings of the pion and nucleons

$$I = 0 : \bar{N}\underline{\tau}.\underline{\pi} N \propto \bar{n}p\,\pi^- + \bar{p}n\,\pi^+ + \frac{1}{\sqrt{2}}(\bar{p}p - \bar{n}n)\,\pi^o \quad , \tag{5.139a}$$

$$I = 1 : \bar{N}N\,\pi_3 \propto (\bar{p}p + \bar{n}n)\,\pi^o \quad , \tag{5.139b}$$

$$I = 1 : \bar{N}(\underline{\tau} \times \underline{\pi})_3 N \propto \bar{n}p\,\pi^- - \bar{p}n\,\pi^+ \quad , \tag{5.139c}$$

$$I = 2 : P^{ij}(\bar{N}\tau^i N)\,\pi^j \propto \bar{n}p\,\pi^- + \bar{p}n\,\pi^+ - \sqrt{2}\,(\bar{p}p - \bar{n}n)\,\pi^o \quad , \tag{5.139d}$$

where P^{ij} is the $(I, I^3) = (2,0)$ projection operator. Notice (as claimed) that under the operation of $\mathcal{C}$ only the second of the $I = 1$ combinations (5.139c) has the required C-parity -1. Notice further that under the charge symmetry operation

$$e^{i\pi v^2}|p> = |n> \qquad e^{i\pi v^2}|n> = -|p>$$

$$e^{i\pi v^2}|\bar{n}> = -|\bar{p}> \qquad e^{i\pi v^2}|\bar{p}> = |\bar{n}>$$

$$e^{i\pi v^2}|\pi^+> = -|\pi^-> \qquad e^{i\pi v^2}|\pi^o> = -|\pi^o> \qquad e^{i\pi v^2}|\pi^-> = -|\pi^+> \, ,$$

defined in (4.2.6), the $I = 0, 2$ combinations are invariant, while both of the $I = 1$ combinations change sign. Now recall that under this operation $J^{\pm} \rightarrow -J^{\mp}$ (see 4.247), so that the term $\cos^2\theta_c\{J^+, J^-\}$ in $\mathcal{H}_W^H$ is invariant. It therefore cannot give rise to <u>either</u> of the $I = 1$ couplings (5.139b, c). The only way the $\cos^2\theta_c$ term could contribute to these couplings would be if J^+ and J^- did not belong to the same isomultiplet. In principle, therefore, observation of V_{pv} provides a sensitive test of our hypothesis.

Since single pion exchange can arise only from pions and nucleons coupled as in (5.139c), we see that only charged pion exchange is allowed. Thus V_{pv} is determined by the processes

$$(n_-) : n \rightarrow p\pi^- \tag{5.140a}$$

$$(p_+) : p \rightarrow n\pi^+ \, , \tag{5.140b}$$

whose s-wave amplitudes satisfy

$$A(n_-) = - A(p_+) \; , \qquad (5.141)$$

as is apparent from (5.139c). We may estimate the size of these amplitudes by using SU(3) symmetry and the properties (5.135) of $\mathcal{H}_W^H(\Delta Y = 0)$; the SU(3) singlet part cannot contribute because of the C-invariance requirement of a $|\Delta \underline{I}| = 1$ rule. The $\underline{8}$ and $\underline{27}$ pieces are different components of the <u>same</u> representations as contributed in the hyperon decays. These new amplitudes can therefore also be written in terms of the five reduced amplitudes encountered there. We are thus able to express the $\Delta Y = 0$ amplitudes in terms of the $|\Delta Y| = 1$ amplitudes. We leave it as an exercise (which parallels that for the hyperon decays) to show that[20]

$$A(n_-) = - A(p_+) = \frac{1}{\sqrt{6}} \tan\theta_c \left\{ 4\,A(\Lambda^o_-) + 2\,A(\Xi^-_-) + A(\delta_\Lambda) - (\tfrac{3}{2})^{\frac{1}{2}} A(\delta_\Sigma) \right\} . \qquad (5.142)$$

Since $A(\delta_\Lambda)$ and $A(\delta_\Sigma)$ are experimentally found to be very small, to good approximation this gives

$$A(n_-) = - A(p_+) = (\tfrac{2}{3})^{\frac{1}{2}} \tan\theta_c \, [\, 2\,A(\Lambda^o_-) + A(\Xi^-_-) \,] \; . \qquad (5.143)$$

This expression can also be obtained by using the soft pion programme[21]. Note that, since the hyperon decays are proportional to $\sin\theta_c \cos\theta_c$, the amplitudes above are proportional to $\sin^2\theta_c$, as they have to be.

Because of this damping of the contribution $V_{pv}^{(\pi)}$ to V_{pv} from pion exchange, we must also consider the contributions from the exchange of higher mass bosons. CP-invariance again rules out the exchange of any neutral pseudoscalar such as the η or X^o mesons; the only possible combinations are $\bar{n}n\eta$ and $\bar{p}p\eta$, and both of these have C-parity +1, which implies they are forbidden. The same argument applies to X^o. The next contributions to worry about are the exchange of the vector mesons ρ, ω, φ.

The ω and φ mesons are neutral isoscalars with C-parity -1. Thus there are two possible (non-derivative parity violating) couplings to the nucleons

$$I = 0 : \; \bar{N}\gamma_\mu \gamma_5 N \omega^\mu \propto (\bar{p}\,\gamma_\mu \gamma_5\, p + \bar{n}\,\gamma_\mu \gamma_5\, n)\; \omega^\mu \qquad (5.144a)$$

$$I = 1 : \bar{N}\gamma_\mu\gamma_5\tau^3 N\omega^\mu \propto (\bar{p}\gamma_\mu\gamma_5 p - \bar{n}\gamma_\mu\gamma_5 n)\,\omega^\mu \;. \tag{5.144b}$$

The $\gamma_\mu\gamma_5$ is necessary to ensure that the resultant amplitude is a pseudoscalar and therefore parity violating. Both of these couplings have C-parity -1 as required; recall that the axial vector $\bar{p}\gamma_\mu\gamma_5 p$ has opposite C-parity to the vector $\bar{p}\gamma_\mu p$, which has C-parity -1. Under the charge symmetry operation the ω is invariant, since it is an isoscalar, so the $I = 0$ combination is invariant, while the $I = 1$ changes sign. Thus the $\cos^2\theta_c$ part of $\mathcal{H}_W^H$, which does have an $I = 0$ part, can contribute to ω-exchange; the $\sin^2\theta_c$ term, which is <u>not</u> an eigenstate of charge symmetry, can contribute to both $I = 0, 1$ combinations. Precisely analogous remarks apply to φ-exchange.

Finally, we consider ρ-exchange. The ρ is an isovector, like the pion but has intrinsic C-parity -1, i.e.

$$\mathcal{C}\,\rho_\mu^o\,\mathcal{C}^{-1} = -\rho_\mu^o \tag{5.145a}$$

$$\mathcal{C}\,\rho_\mu^\pm\,\mathcal{C}^{-1} = -\rho_\mu^\mp \;. \tag{5.145b}$$

The isospin analysis therefore follows (5.139) but the conclusions are reversed. The four possible couplings are

$$I = 0: \bar{N}\gamma_\mu\gamma_5 \underline{\tau}\cdot\underline{\rho}^\mu N \sim \bar{n}p\rho^- + \bar{p}n\rho^+ + \frac{1}{\sqrt{2}}(\bar{p}p - \bar{n}n)\,\rho^o \tag{5.146a}$$

$$I = 1: \bar{N}\gamma_\mu\gamma_5 N\rho^{3\mu} \sim (\bar{p}p + \bar{n}n)\,\rho^o \tag{5.146b}$$

$$I = 1: \bar{N}\gamma_\mu\gamma_5 (\underline{\tau}\times\underline{\rho}^\mu)_3 N \sim \bar{n}p\rho^- - \bar{p}n\rho^+ \tag{5.146c}$$

$$I = 2: P^{ij}\bar{N}\gamma^\mu\gamma_5\tau^i N\rho_\mu^j \sim \bar{n}p\rho^- + \bar{p}n\rho^+ - \sqrt{2}(\bar{p}p - \bar{n}n)\,\rho^o \;. \tag{5.146d}$$

Because the ρ has intrinsic C-parity opposite to that of the pion, our previous conclusions are reversed: with the exception of the $I = 1$ combination (5.146c) all of the above combinations have $C = -1$. On the other hand the properties under the charge symmetry operation are unaltered, so, as before, the $\cos^2\theta_c$ part of $\mathcal{H}_W^H(\Delta Y = 0)$ can contribute only to the $I = 0, 2$ combinations. Also the charged $\rho^\pm$ are involved only in the $I = 0, 2$ combinations to which this $\cos^2\theta_c$ term contributes.

The estimation of the magnitude of these vector meson matrix elements is much less clear cut than the pion amplitudes. The sophisticated techniques used are beyond the scope of this book. And in any case there is disagreement as to the result, which depends upon assumptions made about the Schwinger terms. What we can do is to estimate the size using the factorization method[22] mentioned earlier (cf. (5.125)). We only retain the $\cos^2\theta_c$ term, which is presumably the dominant contribution. Then

$$<p\rho^-|\mathcal{H}_W^H(\Delta Y=0)|n> \sim \frac{G}{\sqrt{2}}\cos^2\theta_c <p\rho^-|V_\lambda^+ A^{-\lambda} + A_\lambda^+ V^{-\lambda}|n>$$

$$\sim \frac{G}{\sqrt{2}}\cos^2\theta_c <p|A_\lambda^+|n><\rho^-|V^{-\lambda}|0> , \qquad (5.147)$$

since the ρ is a vector (rather than axial vector) particle. Now using isospin symmetry we may relate the ρ matrix element to an electromagnetic one

$$<\rho^-|V_\lambda^-|0> = <\rho^-|[v^1 - iv^2, j_\lambda]|0>$$

$$= \sqrt{2} <\rho^0|j_\lambda|0>$$

$$= \sqrt{2}\, m_\rho^2 f_\rho^{-1} \epsilon_\lambda ,$$

where ϵ_λ is the polarization of the ρ and f_ρ is a constant determined from experiments on $\rho \to e^+ e^-$

$$f_\rho^2/4\pi \simeq 2.4 .$$

The matrix element of A_λ^+ is of course determined by g_1^A, so this estimate yields for the invariant matrix element

$$M(n \to p\rho^-) = G\cos^2\theta_c f_\rho^{-1} m_\rho^2 g_1^A(0)\, \bar{p}\gamma_\lambda\gamma_5 n \epsilon^\lambda . \qquad (5.148)$$

In fact, this estimate coincides with one of the results of more sophisticated analyses[23]. The other treatment[24] yields the value $M \simeq 0$. In either case it is evident from (5.146) that

$$M(n \to p\rho^-) = M(p \to n\rho^+) . \qquad (5.149)$$

The above method does not allow us to calculate the contributions from the exchange of the neutral vector mesons ρ^0, ω, φ, because the weak currents in (5.5b) are charged. It is customary therefore to neglect these contributions.

Once the magnitude of the parity violating amplitudes is known, it is straightforward to calculate the corresponding parity violating potential V_{pv} [25]. The precise form of these potentials is not particularly instructive or even very important for the predictions about observable effects. What is important is the overall magnitudes of the pieces $V_{pv}^{(\pi)}$ and $V_{pv}^{(\rho)}$. We repeat that $V_{pv}^{(\pi)}$ is sensitive to the precise structure of $\mathcal{H}_W^H$ while $V_{vp}^{(\rho)}$ is not. However the actual magnitude of $V_{pv}^{(\rho)}$ is uncertain because of the theoretical problems.

With the inclusion of V_{pv} the total Hamiltonian has the form

$$H = H_o + V_{pv} \quad ,$$

where H_o is the usual nucleon Hamiltonian and is of course dominated by the strong internucleon interaction, which is parity conserving. The eigenstates ψ_i of H_o satisfy

$$H_o \psi_i = E_i \psi_i \qquad \text{(no summation)} \ .$$

Since V_{pv} is small, we may use perturbation theory to find the perturbed eigenstates Ψ_i of H

$$\Psi_i = \psi_i + \sum_{j \neq i} (E_i - E_j)^{-1} \langle j | V_{pv} | i \rangle \psi_j \ . \qquad (5.150)$$

The only states ψ_j contributing to the above sum have parity opposite to that of ψ_i since V_{pv} is a pseudoscalar. Symbolically we rewrite (5.150) as

$$\Psi_i = \psi_i + \mathcal{F} \varphi_i \quad ,$$

where φ_i has opposite parity to ψ_i and $\mathcal{F}$ is a measure of the parity impurity in the state Ψ_i; we expect $\mathcal{F} \simeq 10^{-6}$. There are two types of experiment to detect the presence of this parity violating admixture[26]. Firstly, one can look for a transition which is forbidden if $\mathcal{F} = 0$. For example, an energetically allowed α transition between 2^- and 0^+ states is forbidden if $\mathcal{F} = 0$. This is because the α carries away angular momentum $\ell = 2$ but conserves parity. If the 2^- has an admixture of 2^+, the decay is allowed with an amplitude proportional to $\mathcal{F}$. But then the rate is proportional to $\mathcal{F}^2$, which makes the effect difficult, but not impossible, to detect. Secondly, one can search for observable pseudoscalar quantities arising from interference between the (strong) parity conserving and (weak) parity violating pieces of the Hamiltonian; this is just the way that parity violation was originally discovered in β-decay. For example, one could look for asymmetry A_γ in the emission of γ-rays relative to the spin direction of a polarized nucleus. The presence of such a term signals parity violation, just as the electron asymmetry from polarised cobalt signalled it in the classical experiment. Alternatively, one can measure the circular polarization P_γ of γ-rays emitted from unpolarized nuclei. The important thing is that the effect is caused by interference, and is therefore of order $\mathscr{F}$ rather than $\mathscr{F}^2$. We shall not repeat here the processes by which the parity violating potential V_{pv} is converted into predictions for the experimentally observed quantities. The interested reader is referred to Blin-Stoyle[27], and references cited there.

Instead we describe briefly the experimental situation. This has been fully reviewed recently by Gari[28]. Experiments of the first type described above have all been directed to the search for an α-transition between the 2^- state of ^{16}O at 8.88 MeV and the 0^+ ground state of ^{12}C. Parity impurities in the 2^- ^{16}O state arise primarily from the nearby 2^+ states at 6.9 MeV, 9.8 MeV and 11.5 MeV. Other more distant states and parity impurities in the 0^+ ^{12}C state are expected to be negligible. Since all of these states have $I = 0$, only the part of V_{pv} having $I = 0$ can contribute. Thus this experiment is essentially a measure of $V_{pv}^{(\rho)}$, since $V_{pv}^{(\pi)}$ has $I = 1$, as we have seen. Using the value value[23] (5.148) leads to a predicted decay width[29] of around 10^{-10} eV. A number of experiments have sought this decay but only one[30] has a positive result - a decay width of $(1.0 \pm 0.2)10^{-10}$ eV. Plainly this provides strong evidence of parity mixing of nuclear states, but independent verifications would be most welcome. At face value this experiment strongly indicates that $V_{pv}^{(\rho)}$ is non-zero, and technique refinements might enable this method to give information on the isospin structure of $\mathcal{H}_W^H(\Delta Y = 0)$.

Experiments of the second type have been performed on at least 17 nuclei, and many of these yield non-zero measurements of A_γ and P_γ. The idea behind all of these is to select states for which the "regular" transition

$\psi_1 \rightarrow \psi_2$ is hindered for some reason. The observed asymmetry or circular polarization is determined by the ratio of the amplitudes of the "irregular" transitions $\varphi_1 \rightarrow \psi_2$, $\psi_1 \rightarrow \varphi_2$ and the regular $\psi_1 \rightarrow \psi_2$. Thus by selecting a hindered regular transition the observed effect is enhanced. The largest effect so far observed is in the 501 keV line of ^{180}Hf, in which the observed circular polarization $P_\gamma \sim 10^{-3}$ and all experiments are consistent. Unfortunately the theoretical difficulties in estimating the effect for this particular transition make it unsuitable for testing the consistency of theory with experiment. The most popular transition is the 482 keV line of ^{181}Ta; the regular M1 transition is hindered by a factor of 3×10^{-6}, and the E2 by a factor of 35. The successful experimental measurements of P_γ divide into two classes, each of which are consistent within themselves, but which are mutually inconsistent, although they agree that P_γ is negative. Both $V_{pv}^{(\pi)}$ and $V_{pv}^{(\rho)}$ contribute to this transition. Although the sign of $V_{pv}^{(\pi)}$ is not known, the resulting theoretical estimate is not crucially affected.(27) Our hypothesis leads to a predicted P_γ of $(2-3)\ 10^{-5}$, which agrees with the group of larger magnitude observations. One of the major problems facing the theorist in calculating the observable effects of parity nonconservation is the lack of information on nuclear wave functions. For this reason data obtained from few-nucleon systems are particularly valuable. In the case of the process $np \rightarrow d\gamma$ this is even more true; Danilov(31) has shown that for unpolarized neutrons P_γ depends only on $V_{pv}^{(\rho)}$, while the asymmetry A_γ with respect to the captured neutron's spin polarization is determined by $V_{pv}^{(\pi)}$. So in principle the relative magnitude of these two important contributions may be measured. However, the predicted effects turn out to be of order 10^{-7} to 10^{-9}, which are too small to be observed at present.

The problems facing both theoretical and experimental physicists in this area are truly formidable. It is therefore not surprising that we are at present unable to draw firm conclusions as to the consistency of our hypothesis with the experimental data. Nevertheless, it is probably true to say that the large parity violating effects presently being observed leave most theorists with some doubt that the hypothesis we have advanced will in the end fit the data.

Chapter 6

UNIFIED GAUGE THEORIES OF WEAK AND ELECTROMAGNETIC INTERACTIONS

6.1 INTRODUCTION

We have seen in Chapter 3 that the current-current hypothesis (3.1) leads to the prediction that the cross section for neutrino-lepton scattering increases linearly with energy in the laboratory frame. In general we have

$$\sigma(\nu_\ell \ell \to \nu_\ell \ell) = \frac{G^2}{\pi} W^2 , \tag{6.1}$$

where W is the Lorentz invariant total c.m. energy. In the approximation which we have made of neglecting the lepton mass, this process is purely $J = 0$ in the centre-of-mass frame, since both ν_ℓ and ℓ have helicity -1. The unitarity bound for such a process is $4\pi W^{-2}$. Thus the cross section (6.1) cannot possibly be correct when it exceeds this limit. This occurs when

$$W^2 > W_c^2 \equiv 2\pi G^{-1} . \tag{6.2}$$

Using the value (3.63) for G, we find that the critical centre-of-mass energy is given by

$$W_c \simeq 350 \text{ GeV} , \tag{6.3}$$

at which energy the cross section has the value

$$\sigma(\nu_\ell \ell \to \nu_\ell \ell) = 2G \sim 10^{-31} \text{ cm}^2 . \tag{6.4}$$

Thus beyond 350 GeV lowest order perturbation theory, which we used to derive (6.1), cannot give an adequate description of weak interactions. Evidently we must include higher order contributions; at the energy in question, $GW^2 \sim 1$, from (6.2), so all higher order contributions are of comparable size, as is the case for strong interactions.

One may feel that, since presently available energies are so far removed from this critical value, we do not need to worry yet about these higher order terms; this, indeed, is what we have assumed hitherto. The imp'ication has been that, if we had calculated the order G^2 contributions to the weak matrix elements, then they would have been small compared with the leading order G term, at least for the energies we have been considering. However, as Heisenberg[1] observed, the higher order corrections to $\nu_\ell - \ell$ scattering, for example, diverge. The order G^2 contribution is quadratically divergent, as can easily be seen by considering the Feynman diagram in Fig. 6.1.

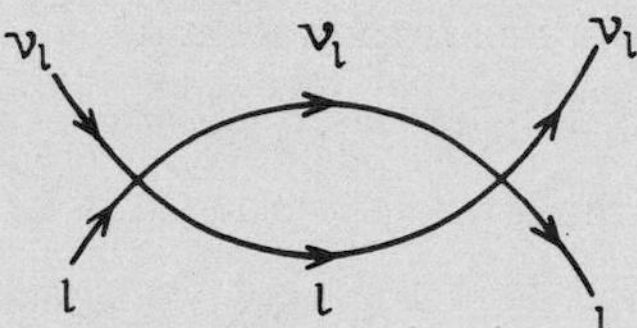

Fig. 6.1 Order G^2 contribution to $\nu_\ell - \ell$ scattering.

The closed loop implies we must integrate over the loop four-momentum k; each of the fermion propagators contributes essentially one inverse power of k to the integral, so we are left with an integral looking like

$$\int d^4k \, (k^2)^{-1}$$

for large k. Clearly this is quadratically divergent. If we cut off the integral at large k with a cut-off Λ, as described in 4.2.12, we shall obtain a contribution to the matrix element proportional to $G^2\Lambda^2$; what is happening is that the extra power of G is being balanced by a factor Λ^2 to keep the dimensions right. This, of course, is just the argument we used to understand the linear energy dependence of the lowest order matrix element. So the unitarity problem and the divergence problem are intimately related.

But do we have to worry about these problems ? After all, it is well known that, if the perturbation series is terminated at any finite order, the theory is non-unitary whatever the interaction. And of course the radiative corrections in QED are divergent. The answer is "unfortunately, yes".

The QED divergences are only logarithmic, i.e. proportional to $\ln \frac{\Lambda}{m}$ to some power; this also is obvious on dimensional grounds - the fine structure constant α is dimensionless, so higher powers are at most logarithmically divergent. These logarithmic divergences may be collected together and the theory reformulated in terms of finite "renormalized" electron charge and mass. Unfortunately, the current-current weak interaction theory is not renormalizable. The order G^n contributions diverge as $\left(\Lambda^2\right)^{n-1}$, and the upshot is that we have to introduce more and more renormalized quantities at each order. Consequently we just do not know how to calculate with this theory, and the neglect of higher order contributions is pure faith on our part.

One might wonder whether the formulation of our theory in terms of an intermediate vector boson W, as was done in section 3.3, might solve some of the problems to which we have alluded. After all, the theory then looks quite like QED. Instead of Fig. 6.1 we should have the order f^4 diagram shown in Fig. 6.2.

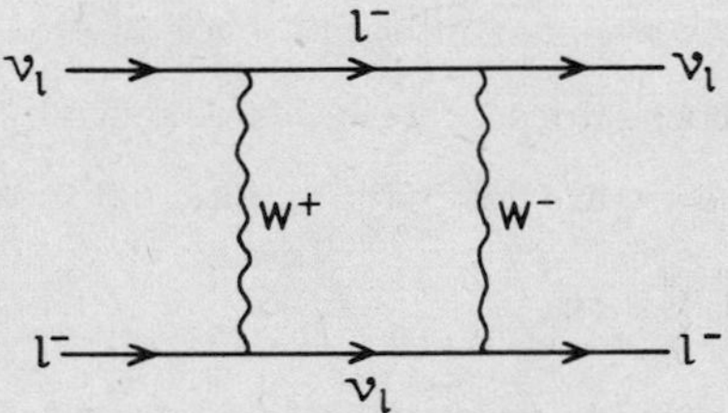

Fig. 6.2 Order f^4 contribution to $\nu_\ell - \ell$ scattering.

It is easy to see that this diagram is quadratically divergent, just as that in Fig. 6.1. The reason is that we get no extra convergence from the W-propagators; as shown in (3.103), each of them contributes something like

$$(g_{\alpha\beta} - k_\alpha k_\beta / m_w^2)\,(k^2 - m_w^2)^{-1} \quad , \tag{6.5}$$

and we see that both numerator and denominator have two powers of k. The analogous diagram in QED is shown in Fig. 6.3. This diagram is convergent, since each of the photon propagators contributes something like

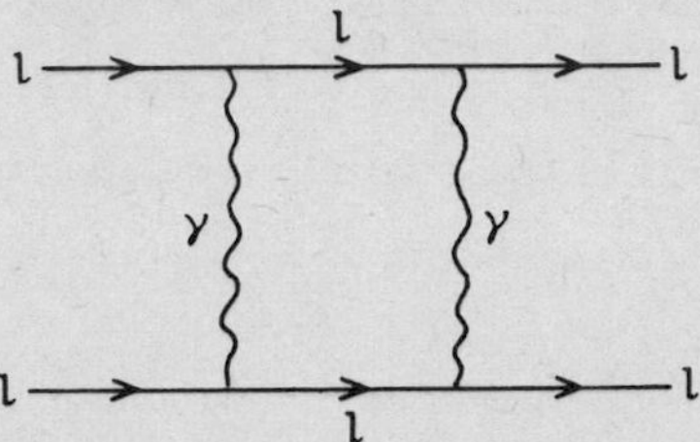

Fig. 6.3 Order e^4 contribution to ℓ-ℓ scattering.

$$\frac{g_{\alpha\beta}}{k^2} \tag{6.6}$$

in the Feynman gauge. Overall we have an integral like

$$\int d^4k \ (k^2)^{-3} \ ,$$

for large k, which is convergent. Evidently the difference between the two theories arises from the $k_\alpha k_\beta / m_W^2$ in the numerator of the W-propagator. Eventually these terms, arising from the longitudinal polarization vectors, yield the anticipated quadratic divergence proportional to $f^4 \Lambda^2 m_W^{-2}$, again, essentially from dimensional considerations, once it is recognized that the m_W^{-4} terms are present. We can always add terms proportional to $k_\alpha k_\beta$ in the numerator of the <u>photon</u> propagator, but gauge invariance ensures that they cannot contribute. Thus, if we want a renormalizable weak theory, this suggests that we should start from a gauge invariant Lagrangian involving charged vector bosons.

Before doing this, let us first review the gauge invariance of QED. We denote the electron's field by ψ and the photon's by A_μ. A U(1) gauge group transformation is characterized by a single gauge function $\Lambda(x)$; we write

$$\psi'(x) = \exp[ie\,\Lambda(x)]\,\psi(x) \ , \tag{6.7}$$

so, if Λ is infinitesimal, we have

$$\psi'(x) \simeq [1 + ie\,\Lambda(x)]\,\psi(x) \ , \tag{6.8}$$

neglecting $O(\Lambda^2)$. The electromagnetic field transforms as follows:

$$A'_\mu(x) = A_\mu(x) + \partial_\mu \Lambda(x) \quad . \tag{6.9}$$

Then it follows trivially that

$$[\partial_\mu - ie A'_\mu(x)]\, \psi'(x) \simeq [1 + ie\Lambda(x)][\partial_\mu - ie A_\mu(x)]\, \psi(x) \quad . \tag{6.10}$$

Now

$$\bar{\psi}'(x) \simeq \bar{\psi}(x)\, [1 - ie\Lambda(x)] \quad , \tag{6.11}$$

so

$$\bar{\psi}'(x)\, \gamma^\mu \left(\partial_\mu - ie\, A'_\mu(x)\right) \psi'(x) = \bar{\psi}(x)\, \gamma^\mu \left(\partial_\mu - ie\, A_\mu(x)\right) \psi(x) \quad , \tag{6.12}$$

and $\bar{\psi}\gamma^\mu(\partial_\mu - ie A_\mu)\psi$ is gauge <u>invariant</u>.

It is trivial to see that $\bar{\psi}\psi$ and $F_{\mu\nu} \equiv \partial_\mu A_\nu - \partial_\nu A_\mu$ are both gauge invariant, so

$$\mathcal{L} = -\tfrac{1}{4} F^{\mu\nu} F_{\mu\nu} + i\psi\gamma^\mu(\partial_\mu - ie A_\mu)\psi - m\bar{\psi}\psi \tag{6.13}$$

is also gauge invariant. This is just the QED Lagrangian, whose interaction part we wrote down in (2.220).

6.2 GAUGE INVARIANT THEORY FOR WEAK AND ELECTROMAGNETIC INTERACTIONS

We shall now do something analogous for the weak interactions. We suppose, initially, that the only leptonic fields are the known ones; we shall work with e and ν_e-everything can be paralleled for the muonic leptons. First define a doublet of fields, as in 4.1.5 ,

$$L \equiv \begin{pmatrix} a\,\nu_e \\ a\,e \end{pmatrix} \quad , \tag{6.14}$$

where as before $a = \frac{1}{2}(1 - \gamma_5)$ is the left-handed projection operator. Using the 2 x 2 Pauli matrices τ_i (i = 1,2,3), it is easy to write down the leptonic weak current in terms of L. Defining

$$\tau_\pm = \tfrac{1}{2}(\tau_1 \pm i\,\tau_2) \quad ,$$

$$\tau_+ = \begin{pmatrix} 0 & 1 \\ 0 & 0 \end{pmatrix} \quad ,$$

so

$$\tau_+ L = \begin{pmatrix} a\,e \\ 0 \end{pmatrix} \quad .$$

Now

$\bar{L} = (\bar{\nu}_e a', \bar{e} a')$ and $a' \equiv \frac{1}{2}(1+\gamma_5)$.

Thus

$$\bar{L}\gamma_\lambda \tau_+ L = \bar{\nu}_e a' \gamma_\lambda a e \tag{6.15a}$$

$$= \bar{\nu}_e \gamma_\lambda \tfrac{1}{2}(1-\gamma_5) e . \tag{6.15b}$$

The hermitian conjugate is

$$\bar{e}\gamma_\lambda \tfrac{1}{2}(1-\gamma_5)\nu_e = \bar{L}\gamma_\lambda \tau_- L . \tag{6.16}$$

The simplest group containing these two currents is obviously $SU(2)_L$ - leptonic isospin. We could write down a $SU(2)_L$ gauge invariant Lagrangian coupling the three currents $\bar{L}\gamma_\lambda \tau_i L$ to gauge vector bosons A_λ^i . This would involve the neutral current

$$\bar{L}\gamma_\lambda \tau_3 L = \bar{\nu}_e \gamma_\lambda a \nu_e - \bar{e}\gamma_\lambda a e . \tag{6.17}$$

Plainly this is not the electromagnetic current, since it is parity violating and involves the neutrino. If we insist on having the electromagnetic current within the theory, we must at some stage involve the right-hand component of the electron's field

$$R = a' e . \tag{6.18}$$

Now,

$$\bar{L}\gamma_\lambda I_2 L = \bar{\nu}_e \gamma_\lambda a \nu_e + \bar{e}\gamma_\lambda a e \tag{6.19a}$$

$$\bar{R}\gamma_\lambda R = \bar{e}\gamma_\lambda a' e , \tag{6.19b}$$

hence

$$\bar{e}\gamma_\lambda e = -\tfrac{1}{2}\bar{L}\gamma_\lambda \tau_3 L + \{\tfrac{1}{2}\bar{L}\gamma_\lambda I_2 L + \bar{R}\gamma_\lambda R\} . \tag{6.20}$$

Thus the simplest way to ensure the electromagnetic current is contained in the theory is to associate the current { } with a one parameter group U(1) - leptonic hypercharge. Thus we are led to write down a Lagrangian which is invariant under an $SU(2)_L \times U(1)$ gauge transformation. We denote the three gauge fields associated with $SU(2)_L$ by $\underline{A}_\mu$, and that associated with U(1) by B_μ. Then under an infinitesimal $SU(2)_L \times U(1)$ gauge transformation the fields transform as

$$L' = (1 - i g \underline{\Lambda}.\underline{T} + i \ell \Lambda) \, L \,, \tag{6.21}$$

where $\underline{\Lambda}$, Λ are the associated gauge functions and g, ℓ are constants. The matrices $T^i = \frac{1}{2}\tau_i$ constitute the two-dimensional representation of $SU(2)_L$. The remaining fields transform as

$$R' = (1 + ir\,\Lambda)\, R \tag{6.22a}$$

$$\underline{A}'_\mu = \underline{A}_\mu + \partial_\mu \underline{\Lambda} + g\,\underline{\Lambda} \times \underline{A}_\mu \tag{6.22b}$$

$$B'_\mu = B_\mu + \partial_\mu \Lambda \,. \tag{6.22c}$$

The form for $\underline{A}'_\mu$ follows because the matrices $\underline{t}$, constituting the regular (three-dimensional) representation of the group, are given by

$$(t^i)_{jk} = i\,\epsilon_{jik} \,. \tag{6.23}$$

It then follows, as before, that the following forms are $SU(2)_L \times U(1)$ gauge invariant:

$$\overline{L}\,\gamma^\mu (\partial_\mu + i g\,\underline{T}.\underline{A}_\mu - i \ell B_\mu)\, L \tag{6.24a}$$

$$\overline{R}\,\gamma^\mu (\partial_\mu \qquad\qquad - i r B_\mu)\, R \tag{6.24b}$$

$$B_{\mu\nu} \equiv \partial_\mu B_\nu - \partial_\nu B_\mu \,. \tag{6.24c}$$

Further, if we define

$$\underline{A}_{\mu\nu} \equiv \partial_\mu \underline{A}_\nu - \partial_\nu \underline{A}_\mu - g\,\underline{A}_\mu \times \underline{A}_\nu \,, \tag{6.25a}$$

then

$$\underline{A}'_{\mu\nu} = \underline{A}_{\mu\nu} + g\,\underline{\Lambda} \times \underline{A}_{\mu\nu} \,. \tag{6.25b}$$

Thus $\underline{A}_{\mu\nu} . \underline{A}^{\mu\nu}$ is also gauge invariant, and so is the Lagrangian

$$\mathscr{L}_{(1)} = -\tfrac{1}{4}\,\underline{A}_{\mu\nu}.\underline{A}^{\mu\nu} - \tfrac{1}{4}\,B_{\mu\nu}B^{\mu\nu} + i\overline{R}\,\gamma^\mu(\partial_\mu - irB_\mu)R \, +$$

$$+ \, i\overline{L}\,\gamma^\mu(\partial_\mu + i g\,\underline{T}.\underline{A}_\mu - i\ell B_\mu)\, L \,. \tag{6.26}$$

Now, if we demand that B_μ is coupled to the current $\{\cdot\}$ given in (6.20), then we must have

$$\ell = \tfrac{1}{2} r \equiv \tfrac{1}{2} g' \,. \tag{6.27}$$

This will ensure that the electromagnetic current is coupled by the theory, as are the two charged weak currents, by design. Plainly, however, there are two neutral currents within the theory. The neutrino field is seen to be coupled to the following linear combination of A_μ^3 and B_μ

$$Z_\mu = (g A_\mu^3 - g' B_\mu)(g^2 + g'^2)^{-\frac{1}{2}} \tag{6.28a}$$

$$= \cos\theta_W \, A_\mu^3 - \sin\theta_W \, B_\mu \, , \tag{6.28b}$$

where

$$g'/g \equiv \tan\theta_W \, . \tag{6.28c}$$

The orthogonal combination of the fields is (by construction) coupled to the electromagnetic current, so we define

$$A_\mu \equiv \sin\theta_W \, A_\mu^3 + \cos\theta_W \, B_\mu \, . \tag{6.29}$$

Inverting these we obtain

$$A_\mu^3 = \cos\theta_W \, Z_\mu + \sin\theta_W \, A_\mu \tag{6.30a}$$

$$B_\mu = -\sin\theta_W \, Z_\mu + \cos\theta_W \, A_\mu \, . \tag{6.30b}$$

Then, by inspection, the coupling of A_μ to the leptons in $\mathcal{L}_{(1)}$ is

$$A^\mu \left[-\tfrac{1}{2} g \sin\theta_W \, \overline{L} \gamma_\mu \tau_3 L + g' \cos\theta_W \, (\overline{R} \gamma_\mu R + \tfrac{1}{2} \overline{L} \gamma_\mu I_2 L) \right] \, .$$

To get the charge of the electron correct, we must take (see (6.20))

$$g \sin\theta_W = g' \cos\theta_W = e \, . \tag{6.31}$$

Thus we have constructed a gauge invariant Lagrangian $\mathcal{L}_{(1)}$, which economically incorporates both the weak and electromagnetic currents of the leptons. Of course, we have one additional neutral weak current, which was not sought, but which was unavoidable. Despite these positive features, $\mathcal{L}_{(1)}$ is plainly not a realistic Lagrangian. There are no mass terms either for the leptons or the gauge fields. The terms missing, $\underline{A}_\mu \cdot \underline{A}_\mu$, $B_\mu B^\mu$, $\overline{L} R$ are <u>not</u> gauge invariant, so the gauge invariance is true only for massless leptons and gauge bosons. In fact this masslessness, which arises from the gauge invariance, is sufficient to ensure the renormalizability of the theory, just like QED. The problem remaining, then, is to generate the required mass terms without destroying the renormalizability of the theory.

6.3 SPONTANEOUSLY BROKEN GAUGE THEORIES

The technique used to generate the masses is known as the Higgs-Kibble mechanism[(2)]. We illustrate its workings first on a similar but simpler gauge model, the Higgs Model. Consider the following renormalizable theory

$$\mathcal{L} = -\tfrac{1}{4} F_{\mu\nu} F^{\mu\nu} + |(\partial_\mu - ie A_\mu)\varphi|^2 - \mu^2 |\varphi|^2 - h|\varphi|^4 \quad , \tag{6.32}$$

where $F_{\mu\nu} \equiv \partial_\mu A_\nu - \partial_\nu A_\mu$ and φ is a complex scalar field. We take $h > 0$, so that the corresponding classical Hamiltonian (with φ as a coordinate) is bounded below. $\mathcal{L}$ is invariant under the gauge transformation

$$A'_\mu = A_\mu + \partial_\mu \Lambda \ , \quad \varphi' = \exp(ie\Lambda)\,\varphi \ . \tag{6.33}$$

Now suppose φ has a non-zero vacuum expectation value:

$$<0|\varphi|0> \ \equiv \ <\varphi>_o = \frac{1}{\sqrt{2}} \lambda \ . \tag{6.34}$$

Plainly we should define new fields which <u>do</u> have vanishing vacuum expectation values, so we write

$$\varphi = \frac{1}{\sqrt{2}} (\lambda + \varphi_1 + i\varphi_2) \ , \tag{6.35}$$

where φ_i are real scalar fields and $<\varphi_i>_o = 0$. Equivalently, we may work with the fields θ, χ, where

$$\varphi = \frac{1}{\sqrt{2}} (\lambda + \chi) \ \exp(i\theta/\lambda) \tag{6.36a}$$

and

$$<\chi>_o = 0 = <\theta>_o \ . \tag{6.36b}$$

The fact that we <u>can</u> define <u>fields</u> χ, θ in this way needs a little elaboration. Expanding the exponential above gives

$$\varphi = \frac{1}{\sqrt{2}} (\lambda + \chi + i\theta + \dots) \ , \tag{6.37}$$

where ... indicates quadratic and higher powers of θ, χ. The λ in the exponent must be the same as that in $<\varphi>_o$, so that θ occurs with coefficient i, as above. This is important, because comparing (6.37) and (6.35) we see that

$$\varphi_1 = \chi + \cdots$$

$$\varphi_2 = \theta + \cdots \quad .$$

These can be inverted to give

$$\chi = \varphi_1 + \cdots$$

$$\theta = \varphi_2 + \cdots \quad ,$$

so that χ and θ are properly defined fields satisfying the correct canonical commutation relations.

Under the gauge transformation

$$\chi' = \chi \qquad \text{and} \qquad \theta' = \theta + e\lambda\Lambda \ . \tag{6.38}$$

Suppose we define a new field

$$B_\mu = A_\mu - (e\lambda)^{-1}\partial_\mu\theta \ . \tag{6.39}$$

Then

$$\begin{aligned} B'_\mu &= A'_\mu - (e\lambda)^{-1}\partial_\mu\theta' \\ &= A_\mu + \partial_\mu\Lambda - (e\lambda)^{-1}\partial_\mu\theta - \partial_\mu\Lambda \\ &= B_\mu \ . \end{aligned} \tag{6.40}$$

So B_μ is a gauge invariant field. When we rewrite $\mathcal{L}$ in terms of the two gauge invariant fields B_μ, χ and θ, we find that θ does not enter; explicitly, defining $f_{\mu\nu} \equiv \partial_\mu B_\nu - \partial_\nu B_\mu$, we find

$$\begin{aligned} \mathcal{L} = & \left[-\tfrac{1}{4} f_{\mu\nu} f^{\mu\nu} + \tfrac{1}{2} e^2\lambda^2 B_\mu B^\mu\right] + \left[\tfrac{1}{2}(\partial_\mu\chi)^2 - \tfrac{1}{2}(\mu^2 + 3h\lambda^2)\chi^2\right] + \\ & + \tfrac{1}{2} e^2 B_\mu^2 (\chi^2 + 2\lambda\chi) - \tfrac{1}{4} h(\chi^4 + 4\lambda\chi^3) - \lambda(\mu^2 + h\lambda^2)\chi + \\ & + \text{constant} \ . \end{aligned} \tag{6.41}$$

The term linear in χ must be absent if we require $\langle\chi\rangle_o = 0$. This occurs if

$$\lambda = 0 \tag{6.42a}$$

or

$$\mu^2 + h\lambda^2 = 0 \ . \tag{6.42b}$$

The first possibility is the conventional situation. The second possibility is what concerns us. For this to be true we must have $\mu^2 < 0$, since

$h\lambda^2 > 0$, and then

$$\lambda = (-\mu^2 h^{-1})^{\frac{1}{2}} . \qquad (6.43)$$

Such a choice for λ has a simple interpretation in classical mechanics. The classical potential corresponding to the Lagrangian (6.32) is

$$V(x) = \mu^2 x^2 + hx^4 .$$

So the stationary points are given by

$$V'(x) = 0$$

i.e.

$$x = 0 \quad \text{and} \quad x = \frac{1}{\sqrt{2}}(-\mu^2 h^{-1})^{\frac{1}{2}} .$$

Evaluating the second derivative gives

$$V''(0) = 2\mu^2$$

and

$$V''\left(\sqrt{\frac{-\mu^2}{2h}}\right) = -4\mu^2 .$$

Thus in the conventional case $(\mu^2 > 0)$ the origin is the stable equilibrium point. On the other hand, if $\mu^2 < 0$, the equilibrium point is shifted to $x^2 = -\mu^2/2h$, and we should choose normal coordinates which vanish at <u>this</u> point. At any rate, with the choice (6.43) we obtain

$$\mathcal{L} = \left[-\tfrac{1}{4} f_{\mu\nu} f^{\mu\nu} + \tfrac{1}{2}e^2\lambda^2 B_\mu^2\right] + \tfrac{1}{2}\left[(\partial_\mu \chi)^2 + 2\mu^2\chi^2\right] + $$

$$+ \tfrac{1}{2}e^2 B_\mu^2 (\chi^2 + 2\lambda\chi) - \tfrac{1}{4}h(\chi^4 + 4\lambda\chi^3) + \text{constant.} \qquad (6.44)$$

Thus we now have a Lagrangian describing a vector field B_μ, having mass $e\lambda$, and a real scalar field χ, of mass $(-2\mu^2)^{\frac{1}{2}}$. No gauge symmetry remains, since B_μ and χ are gauge <u>invariant</u> fields, but the relic survives in the trilinear and quadrilinear couplings of the fields; instead of having arbitrary couplings they are related. Thus by breaking the symmetry $(\lambda \neq 0)$ we eliminate one field, and the extra degree of freedom is taken up by the vector field acquiring a non-zero mass; a massive spin-one particle has <u>three</u> polarization states, rather than the <u>two</u> possessed by a massless vector particle. But it is precisely this longitudinal polarization vector which gives the $k_\alpha k_\beta/m_W^2$ term in the W-propagator, and which in turn leads to the non-renormalizability of such field theories usually.

However the field theory (6.44), with the relationships between masses and coupling constants, is plainly no 'usual' theory. In fact it is renormalizable, although the proof of this is tricky. Instead of eliminating the gauge degree of freedom φ_2 in (6.35), as we have done, it is retained. The Lagrangian is thus manifestly renormalizable, as it has only massless vector particles, like the photon. However, the theory is not obviously unitary, as the Feynman rules have poles at $k^2 = 0$ corresponding to the unphysical particle φ_2. The proof[3] thus consists of showing that these $k^2 = 0$ poles do not appear in any physical amplitude.

At any rate, if we assume in general that spontaneously breaking the symmetry of a gauge invariant theory does not destroy its renormalizability, then the way forward in the problem which concerns us is reasonably clear. Evidently we have to break the $SU(2)_L \times U(1)$ symmetry in a way which gives the weak vector bosons a mass. However, the symmetry breaking must preserve the $U(1)$ symmetry, associated with the conservation of charge, so that the photon stays massless. To do this we must introduce some scalar fields, and to break the $SU(2)_L$ we take a doublet

$$\varphi = \begin{pmatrix} \varphi^+ \\ \varphi^o \end{pmatrix}. \tag{6.45}$$

Under the $SU(2)_L \times U(1)$ gauge transformation

$$\varphi' = (1 - ig\underline{\Lambda}.\underline{T} + if\Lambda)\varphi \ , \quad T^i = \tfrac{1}{2}\tau^i \ , \tag{6.46}$$

and, as before, it is easy to verify that $|(\partial_\mu + ig\underline{A}_\mu.\underline{T} - if B_\mu)\varphi|^2$ and $|\varphi|^2$ are gauge invariant $(|\varphi|^2 \equiv \varphi^+\varphi)$. Also, $\overline{L}\varphi R$ is gauge invariant, provided $f = -\frac{1}{2}g'$. Thus

$$\mathcal{L}_\varphi \equiv |(\partial_\mu + ig\underline{A}_\mu.\underline{T} + \tfrac{1}{2}ig'B_\mu)\varphi|^2 - G_e(\overline{L}\varphi R + h.c.) - $$
$$- \mu^2|\varphi|^2 - h(|\varphi|^2)^2 \tag{6.47}$$

is also gauge invariant. So

$$\mathcal{L}_{(2)} \equiv \mathcal{L}_{(1)} + \mathcal{L}_\varphi \tag{6.48}$$

is gauge invariant and still renormalizable, since φ^4 theory is. Now we shall break the gauge symmetry by giving at least one of the scalar fields a non-zero vacuum expectation value. Plainly we must take

$$< \varphi^o >_o = \frac{1}{\sqrt{2}} \lambda \tag{6.49a}$$

but

$$< \varphi^+ >_o = 0 \ . \tag{6.49b}$$

Otherwise we should violate charge conservation. As before, we may use the gauge symmetry to transform away the unphysical fields, and the Lagrangian describing the theory in this unitary gauge is given by making the replacement

$$\begin{pmatrix} \varphi^+ \\ \varphi^o \end{pmatrix} \rightarrow \begin{pmatrix} 0 \\ \frac{1}{\sqrt{2}} (\lambda + \varphi) \end{pmatrix}, \tag{6.50}$$

where φ is a real scalar field. With this substitution

$$\mathcal{L}_\varphi = \tfrac{1}{2} (\partial_\mu \varphi)^2 + \frac{1}{8} g^2 (\lambda + \varphi)^2 |A^1_\mu - i A^2_\mu|^2 + \frac{1}{8} (g^2 + g'^2) Z^2_\mu (\lambda + \varphi)^2 -$$

$$- \frac{G_e}{\sqrt{2}} (\bar{e} e) (\lambda + \varphi) - \tfrac{1}{2} \mu^2 (\lambda + \varphi)^2 - \tfrac{1}{4} h (\lambda + \varphi)^4 \ , \tag{6.51}$$

using the definition (6.28) for the field Z_μ.

As before, we may ignore the constant term, and the term linear in φ is absent if $\mu^2 + h \lambda^2 = 0$; in that case $m^2_\varphi = -2\mu^2 > 0$. If we identify the field operator of the W-boson as

$$W^+_\mu = \frac{1}{\sqrt{2}} (A^1_\mu - i A^2_\mu) \ , \tag{6.52}$$

we see that $\mathcal{L}_\varphi$, given in (6.51), contains mass terms for the W and Z, as anticipated. There is no term proportional to $A_\mu A^\mu$, so the photon stays massless, as required. By inspection we see that

$$m^2_W = \tfrac{1}{4} g^2 \lambda^2 \tag{6.53a}$$

$$m^2_Z = \tfrac{1}{4} (g^2 + g'^2) \lambda^2 \tag{6.53b}$$

$$m_e = \frac{1}{\sqrt{2}} G_e \lambda \ . \tag{6.53c}$$

We may now write out $\mathcal{L}_{(2)}$ in detail to obtain the following interaction Lagrangian for our theory

$$\mathcal{L}_I = -\tfrac{1}{4}\left\{g^2(\underline{A}_\mu \times \underline{A}_\nu)\cdot(\underline{A}^\mu \times \underline{A}^\nu) - 2g(\underline{A}_\mu \times \underline{A}_\nu)\cdot(\partial^\mu \underline{A}^\nu - \partial^\nu \underline{A}^\mu)\right\} +$$

$$+ \; e(\bar{e}\gamma^\mu e)A_\mu + e\tan\theta_W Z^\mu\left[-\bar{e}\gamma_\mu e - \operatorname{cosec}^2\theta_W \bar{L}\gamma_\mu T_3 L\right] -$$

$$- \frac{1}{\sqrt{2}} g\left(\left[\bar{\nu}_e \gamma^\mu \tfrac{1}{2}(1-\gamma_5)e\right]W^+_\mu + \text{h.c.}\right) + \tfrac{1}{4}g^2|W_\mu|^2(\varphi^2 + 2\lambda\varphi) +$$

$$+ \frac{1}{8}(g^2 + g'^2)Z^2_\mu(\varphi^2 + 2\lambda\varphi) - \frac{m_e}{\lambda}(\bar{e}e)\varphi - \tfrac{1}{4}h(\varphi^4 + 4\lambda\varphi^3) \; . \qquad (6.54)$$

This is the unified field theory which was written down by Weinberg and Salam[(4)]. Just like the Higgs Model, discussed previously, it is renormalizable, but the proof of this was not given until several years after the formulation of the theory. The proof, which is highly complicated by the various relations between the coupling constants in (6.54) and the masses (6.53), was given by 't Hooft[(5)].

Comparing the coupling of the leptonic current to the W-boson, in (6.54), with that discussed previously in (3.91), we see that

$$f = \frac{g}{2\sqrt{2}} \; . \qquad (6.55)$$

Thus from (3.105) we may relate g to the known weak coupling constant G:

$$\frac{g^2}{8\,m_W^2} = \frac{G}{\sqrt{2}} \; . \qquad (6.56)$$

It follows, using (6.31), that

$$m_W^2 = \frac{g^2\sqrt{2}}{8\,G} = \frac{e^2\sqrt{2}}{8\,G\sin^2\theta_W} \geq \frac{\alpha\,\pi}{G\sqrt{2}} \; . \qquad (6.57)$$

Substituting (3.63) for G gives

$$m_W \geq 37.3 \text{ GeV} \; . \qquad (6.58)$$

In the same way

$$m_Z^2 = m_W^2 \sec^2\theta_W = \frac{e^2\sqrt{2}}{2\,G\sin^2 2\theta_W} \; , \qquad (6.59)$$

so that

$$m_Z \geq 74.6 \text{ GeV} \; . \qquad (6.60)$$

6.3.1 LEPTON-LEPTON SCATTERING

Before going any further with this particular model, it is as well to check that its predictions are at least consistent with the extremely limited data available on purely leptonic processes. The most welcome experimental result would be the direct production of the W, but, given the lower bound (6.58), this is not possible at presently available energies (see for example the lower bound given at the end of Chapter 3). The most obvious feature of the model, which conflicts with the hypothesis used in most of this book, is that it contains a neutral parity violating current coupled to the Z-boson; this of course was inevitable from the outset, as we noted. Notice that we cannot get rid of the neutral current effects by making the mass of the Z arbitrarily large. From (6.59) we see that this would involve taking $\theta_W \to 0$, and then $m_Z^2 \sim \theta_W^{-2}$. On the other hand, the coupling constant g_Z for the lepton-Z interaction is seen from (6.54) to increase as $\theta_W \to 0$ ($g_Z \sim \theta_W^{-1}$). Thus the contribution to the lepton-lepton scattering amplitude from Z exchange approaches a constant as $\theta_W \to 0$, since $M \sim g_Z^2 / m_Z^2$.

As far as the purely leptonic processes are concerned, there is no experimental reason why we should "get rid" of the neutral currents, since it is barely possible to obtain accuracies sufficient to test the conventional charged current-current theory. Since muon targets are not available, the only processes which we might study are

$$\nu_e e \to \nu_e e \tag{6.61a}$$

$$\bar{\nu}_e e \to \bar{\nu}_e e \tag{6.61b}$$

$$\nu_\mu e \to \nu_\mu e \tag{6.61c}$$

$$\bar{\nu}_\mu e \to \bar{\nu}_\mu e \;. \tag{6.61d}$$

Colliding beam machines enable us to study

$$e^+ e^- \to \mu^+ \mu^- \tag{6.62a}$$

$$e^+ e^- \to \nu_\ell \bar{\nu}_\ell \qquad (\ell = e, \mu) \;. \tag{6.62b}$$

The first two processes in (6.61) are allowed in the conventional current-current theory, as we have seen. The last two, however, as noted in

(3.89), can proceed only via neutral currents (assuming lepton conservation), and their observation would correspondingly provide direct experimental evidence for the existence of such currents. The lowest order Feynman diagrams for the process (6.61a) are shown in Fig. 6.4.

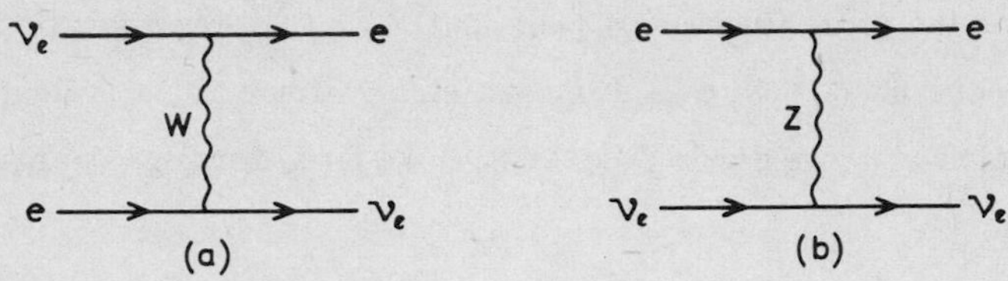

Fig. 6.4 Feynman diagrams for ν_e - e scattering

The invariant matrix element arising from diagram (a) may be written down directly from (6.54). It is

$$M_W = -\frac{g^2}{8}\,\frac{g_{\alpha\beta} - k_\alpha k_\beta/m_W^2}{k^2 - m_W^2}\left[\bar{e}_2\gamma^\alpha(1-\gamma_5)\nu_1\right]\left[\bar{\nu}_2\gamma^\beta(1-\gamma_5)e_1\right] , \quad (6.63)$$

where the suffices 1,2 label the initial and final lepton spinors respectively, and

$$k = \nu_1 - e_2 = \nu_2 - e_1 .$$

Even for quite high energies $k^2 << m_W^2$, and to good approximation

$$M_W = -\frac{G}{\sqrt{2}}\left[\bar{e}_2\gamma^\alpha(1-\gamma_5)e_1\right]\left[\bar{\nu}_2\gamma_\alpha(1-\gamma_5)\nu_1\right] , \quad (6.64)$$

using (6.56) and the Fierz Identities. In the same way, the contribution from diagram (b) is

$$M_Z = -\frac{G}{\sqrt{2}}\Big\{(2\sin^2\theta_W - \tfrac{1}{2})(\bar{e}_2\gamma^\alpha e_1)[\bar{\nu}_2\gamma_\alpha(1-\gamma_5)\nu_1] + \tfrac{1}{2}(\bar{e}_2\gamma^\alpha\gamma_5 e_1)[\bar{\nu}_2\gamma_\alpha(1-\gamma_5)\nu_1]\Big\} . \quad (6.65)$$

In deriving this expression an additional factor of -1 has been introduced (arising from the Fermi statistics of the particles involved), so that initial and final states are defined in the same order for both diagrams. We see that M_Z is never zero, as already noted. Combining (6.64) and (6.65)

$$M \equiv M_W + M_Z = -\frac{G}{\sqrt{2}}\left[\bar{e}_2\gamma^\alpha(C_V - C_A\gamma_5)e_1\right]\left[\bar{\nu}_2\gamma_\alpha(1-\gamma_5)\nu_1\right] , \quad (6.66a)$$

where

$$C_V = 1 + (2\sin^2\theta_W - \tfrac{1}{2}) \tag{6.66b}$$

$$C_A = 1 + (-\tfrac{1}{2}) \ . \tag{6.66c}$$

For the process (6.61c) only diagram (b) contributes. The matrix element has the same form as (6.66a), but C_V, C_A are given by the terms in parentheses in (6.66b, c). It is now straightforward to evaluate the cross section[6], as we did in Chapter 3. We find, for high energies ($\omega \equiv E_\nu m_e^{-1} \gg 1$),

$$\sigma_L(\nu e \to \nu e) = \tfrac{1}{4}\sigma_o\omega\left[(C_V + C_A)^2 + \tfrac{1}{3}(C_V - C_A)^2\right] .$$

In the same way the cross section for $\bar{\nu}$ e scattering is given by

$$\sigma_L(\bar{\nu} e \to \bar{\nu} e) = \tfrac{1}{4}\sigma_o\omega\left[(C_V - C_A)^2 + \tfrac{1}{3}(C_V + C_A)^2\right] .$$

Explicitly, using (6.66b, c), we find

$$\sigma_L(\nu_e e \to \nu_e e) = \tfrac{1}{4}\sigma_o\omega\left[1 + 4\sin^2\theta_W + \frac{16}{3}\sin^4\theta_W\right] \tag{6.67a}$$

$$\sigma_L(\bar{\nu}_e e \to \bar{\nu}_e e) = \tfrac{1}{4}\sigma_o\omega\left[\tfrac{1}{3} + \frac{4}{3}\sin^2\theta_W + \frac{16}{3}\sin^4\theta_W\right] . \tag{6.67b}$$

These should be compared with the predictions of the conventional V-A theory given in (3.81)

$$\sigma_L(\nu_e e \to \nu_e e) = 3\sigma_L(\bar{\nu}_e e \to \bar{\nu}_e e) = \sigma_o\omega \ . \tag{3.81}$$

Likewise, the predictions for the muonic neutrino cross sections are

$$\sigma_L(\nu_\mu e \to \nu_\mu e) = \tfrac{1}{4}\sigma_o\omega\left[1 - 4\sin^2\theta_W + \frac{16}{3}\sin^4\theta_W\right] \tag{6.68a}$$

$$\sigma_L(\bar{\nu}_\mu e \to \bar{\nu}_\mu e) = \tfrac{1}{4}\sigma_o\omega\left[\tfrac{1}{3} - \frac{4}{3}\sin^2\theta_W + \frac{16}{3}\sin^4\theta_W\right] . \tag{6.68b}$$

The conventional V-A theory, of course, predicts both of these to be zero. But in the Weinberg-Salam model these cross sections do not vanish for any value of θ_W. The minimum value of both is in fact $\frac{1}{16}\sigma_o\omega$, although these cannot both be realized for the same value θ_W. Thus the model can be eliminated, if not enough events are seen (and, of course, if too many are seen).

During the last year experiments using the neutrino beam at CERN and the Gargamelle bubble chamber indicate that the process (6.61d) does actually occur, see 3.2.2. Just three events have been identified as this process[7], but, since the background is minute, there is little doubt that neutral currents have now been seen. Of course, there is no really solid reason why the Weinberg-Salam model we have been discussing should in fact be responsible for this process. However it was the first of the unified field theories, and its aesthetic appeal is (I hope) self-evident. So it plainly is worth checking whether, for some value of θ_W, it is consistent with the $\bar{\nu}_\mu e \to \bar{\nu}_\mu e$ data. In fact, it turns out that the cross section corresponding to the three events so far observed is above the minimum value predicted by the model, and well below the maximum which could be accommodated. The cross sections measured satisfy

$$\sigma(\nu_\mu e \to \nu_\mu e) < 0.26 \times 10^{-41} \left(\frac{E_\nu}{1\,\text{GeV}}\right) \text{cm}^2 \tag{6.69a}$$

$$0.03 \times 10^{-41}\left(\frac{E_\nu}{1\,\text{GeV}}\right)\text{cm}^2 < \sigma(\bar{\nu}_\mu e \to \bar{\nu}_\mu e) < 0.29 \times 10^{-41}\left(\frac{E_\nu}{1\,\text{GeV}}\right)\text{cm}^2 . \tag{6.69b}$$

If the Weinberg-Salam model describes reality, we may calculate the probability that these two results correspond to the same value of θ_W. This leads to the conclusion[7]

$$0.1 < \sin^2\theta_W < 0.45 \quad (90\%\ \text{C.L.}) . \tag{6.70}$$

Further information on θ_W can be derived from the searches for the process (6.61b) using the Savannah River reactor. These concluded that the measured cross section σ and the conventional V-A cross section σ_{V-A} given in (3.80) satify[8]

$$\frac{\sigma}{\sigma_{V-A}} = 1.1 \pm 0.8 .$$

From this result and (6.67) we find θ_W satisfies

$$\sin^2\theta_W < 0.3 \quad (90\%\ \text{C.L.}) .. \tag{6.71}$$

It is plainly most important that neutral current effects should be searched for in as many processes as possible, thereby checking the self-consistency of the model. With the advent of high energy colliding beam machines the processes (6.62) will be the subject of intense study. The process (6.62a) is, of course, allowed by single photon exchange, as shown in diagram (a) of Fig. 6.5.

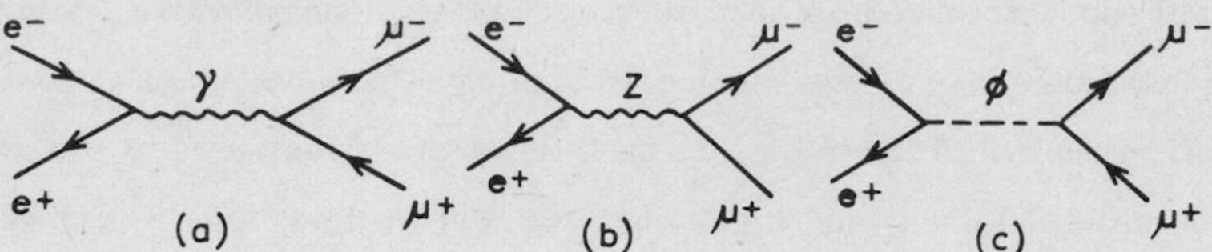

Fig. 6.5. Contributions to $e^+e^- \rightarrow \mu^+\mu^-$ in Weinberg-Salam model.

In the unified model under discussion, we see, from (6.54), that Z and φ exchange can also contribute, as shown in diagrams (b), (c) of Fig. 6.5. At low energies the purely electromagnetic contribution (a) will dominate, because (b), for example, like M_Z in (6.65) is proportional to G, the weak coupling constant. At higher energies the electromagnetic contribution is damped by the $(k^2)^{-1}$ in the photon propagator, while the Z-contribution is enhanced as its propagator denominator $k^2 - m_Z^2$ gets smaller. As a result, measurable interference effects between the γ- and Z-exchange contributions may be obtained[9]. Since the interaction of the Z boson is parity violating, these interference effects show up in the form of non-zero muon polarizations, and asymmetries in the differential cross-section. For example, Budny[9] has shown that Z-γ interference effects lead to an asymmetry in the cross sections, as a function of the scattering angle, of at least 12%. Similar effects are predicted for the average helicity of the μ^-, and in the total rate correction arising from the new exchanges. The effects of φ exchange are very small (because of the proportionality of the coupling to m_e), except possibly right at the pole, if its width is small.

It is important to recognize that the existence of neutral currents does not affect the mechanism for muon decay, which is the one purely leptonic process to be subjected to really intense experimental investigation over many years. This is because the Weinberg model carefully separates the electronic and muonic sectors, so there are no (neutral <u>or</u> charged) currents which mix electronic and muonic fields. However, the unified theory will affect the corrections to muon decay. In addition to the purely electromagnetic corrections, arising from the exchange of photons between the charged particles participating in the decay, we must now include those arising from Z and φ exchange. In principle, all of these are of order α in a unified theory, since the coupling constants are all of order e. It turns out, however, that the truly order α corrections affect only the total decay rate and not the spectrum shape[10]; the new corrections to the shape are all of order α/m_W^2, and so are quite negligible. Of course, the corrections to the total rate are important when we compare the muon and neutron decay rates with a view to evaluating the universality postulates, as we saw in 4.2.12. We shall therefore return to this topic after we have seen how the hadrons are to be included in the model.

6.3.2 GENERAL STRUCTURE OF UNIFIED THEORIES

Having constructed one renormalizable field theory of the weak and electromagnetic interactions, it is easy to see how the method could be generalized. While there is no objection to the Weinberg-Salam model, there is also very little positive evidence in its support. It is therefore only prudent not to prejudge the issue, and to this end we present a general formalism for the construction of such theories[11].

Our experience so far indicates that we should start with a manifestly renormalizable gauge invariant Lagrangian, which in general will involve gauge fields A_μ^α, spinor fields ψ^n and scalar fields φ^i

$$\mathcal{L} = -\tfrac{1}{4} F_{\mu\nu}^{\alpha} F^{\alpha\mu\nu} + \tfrac{1}{2} (D_\mu\varphi)^i (D^\mu\varphi)^i + i\bar{\psi}\gamma^\mu D_\mu \psi - \\ - \bar{\psi}\Gamma^i\psi\varphi^i - P(\varphi) - \bar{\psi} m_o \psi \ . \qquad (6.72)$$

$F_{\mu\nu}^\alpha$ is the gauge covariant curl analogous to (6.25a)

$$F_{\mu\nu}^\alpha \equiv \partial_\mu A_\nu^\alpha - \partial_\nu A_\mu^\alpha - g\, C^{\alpha\beta\gamma} A_\mu^\beta A_\nu^\gamma \ , \qquad (6.73)$$

where $C^{\alpha\beta\gamma}$ are the structure constants of the gauge group G. $(D_\mu\varphi)^i$ and $D_\mu\psi$ are the gauge covariant derivatives of the scalar and spinor fields analogous to (6.10)

$$(D_\mu \varphi)^i \equiv \partial_\mu \varphi^i + i g A^\alpha_\mu (t^\alpha)^{ij} \varphi^j \tag{6.74a}$$

$$(D_\mu \psi)^n \equiv \partial_\mu \psi^n + i g A^\alpha_\mu (T^\alpha)^{nm} \psi^m \quad , \tag{6.74b}$$

where the matrices t^α, T^α are representations of the group G

$$[t^\alpha, t^\beta] = i C^{\alpha\beta\gamma} t^\gamma \tag{6.75a}$$

$$[T^\alpha, T^\beta] = i C^{\alpha\beta\gamma} T^\gamma \quad . \tag{6.75b}$$

These definitions guarantee the gauge invariance of the first three terms of $\mathcal{L}$ under the gauge transformations

$$A'^\alpha_\mu = A^\alpha_\mu + \partial_\mu \Lambda^\alpha + g C^{\alpha\beta\gamma} \Lambda^\beta A^\gamma_\mu \tag{6.76a}$$

$$\psi'^n = \psi^n - i g \Lambda^\alpha (T^\alpha)^{nm} \psi^m \tag{6.76b}$$

$$\varphi'^i = \varphi^i - i g \Lambda^\alpha (t^\alpha)^{ij} \varphi^j \quad . \tag{6.76c}$$

The Yukawa coupling must also be gauge invariant, which requires that Γ^i satisfies

$$[T^\alpha, \Gamma^i] = (t^\alpha)^{ji} \Gamma^j \quad . \tag{6.77}$$

Similarly, the fermion mass term must be gauge invariant, so

$$[T^\alpha, m_o] = 0 \quad . \tag{6.78}$$

$P(\varphi)$ is an arbitrary quartic polynomial in the scalar fields, and it too must be gauge invariant, which requires

$$\frac{\partial P(\varphi)}{\partial \varphi^i} (t^\alpha)^{ij} \varphi^j = 0 \quad . \tag{6.79}$$

Next we break the symmetry by allowing φ^i to have a non-zero vacuum expectation value λ^i. The value of λ^i is determined, as before, by the requirement that $P(\varphi)$ is stationary when $\varphi^i = \lambda^i$, i.e.

$$\left. \frac{\partial P}{\partial \varphi^i} \right|_{\varphi^j = \lambda^j} = 0 \quad . \tag{6.80}$$

So we must rewrite the Lagrangian in terms of the shifted field

$$\chi^i = \varphi^i - \lambda^i \quad ,$$

which <u>does</u> have zero vacuum expectation value. Then

$$P(\varphi) = P(\lambda + \chi) = P(\lambda) + \frac{\partial P}{\partial \varphi^i}\bigg|_{\varphi^j = \lambda^j} \chi^i + \tfrac{1}{2} \frac{\partial^2 P}{\partial \varphi^i \partial \varphi^k}\bigg|_{\varphi^j = \lambda^j} \chi^i \chi^k + \cdots .$$

$P(\lambda)$ is an irrelevant constant, the second term vanishes by virtue of (6.80), and the third term defines the mass matrix for the new scalar fields

$$(\mu^2)^{ij} \equiv \frac{\partial^2 P}{\partial \varphi^i \partial \varphi^j}\bigg|_{\varphi = \lambda} . \tag{6.81}$$

Now differentiating (6.79) gives

$$\frac{\partial^2 P}{\partial \varphi^k \partial \varphi^i} (t^\alpha)^{ij} \varphi^j + \frac{\partial P}{\partial \varphi^i} (t^\alpha)^{ij} \delta^{jk} = 0 ,$$

so, putting $\varphi = \lambda$ and using (6.80) and (6.81), we see that μ^2 satisfies

$$(\mu^2)^{ki} (t^\alpha \lambda)^i = 0 . \tag{6.82}$$

This shows that in general the matrix μ^2 has zero eigenvalues corresponding to certain combinations of the fields χ. As before, we want to work in the unitary gauge, in which these massless scalar modes have been transformed away and used to give the hitherto massless vector bosons the extra degree of freedom required to make them massive. In this unitary gauge the fields are perpendicular to the above eigenvectors

$$\chi^i (t^\alpha \lambda)^i = 0 . \tag{6.83}$$

As before, we rewrite $\left(D_\mu \varphi^i\right)^2$ in terms of χ^i and λ^i and read off the vector meson mass matrix

$$(M^2)^{\alpha\beta} = g^2 (t^\alpha t^\beta)^{ij} \lambda^i \lambda^j . \tag{6.84}$$

Thus, if C^β is such that

$$(C^\beta t^\beta)^{ij} \lambda^j = 0 , \tag{6.85a}$$

then

$$(M^2)^{\alpha\beta} C^\beta = 0 , \tag{6.85b}$$

which shows that the linear combination of gauge fields $C^\beta A^\beta$ has zero mass, corresponding to an unbroken symmetry. Finally, we note that the fermion mass matrix is given by

$$m = m_o + \Gamma^i \lambda^i . \tag{6.86}$$

The Georgi Glashow Model

We may illustrate the above general notation by considering another unified model constructed by Georgi and Glashow[12]. This is based on the three-dimensional rotation group $SO(3) \cong SU(2)$, with three generators and therefore three gauge fields A_μ^α $(\alpha = 1,2,3)$. The structure constants are

$$C^{\alpha\beta\gamma} = \epsilon^{\alpha\beta\gamma} \quad . \tag{6.87}$$

We wish to identify A_μ^3 with the electromagnetic field, and $A_\mu^1 \pm i A_\mu^2$ with the charged W-fields. Thus there are no new neutral currents. In such a model the charge must be the third component of (leptonic) isospin, so the lowest representation to which the leptons can be assigned is the (regular) three-dimensional representation. Since the known leptons (e^-, ν_e) have charges -1, 0, we must introduce a new lepton (X^+) having charge +1 to complete such a representation. In fact, we must introduce two triplet representations, one left-handed and the other right-handed. Then, since only the left-handed neutrino field exists, we must introduce a new neutral lepton (X^o) to complete the right-handed triplet; in general, its left-handed component will also appear in the left-handed triplet. Thus we have X^+, X^o, e^- each with two components and ν_e with one. These seven components are assigned as follows

$$\underline{\psi}_L = \tfrac{1}{2}(1-\gamma_5) \begin{pmatrix} X^+ \\ \sin\beta\,\nu_e + \cos\beta\, X^o \\ e^- \end{pmatrix} \tag{6.88a}$$

$$\underline{\psi}_R = \tfrac{1}{2}(1-\gamma_5) \begin{pmatrix} X^+ \\ X^o \\ e^- \end{pmatrix} \tag{6.88b}$$

$$S_L = \tfrac{1}{2}(1-\gamma_5)\,(\sin\beta\, X^o - \cos\beta\,\nu_e) \quad . \tag{6.88c}$$

X^+, X^o are presumably 'heavy' leptons since they have not (yet ?) been observed. The above assignments are in the spherical basis in which T^3 is diagonal. We can, of course, write $X^+ = \frac{1}{\sqrt{2}}(X^1 - iX^2)$, $e^- = \frac{1}{\sqrt{2}}(X^1 + iX^2)$, thereby defining an hermitian basis. In this hermitian basis the three-dimensional representation of the group is given by

$$(T^\alpha)^{nm} = i\,\epsilon^{n\alpha m} \quad . \tag{6.89}$$

(In the singlet representation $T^\alpha = 0.$) Thus the fermion kinetic energy term in (6.72) is

$$i\bar{\psi}\gamma^\mu D_\mu \psi = i\underline{\bar{\psi}}_L \cdot \gamma^\mu \partial_\mu \underline{\psi}_L + i\underline{\bar{\psi}}_R \cdot \gamma^\mu \partial_\mu \underline{\bar{\psi}}_R + $$

$$+ i g \underline{A}_\mu \cdot \underline{\bar{\psi}}_L \times \gamma^\mu \underline{\psi}_L + ig \underline{A}_\mu \cdot \underline{\bar{\psi}}_R \times \gamma^\mu \underline{\bar{\psi}}_R + $$

$$+ i\bar{S}_L \gamma^\mu \partial_\mu S_L , \tag{6.90}$$

using vector notation to denote the (hermitian) triplets of fields. Writing out the interaction term explicitly, we find that it is

$$g A^3_\mu [\bar{e}\gamma^\mu e - \bar{X}^+ \gamma^\mu X^+] -$$

$$- \tfrac{1}{2} g \sin\beta [\bar{e}\gamma^\mu (1-\gamma_5)\nu_e W^-_\mu + \bar{\nu}_e \gamma^\mu (1-\gamma_5) e W^+_\mu + \cdots] , \tag{6.91}$$

where .... are additional terms all involving the fields X^+ and/or X^o. Plainly to get the charge on the electron right we must take

$$g = e . \tag{6.92}$$

The semi-weak coupling constant is then $f = -\frac{1}{2} e \sin\beta$. Thus, in this model (c.f. (6.56)), we will require

$$\frac{G}{\sqrt{2}} = \frac{f^2}{m^2_W} = \frac{\pi\alpha \sin^2\beta}{m^2_W} .$$

Hence (when we have arranged for m_W to have a mass)

$$m^2_W \leq \frac{\sqrt{2}\,\alpha\pi}{G} \simeq (53 \text{ GeV})^2 . \tag{6.93}$$

From (6.78) and (6.89), m_o must be a multiple of the unit matrix and

$$\bar{\psi} m_o \psi = m_o \underline{\bar{\psi}}_L \cdot \underline{\psi}_R + \text{h.c.}$$

$$= m_o \left\{ \bar{e} e + \bar{X}^+ X^+ + \cos\beta \bar{X}^o X^o + \tfrac{1}{2}\sin\beta [\bar{X}^o (1-\gamma_5)\nu_e + \text{h.c.}] \right\} \tag{6.94}$$

Thus, until we break the symmetry, e and X^+ have the same mass, and the matrix is not diagonal in the X^o and ν_e fields. We wish to break the symmetry, so that $W^\pm$ acquires a mass. This requires us to put the scalar fields φ^i in some non-singlet representation of the group. The simplest choice is that the φ^i also transform as the vector representation. In this case $t^\alpha = T^\alpha$, and the most general gauge invariant quartic $P(\varphi)$ has the form

$$P(\varphi) = a + b\,\underline{\varphi}^2 + c\,(\underline{\varphi}^2)^2 \,, \tag{6.95a}$$

where

$$\underline{\varphi}^2 = (\varphi^1)^2 + (\varphi^2)^2 + (\varphi^3)^2 \,. \tag{6.95b}$$

It is easy to see that this satisfies (6.79). Thus, when we break the symmetry, (6.80) requires that λ^i satisfies

$$b\,\lambda^i + 2c\lambda^i\,(\lambda^j\lambda^j) = 0 \,. \tag{6.96}$$

Since we want to retain charge conservation, we must take

$$\lambda^1 = \lambda^2 = 0 \tag{6.97a}$$

$$\lambda^3 = (-b/2c)^{\frac{1}{2}} \equiv \lambda \,. \tag{6.97b}$$

c must be positive, as before, so b must be negative. With this choice, we see from (6.81) that the scalar mass matrix is

$$(\mu^2)^{ij} = -4b\,\delta^{i3}\,\delta^{j3} \,. \tag{6.98}$$

Thus the zero mass (Goldstone bosons) states are χ^1 and χ^2, as may also be seen from (6.82). In the unitary gauge specified in (6.83),

$$\chi^i\,(t^\alpha\lambda)^i = \chi^i\,i\epsilon^{i\alpha 3}\,\lambda^3 = 0 \,. \tag{6.99}$$

Thus only χ^3 survives, and

$$\varphi^i = \chi^i + \lambda^i = (\chi^3 + \lambda^3)\,\delta^{i3} \,. \tag{6.100}$$

As a result the gauge fields' mass matrix is found from (6.84) to be

$$(M^2)^{\alpha\beta} = g^2\lambda^2\,(\delta^{\alpha\beta} - \delta^{\alpha 3}\,\delta^{\beta 3}) \,. \tag{6.101}$$

Thus A^3_μ has zero mass, as expected, while $A^{1,2}_\mu$ have mass $|g\lambda|$. Hence, using (6.92),

$$m_W = |e\lambda| \,. \tag{6.102}$$

Finally, we look at the Yukawa coupling term of (6.72). Since

$$(t^\alpha)^{ij} = (T^\alpha)^{ij} = i\epsilon^{i\alpha j} \,,$$

(6.77) allows Γ^i to be 3 x 3 or 3 x 1 :

$$(\Gamma^i)^{mn} = g_V\,i\epsilon^{min} \tag{6.103a}$$

$$(\Gamma^i)^n = g_S\,\delta^{in} \,, \tag{6.103b}$$

where g_V and g_S are constants. Thus the general Yukawa term is

$$\bar{\psi}\Gamma^i \psi \varphi^i = -ig_V \bar{\underline{\psi}}_R \times \underline{\psi}_L \cdot \underline{\varphi} + g_S \bar{\underline{\psi}}_R S_L \cdot \underline{\varphi} + \text{h.c.} \quad .$$

When we break the symmetry using (6.100), this term generates mass terms, in addition to the Yukawa couplings to χ^3; these are

$$\lambda g_V (\bar{X}^+ X^+ - \bar{e} e) + \lambda g_S \sin\beta \bar{X}^o X^o - \tfrac{1}{2}\lambda g_S \cos\beta [\bar{X}^o (1-\gamma_5)\nu_e + \text{h.c.}] \tag{6.104}$$

Thus, combining with (6.94), we may eliminate the off-diagonal terms by the choice

$$m_o \sin\beta = \lambda g_S \cos\beta \quad . \tag{6.105}$$

Then the masses of the remaining states are

$$m_e = m_o - \lambda g_V \tag{6.106a}$$

$$m_{X^+} = m_o + \lambda g_V \tag{6.106b}$$

$$m_{X^o} = m_o \cos\beta + \lambda g_S \sin\beta = m_o \sec\beta \quad . \tag{6.106c}$$

Hence

$$m_e + m_{X^+} = 2 \cos\beta \, m_{X^o} \quad . \tag{6.107}$$

The calculation of the Yukawa couplings to χ^3 is left as an exercise. Plainly the general formalism (6.72) admits of innumerable models, which may be more or less attractive according to one's prejudices. The original attraction of the model just described was that it contained no new neutral currents, although the price paid was the introduction of new leptons. Since neutral currents do exist, the model is no longer so popular. However, it does have the attraction that it is formulated in terms of one gauge coupling constant: $g = e$; this is because the underlying group O(3) is simple, while the Weinberg model's group SU(2) x U(1) is not.

6.3.3 INCLUSION OF HADRONS IN THE WEINBERG-SALAM MODEL

We have seen that the possibilities of testing the Weinberg model are restricted by the difficulty (and expense) of performing neutrino scattering or colliding beam experiments. Since most of the original data in weak interactions was collected from semileptonic decays, it is natural to expect that the inclusion of hadrons in the model will provide us with more opportunities for testing the theory. In addition, we may hope that in doing this we shall be led to some additional constraints, so as to reduce the number of acceptable models which can be constructed according to the prescriptions of the previous section.

The obvious way to include the hadrons in any model is to assign the quark fields to some representation of the gauge group in such a way as to generate the known hadronic currents $\mathcal{J}_\alpha^\pm$. The expression for $\mathcal{J}_\alpha^+$ in terms of the quark field operators p, n, λ is, as we have seen,

$$\mathcal{J}_\alpha^+ = \cos\theta_c\, \bar{p}\gamma_\alpha(1-\gamma_5)n + \sin\theta_c\, \bar{p}\gamma_\alpha(1-\gamma_5)n$$

$$= \bar{p}\gamma_\alpha(1-\gamma_5)\,[\,n\cos\theta_c + \lambda\sin\theta_c\,]\,. \tag{6.108}$$

We confine our attention henceforth to the Weinberg Model. Then the simplest way of ensuring that $\mathcal{J}_\alpha^\pm$ are coupled to $W_\alpha^\pm$ is to place the quark fields in a doublet (c.f. (6.14) and (6.15))

$$N_L = a\begin{pmatrix} p \\ n\cos\theta_c + \lambda\sin\theta_c \end{pmatrix}\,. \tag{6.109}$$

The SU(2) group "ladder" operators $\tau^\pm$ will then generate $\mathcal{J}_\alpha^\pm$, just as we obtained L_α, $L_\alpha^\dagger$. The remaining left-handed component and the three right-handed components

$$a\,(-n\sin\theta_c + \lambda\cos\theta_c)$$

$$a'p,\ a'(n\cos\theta_c + \lambda\sin\theta_c),\ a'(-n\sin\theta_c + \lambda\cos\theta_c)\,, \tag{6.110}$$

may be assigned to singlet representations of the $SU(2)_L$. The couplings to the U(1) gauge vector meson B_μ are adjusted, as before, to get the correct charges. The coupling to the Z-boson is then completely determined. As in (6.54) it has the form

$$e\tan\theta_W\, Z^\mu\,[\,j_\mu - \mathrm{cosec}^2\theta_W\, \bar{N}_L\gamma_\mu T^3 N_L\,]\quad, \tag{6.111}$$

where j_μ is the electromagnetic current, and $T^3 = \frac{1}{2}\tau^3$, as before. We consider the structure of the last term, suppressing the γ_μ a factor

$$2\bar{N}_L\gamma_\mu T^3 N_L \sim \bar{p}p - \cos^2\theta_c\,\bar{n}n - \sin^2\theta_c\,\bar{\lambda}\lambda -$$

$$- \sin\theta_c\cos\theta_c\,(\bar{n}\lambda + \bar{\lambda}n)\,. \tag{6.112}$$

The last term of this expression $(\bar{n}\lambda + \bar{\lambda} n)$ is a strangeness changing neutral current, and is completely unacceptable. It will lead to transitions of the form

$$n\bar{\lambda} \rightarrow \mu^+\mu^-$$

via Z-exchange, as shown in Fig. 6.6. Using the Z-coupling to the muon, as given in (6.54), the invariant matrix element for the above process has the form (for $k^2 << m_Z^2$)

$$M(n\bar{\lambda} \rightarrow \mu^+\mu^-) = -\frac{G}{2\sqrt{2}} \sin\theta_c \cos\theta_c \bar{u}(\mu)\gamma_\alpha(1-4\sin^2\theta_W - \gamma_5)v(\mu) \times$$

$$\times \bar{v}(\lambda)\gamma^\alpha(1-\gamma_5)u(n) \ . \qquad (6.113)$$

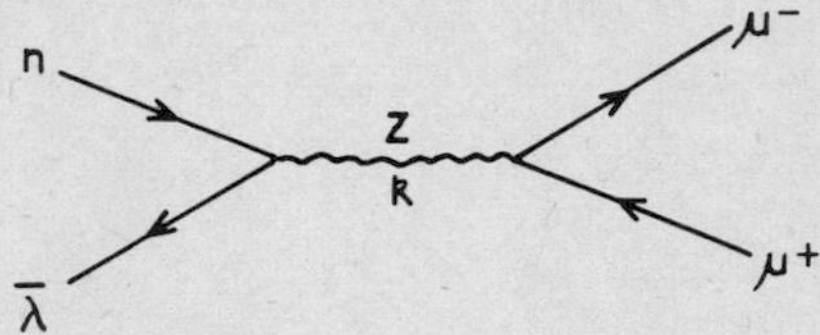

Fig. 6.6 Feynman diagram for $n\bar{\lambda} \rightarrow \mu^+\mu^-$

The $n\bar{\lambda}$ has the same quantum numbers as the kaon K^o, and presumably

$$<0|\bar{\lambda}\gamma_\alpha(1-\gamma_5)n|K^o> \simeq if_K K_\alpha \ . \qquad (6.114)$$

Thus the existence of the strangeness changing neutral current will give rise to the decay

$$K^o \rightarrow \mu^+\mu^- \ ,$$

and, from (6.113) and (6.114), we expect its matrix element to be of the same order of magnitude as that for $K^+ \rightarrow \mu^+\nu_\mu$. Thus we shall predict a branching ratio $\Gamma(K_L \rightarrow \mu^+\mu^-)/\Gamma(K_L)$ of the order of unity. Experimentally this branching ratio is less than 10^{-5}, as noted in Chapter 2. Further, the same mechanism (Z-exchange) will give rise to the transition

$$n\bar{\lambda} \leftrightarrow \lambda\bar{n}$$

or

$$K^o \leftrightarrow \bar{K}^o$$

as a first order (in G) weak effect. This will lead to a mass difference of K_S^o and K_L^o of order G. We shall see in the next chapter that this mass difference is minute and entirely compatible with its being an order G^2 effect, as predicted by the conventional theory. This is why the presence of a strangeness changing neutral current is unacceptable. Plainly we must find some mechanism which removes this unwanted term.

The cheapest way out is to invent one new quark, p', with the same charge as the p quark[13]. Then we may construct a second doublet N_L' using the remaining left-handed component in (6.110)

$$N_L' = a \begin{pmatrix} p' \\ -n\sin\theta_c + \lambda\cos\theta_c \end{pmatrix} . \qquad (6.115)$$

The right-handed $a'p'$ is then left as a singlet in place of the term just placed in the doublet N_L'. With this new doublet we acquire an additional contribution to the Z-interaction arising, as in (6.111), in the combination $\bar{N}_L' \gamma_\mu T^3 N_L'$. In the notation of (6.112)

$$2 N_L' \gamma_\mu T^3 N_L' \sim \bar{p}'p' - \sin^2\theta_c \bar{n}n - \cos^2\theta_c \bar{\lambda}\lambda + \sin\theta_c \cos\theta_c (\bar{n}\lambda + \bar{\lambda}n) ,$$

so

$$2\bar{N}_L \gamma_\mu T^3 N_L + 2\bar{N}_L' \gamma_\mu T^3 N_L' \sim \bar{p}'p' + \bar{p}p - \bar{n}n - \bar{\lambda}\lambda . \qquad (6.116)$$

Thus, although the new quark p' does not contribute to the strangeness changing neutral current, its existence enables us to eliminate it. Of course the <u>known</u> hadrons are reasonably described in terms of just the Gell-Mann Zweig quarks p, n, λ. This non-appearance of p' in the known hadrons is therefore explained by assigning the p' a new quantum number - "charm" - which is conserved by strong interactions. p, n, λ have zero charm, as do all the known hadrons. The assumption is that the charmed hadrons have masses higher than those accessible to present day accelerators.

Of course, the p' also contributes to the charged hadronic current via the SU(2) ladder operators, and the full interaction with the charged $W^\pm$ is

$$- \frac{g}{2\sqrt{2}} W_\mu^+ \left\{ \bar{p} \gamma^\mu (1-\gamma_5) [n \cos\theta_c + \lambda \sin\theta_c] + \right.$$

$$\left. + \bar{p}' \gamma^\mu (1-\gamma_5) [-n \sin\theta_c + \lambda \cos\theta_c] \right\} + \text{h.c.} \quad . \qquad (6.117)$$

Thus the $\Delta S = 2$ transition $n\bar{\lambda} \leftrightarrow \lambda\bar{n}$ can still go, as in the conventional theory, via two W exchange as shown in Fig. 6.7. In addition there are two similar diagrams with a $\bar{p}'$ intermediate state in the $\bar{\lambda}\bar{n}$ line, and

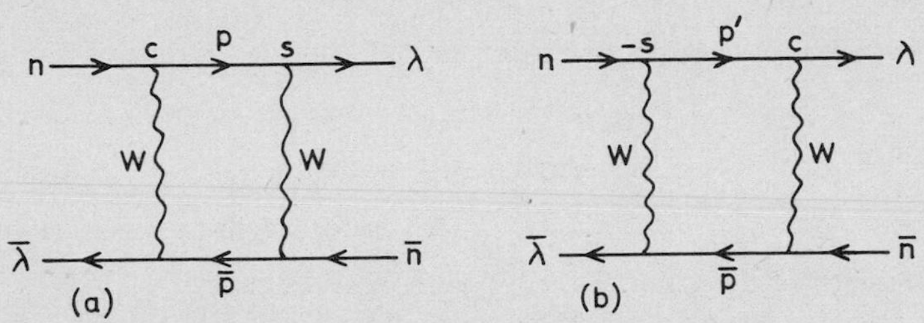

Fig. 6.7 Two W exchange contributions to $n\bar{\lambda} \leftrightarrow \lambda\bar{n}$.

four similar diagrams in which the W's are exchanged in the s-channel. Now, in a unified gauge theory these diagrams are expected to be of order $G\alpha$, rather than the order G^2 indicated by experiment. So some further suppression of these contributions is still required. One way in which this might occur is as follows. We notice from (6.117) that the couplings of the intermediate p and p' are, as indicated in the diagrams, $c = \cos\theta_c$ and $s = \sin\theta_c$. Thus the diagrams (a) and (b) have opposite sign and, <u>if</u> $m_p = m_{p'}$, equal magnitude; the same applies to the other diagrams taken in pairs. The smallness of the observed $K_L^o - K_S^o$ mass difference is therefore explained by making $(m_p - m_{p'})/m_{p'} \ll 1$. Alternatively, if one takes this free quark model seriously, the desired suppression may also be obtained by taking $m_p \ll m_{p'} \simeq 1$ GeV; see M.K. Gaillard and B.W. Lee, Phys. Rev. <u>D10</u>, 897 (1974). Plainly neither explanation is very satisfying, but at the present time it is the conventional wisdom. It may be that some dynamical (rather than kinematical) mechanism, such as will be discussed at the end of this chapter, is "really" responsible for the inhibition; or it may also be that this theory is simply not true.

With the hadrons included in the unified renormalizable field theory, it is now straightforward, in principle, (but technically quite difficult) to evaluate the radiative corrections to beta decay. Of course, we still don't really know how to include strong interaction effects, and the calculations for the most part assume point electromagnetic structure for the hadrons. However, the answer is guaranteed to be finite, and the neglect of higher order terms really is justified. The result is [14]

$$\Gamma(n)/\Gamma(\mu) \equiv R = R_o \left(1 + \frac{3\alpha}{\pi} \ln \frac{m_Z}{m_N}\right) \quad \text{for } m_Z >> m_N \quad . \tag{6.118}$$

Comparing with the results discussed in 4.2.12 , we see that the old cut-off Λ , which became m_W in the W-meson theories, see (4.456), has now been replaced by m_Z. Further, with m_Z given by (6.59) and (6.60), it is of just the order of magnitude (4.546) which seems to be required for the consistency of the Cabibbo theory with experiment.

6.3.4 DEEP INELASTIC NEUTRINO SCATTERING

Having abolished the strangeness changing neutral currents, there still remains a neutral current which conserves hypercharge. In view of the (unexpectedly) large cross-sections obtained in deep inelastic scattering of neutrinos via the charged current, we naturally anticipate that similar results will be forthcoming from scattering by neutral currents, if they actually exist. Further, the success of the quark parton model in both weak and electromagnetic processes suggests that a similar analysis of the neutral current processes will lead to very precise information on the structure and couplings of this neutral current.

Thus we consider the neutral current analogues of (4.369)

$$\nu_\ell N \to \nu_\ell X \tag{6.119a}$$

$$\bar{\nu}_\ell N \to \bar{\nu}_\ell X \ . \tag{6.119b}$$

We use the notation used in 4.2.11 for the momenta of the leptons and hadrons. For $|q^2| << m_Z^2$ the effective interaction between the neutrinos and the hadrons, arising from Z-exchange, has a V, A structure, although not, of course, a V-A structure. Combining the ν-Z interaction of (6.54) with the hadron-Z interaction of (6.111) and (6.116), we obtain an effective interaction

$$\frac{e^2}{4m_Z^2} \operatorname{cosec}^2 2\theta_W K^\mu \bar{\nu}_\ell \gamma_\mu (1-\gamma_5) \nu_\ell , \qquad (6.120)$$

where

$$K_\mu = \bar{p} \gamma_\mu \left(\frac{8}{3} \sin^2 \theta_W - 1 + \gamma_5 \right) p + $$
$$+ \bar{n} \gamma_\mu \left(1 - \frac{4}{3} \sin^2 \theta_W - \gamma_5 \right) n + \cdots . \qquad (6.121)$$

The terms .... above represent the pieces of K_μ involving the p' and λ quarks. We are concerned with the scattering from nucleons, which are constructed only from p and n quarks, so the additional pieces of K_μ cannot contribute, at least when we parallel the naive quark parton model treatment of 4.2.11 .

First we define the hadronic tensor as in (4.370), but with J^+_μ and J^-_λ replaced by K_μ and K_λ ; so both ν and $\bar{\nu}$ scattering are characterized by the <u>same</u> hadronic tensor, since K_μ is hermitian. We may decompose the hadronic tensor as in (4.371), and under the Bjorken scaling hypothesis we find that the deep inelastic scattering processes are described by three structure functions $H_i(x)$. The total ν and $\bar{\nu}$ cross-sections are then given by the analogue of (4.403). So all we need to do is to use the naive quark parton model to evaluate the H_i.

In fact we have already done all of the work necessary for this evaluation, since in (4.410) and thereafter we considered a general V, A hadronic current. The only difference is that the incident ν or $\bar{\nu}$ can now interact with <u>both</u> the p and n quarks inside the nucleon. Then from (4.413) using (6.121) we find, for a proton target,

$$H_2^{(p)}(x) = 2x H_1^{(p)} = \left[\left(\frac{8}{3} \sin^2 \theta_W - 1\right)^2 + 1 \right] x u_p(x) + $$
$$+ \left[\left(\frac{4}{3} \sin^2 \theta_W - 1\right)^2 + 1 \right] x u_n(x) \qquad (6.122a)$$

$$H_3^{(p)}(x) = \left(\frac{8}{3} \sin^2 \theta_W - 1 \right) 2u_p(x) + \left(\frac{4}{3} \sin^2 \theta_W - 1 \right) 2u_n(x) . \qquad (6.122b)$$

Note that the positivity bounds are not all saturated now, since we no longer have a V-A current. Substituting these into (4.403), and replacing $\frac{G}{\sqrt{2}} \cos\theta_c$ by

$$\frac{e^2}{4m_Z^2} \operatorname{cosec}^2 2\theta_W = \frac{G}{2\sqrt{2}}$$

(using (6. 59)), we find that the neutral current cross sections σ_Z are

$$\sigma_Z(\nu p) = \frac{G^2}{4\pi} m_N E_\nu \tfrac{2}{3} \int_0^1 dx \left\{ \left(\frac{64}{9} \sin^4\theta_W - 8\sin^2\theta_W + 3\right) x u_p + \left(\frac{16}{9}\sin^4\theta_W - 4\sin^2\theta_W + 3\right) x u_n \right\} \tag{6.123a}$$

$$\sigma_Z(\bar{\nu} p) = \frac{G^2}{4\pi} m_N E_\nu \tfrac{2}{3} \int_0^1 dx \left\{ \left(\frac{64}{9} \sin^4\theta_W - \frac{8}{3}\sin^2\theta_W + 1\right) x u_p + \left(\frac{16}{9}\sin^4\theta_W - \frac{4}{3}\sin^2\theta_W + 1\right) x u_n \right\} . \tag{6.123b}$$

As before, the cross sections for scattering from a neutron target can be obtained by interchanging u_n and u_p. Thus the isospin averaged cross-sections, defined analogously to those in (4. 430), are

$$\sigma_Z(\nu N) = \frac{G^2}{\pi} m_N E_\nu \left(\frac{20}{27}\sin^4\theta_W - \sin^2\theta_W + \tfrac{1}{2}\right) \int_0^1 dx\, x\,(u_p + u_n)$$

$$\sigma_Z(\bar{\nu} N) = \frac{G^2}{\pi} m_N E_\nu \tfrac{1}{3} \left(\frac{20}{9}\sin^4\theta_W - \sin^2\theta_W + \tfrac{1}{2}\right) \int_0^1 dx\, x\,(u_p + u_n) .$$

We may compare these with the cross sections for the charged current scattering via W-exchange. In the Weinberg-Salam model the result (4. 431a) receives an additional contribution from "charmed" final states, which are excited by the new $\bar{p}'\gamma_\mu\, n \sin\theta_c$ piece of $\mathcal{J}_\mu^+$; the other result (4. 431b) and those for the $|\Delta Y| = 1$ processes, given in 4. 3. 5 , are unaltered. We then find the following branching ratios for the neutral to charged current cross sections[15]

$$\sigma_Z(\nu N)/\sigma_W(\nu N) = \frac{20}{27}\sin^4\theta_W - \sin^2\theta_W + \tfrac{1}{2} \tag{6.124a}$$

$$\sigma_Z(\bar{\nu} N)/\sigma_W(\bar{\nu} N) = \frac{20}{9}\sin^4\theta_W - \sin^2\theta_W + \tfrac{1}{2} . \tag{6.124b}$$

The neutral current cross sections have recently been measured at CERN and NAL. The experiments involved are more difficult than those which measure the charged current cross sections σ_W. This is because in the neutral current processes the final state lepton is a neutrino, rather than a muon, and so cannot be seen directly. The experimenters therefore have to reconstruct the event from measurements of the observed hadrons. The data are[16]

$$\sigma_Z(\nu N)/\sigma_W(\nu N) = 0.21 \pm 0.03 \qquad (6.125a)$$

$$\sigma_Z(\bar{\nu} N)/\sigma_W(\bar{\nu} N) = 0.45 \pm 0.09 \quad , \qquad (6.125b)$$

and[17]

$$(1-\epsilon)\sigma_Z(\nu N)/\sigma_W(\nu N) + \epsilon\sigma_Z(\bar{\nu} N)/\sigma_W(\bar{\nu} N) = 0.29 \pm 0.09 \qquad (6.126a)$$

$$\epsilon = 0.19 \pm 0.05 \ . \qquad (6.126b)$$

These values are consistent with each other, and, <u>if</u> we assume the Weinberg-Salam model is correct, (6.124) gives

$$0.3 < \sin^2\theta_W < 0.4 \ , \qquad (6.127)$$

which is just consistent with the bounds in (6.70) and (6.71) from the purely leptonic processes.

Despite the complexity of these experiments, there is now little doubt that the existence of neutral currents has now definitely been established. The self-consistency of these results and of the leptonic data strongly support this conclusion. Even more remarkable, perhaps, is the fact that the original Weinberg-Salam model is entirely consistent with the data.

Finally we note that the original deep inelastic electron scattering experiments, which established the existence of scaling, will receive contributions from Z-exchange, besides the known photon exchanges. Being weak rather than electromagnetic the Z-exchange contributions will give only small corrections arising from Z-γ interference. However they do lead to apparent deviations from scaling of order q^2/m_Z^2, which should be observable eventually[18].

6.3.5 γ_5 ANOMALIES

We have seen that the observed absence of strangeness changing neutral currents forces us to introduce one extra quark field p'. In fact there are additional technical reasons why we need to include additional fermion fields, if we are to be sure that the field theory we are writing down really makes sense. The technical reasons referred to above arise from a curious feature of "axial-electrodynamics" first noted by Adler and by Bell and Jackiw[19].

Consider the V-V-A diagrams shown in Fig. 6.8a, b. The jagged line

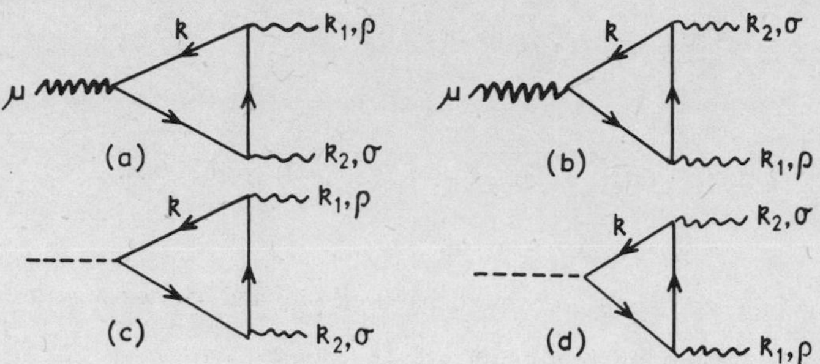

Fig. 6.8 V-V-A and V-V-P diagrams.

is to represent the axial vector vertex $\gamma_\mu \gamma_5$, and the wavy lines vector vertices γ_ρ, γ_σ. The invariant matrix element for diagram (a) is

$$M^{(a)}_{\mu\rho\sigma} = \int dk \, \mathrm{Tr}\,[\gamma_\mu \gamma_5 (\not{k}-m)^{-1} \gamma_\rho (\not{k}+\not{k}_1-m)^{-1} \gamma_\sigma (\not{k}+\not{k}_1+\not{k}_2-m)^{-1}] .$$

Using the trivial identity

$$\not{k}_1 \equiv (\not{k}+\not{k}_1-m) - (\not{k}-m) ,$$

it follows that

$$k_1^\rho M^{(a)}_{\mu\rho\sigma} = \int dk \, \mathrm{Tr}\,[\gamma_\mu \gamma_5 (\not{k}-m)^{-1} \gamma_\sigma (\not{k}+\not{k}_1+\not{k}_2-m)^{-1}]$$
$$- \int dk \, \mathrm{Tr}\,[\gamma_\mu \gamma_5 (\not{k}+\not{k}_1-m)^{-1} \gamma_\sigma (\not{k}+\not{k}_1+\not{k}_2-m)^{-1}] .$$

Both of these integrals are divergent, but formally we may translate the integration variable in the second integral ($k \to k - k_1$), which gives

$$k_1^\rho M^{(a)}_{\mu\rho\sigma} = 0 .$$

The same applies to diagram (b) and to contractions with k_2^σ, so

$$k_1^\rho M_{\mu\rho\sigma} = 0 = k_2^\sigma M_{\mu\rho\sigma} . \tag{6.128}$$

In the same way we may show formally that

$$(k_1 + k_2)^\mu M_{\mu\rho\sigma} = 2m \, M_{\rho\sigma} , \tag{6.129}$$

where $M_{\rho\sigma}$ is the invariant matrix element for the sum of the two diagrams (c), (d) of Fig. 6.8, in which the dashed line represents a pseudoscalar γ_5 interaction. Because $M_{\mu\rho\sigma}$ is a divergent integral, the above arguments are purely formal. One may try and define $M_{\mu\rho\sigma}$ by using a covariant cut-off for example. But in general[19] there is no way of defining $M_{\mu\rho\sigma}$ so that both of the formally derived Ward identities (6.128) and (6.129) are satisfied. In the context of the renormalizable theories under consideration, this is serious, because it indicates that there is no way of renormalizing the divergence of $M_{\mu\rho\sigma}$ while maintaining the Ward identities. Since Ward identities are used in the proof of the renormalizability of the theory, this is an important loophole in the proof. In particular cases both of the Ward identities can be maintained; even though they are not satisfied for any individual fermion triangle, it can happen that the "anomalies" from different loops cancel out. The condition for this to happen is that[20]

$$\sum_f (T_f^3)^2 Q_f = 0 \ ,$$

where Q_f is the charge of a fermion f, T_f^3 is its "weak" isospin, and the summation is over all fermions in the theory. In the Weinberg model the only charged leptons are e^- and μ^-, both with $(T^3)^2 = \frac{1}{4}$. So if all of the hadrons also belong to doublets, the above condition gives

$$\sum_h Q_h = +\ 2 \ , \tag{6.130}$$

where now the summation is over all hadrons (i.e. quarks) in the theory. (6.130) is usually taken as a restriction to be satisfied by any realistic theory. An additional restriction is also imposed, which is related to the $\pi^o \to 2\gamma$ decay. It has been known for many years that the decay rate for this process is reasonably well predicted from the triangle graphs (c) and (d) of Fig. 6.8, in which the fermion is a proton. We want to preserve this feature in our theory, which involves fundamental quark fields. To get the same sign and magnitude we require

$$2 \sum_h I_h^3 Q_h^2 = 1 \ , \tag{6.131}$$

where I_h^3 is the third component of isospin.

It is interesting to note that the decay $\pi^o \to 2\gamma$ only occurs as a result of the anomaly just discussed. Using PCAC we might relate the matrix $M_{\rho\sigma}$

for $\pi^o \to 2\gamma$ to $M_{\mu\rho\sigma}$ using (6.129). Bose symmetry requires

$$M_{\mu\rho\sigma} = (\mu\rho\sigma\, k_1 - k_2)\, C \ ,$$

so using (6.129) we find

$$m\, M_{\rho\sigma} = -\ (\rho\sigma\, k_1\, k_2)\, C \ .$$

But if we also impose (6.128),we deduce $C = 0$, which implies that the $\pi^o \to 2\gamma$ amplitude $M_{\rho\sigma}$ vanishes in the soft pion limit[21].

It is plain that the four quarks p', p, n, λ, so far included in the theory, do not satisfy the restrictions (6.130) and (6.131). Recalling that $Q_{p'} = Q_p$ and taking $I^3_{p'} = 0$,

$$\sum_h Q_h = \tfrac{2}{3}$$

and

$$2 \sum_h I^3_h Q^2_h = \tfrac{1}{3} \ .$$

A model which does satisfy the restrictions is obtained by multiplying the number of quarks by three. We discussed such a model in connection with the nonleptonic hyperon decays in 5.1.4 . The assumption is that there are three "colours" of quarks: p'_i, p_i, n_i, λ_i $(i = 1,2,3)$. For patriotic reasons the three colours $(i = 1,2,3)$ are usually called "red", "white" and "blue"[22], although some more sensitive spectroscopically prefer "red", "yellow" and "blue"[23]. The baryons are made of three quarks, as usual, but there is one of each colour; this enables one to understand the symmetric baryon ground state, since the three fermions are different. The previous results go through unaffected if, for example, we make the weak and electromagnetic interaction colour symmetric. Another way of achieving the desired result is to take Han-Nambu quarks[24]. The three quartets $(p'_i, p_i, n_i, \lambda_i)$ then have integral charges (1,1,0,0), (1,1,0,0) (0,0,-1,-1).

6.3.6 OUTSTANDING PROBLEMS

Plainly it is premature to conclude that the Weinberg-Salam model in fact describes reality, particularly since its gauge vector mesons $W^{\pm}$, Z and the scalar φ have not been observed experimentally. On the other hand, the idea underlying the construction of this and similar models has an appealing simplicity, despite the apparent complexity of the resultant Lagrangian (6.54). We may add to this the facts that these are the first realistic renormalizable field theories of weak and electromagnetic interactions, and that they obey the high energy unitarity constraints noted in the Introduction to this chapter. Indeed, under fairly general assumptions one can invert the argument, and show that <u>only</u> spontaneously broken non-abelian gauge theories (which is what these theories are) satisfy these constraints[(25)]. It is thus hard to resist the belief that the fundamental theory will be something like these theories.

At the moment it seems only too easy (given the basic idea) to write down unified theories. However, none of those we have considered is entirely satisfactory, even if it does not conflict with the data. Surely a unified theory of weak and electromagnetic interactions should be characterized by a single coupling constant. The Weinberg-Salam model does not have this property (θ_W is <u>arbitrary</u>), because SU(2) x U(1) is not a simple group[(11)]; The Georgi-Glashow model does, but <u>it</u> has no neutral currents other than the electromagnetic current. In any case both of the models are "arbitrary" in the sense that the former has to build in the observed charges of the particles by hand (by adjusting the coupling to B_μ), while the latter has to impose (6.107) in every order of perturbation theory. Equally, none of the models <u>requires</u> the existence of the muon and the electron, let alone predicts their masses or their difference. The trouble is the tautology that, if e and μ enter the theory in a fundamentally different way, it is seemingly impossible to obtain from the theory the e-μ universality that nature exhibits. Of course, it is purely a matter of taste whether one feels that these facts need any explanation. Most particle physicists do think an explanation exists; after all $m_e/m_\mu \simeq 2\alpha/\pi$, and written like this, for example, it seems that an explanation <u>must</u> be forthcoming. Similarly, these powerful new models have not explained the observed Cabibbo angle θ_c, which we simply built into the model under discussion. Again, perhaps no explanation is called for, but then we are saying that there are two weak coupling constants.

Another long standing problem, on which we might expect "the" theory to shed some light, is the neutron-proton mass difference. This is "obviously" of order α, so many electromagnetic explanations have been sought. If one tries to calculate the electromagnetic self masses using Feynman diagrams, one soon encounters the familiar logarithmic divergence of the fermion self energy. In a renormalizable theory, such as we are considering, this can be absorbed into a mass renormalization, but then the masses of n and p are in general arbitrary. However, in some spontaneously broken gauge theories it can happen that the most general Lagrangian requires the n and p masses to be equal, even though this does not correspond to an unbroken symmetry of the original gauge group(11). Since the equality is not the result of a symmetry, it will, in general, not survive in higher order calculations. Further, when we include all exchanges, Z, W, φ as well as γ, it must be finite, since if not it would have to be absorbed in a mass renormalization which was not allowed in the most general Lagrangian. So we can see how the n-p mass difference might be calculable, although it would probably involve the masses of the hitherto unobserved vector mesons and scalars.

We shall see in Chapter 7 that CP-violation has been observed in the decays of the neutral kaons. It is certainly conceivable, and some would say likely, that the fundamental CP-violating interaction is weak. If so, we should certainly require that "the" theory predicted CP-violation in a "natural" way. It can be incorporated into the four quark version of Weinberg's model(26). Also Pais(27) and Lee(28) have shown that it can arise as a spontaneous violation of CP-invariance. But it does not appear that the violation is particularly natural, although Georgi and Pais(29) have related it to universality.

6.4 ASYMPTOTIC FREEDOM ?

We have seen how the inclusion of the hadrons into Weinberg's model leads us to postulate the existence of an additional quark and, subsequently, different "coloured" quarks. We have seen also how some of the outstanding problems of the conventional theory of weak interactions seem to require a more detailed understanding of the strong interactions. For example, the success of the naive quark parton model indicates that, in the deep inelastic region under consideration, the nucleon behaves as if it were composed of three free quark constituents, while in other kinematic regions it does not. The development of unified gauge theories of the weak and electromagnetic interactions leads one to wonder whether the strong interactions might not also be included, and possibly even be essential for the overall coherence of the theory. And maybe these theories would enable us to gain some understanding of the problem alluded to above.

Recent developments suggest that the strong interactions might also be described by a non-abelian gauge theory, and that such theories can indeed explain the "switching off" of the interaction in the deep inelastic processes. These developments are very speculative and the reader is warned here, and elsewhere, of the danger of accepting them too readily. My reason for including them in a book such as this is that they do illustrate the sort of way in which some of the outstanding problems might eventually be solved. At present, as we shall see, these theories have obvious defects, which are such as to render them almost unbelievable. Before describing the work we need to consider a property of renormalizable field theories in general.

6.4.1 THE CALLAN-SYMANZIK EQUATION[30]

We consider the renormalizable real scalar field theory described by the bare Lagrangian:

$$\mathcal{L} = \tfrac{1}{2}\,(\partial_\mu \varphi_0)^2 - \tfrac{1}{2}\,\mu_0^2\,\varphi_0^2 - \frac{1}{4!}\,\lambda_0 \varphi_0^4 \;. \tag{6.132}$$

If we calculate the Feynman diagrams (Green's functions) of the field theory, we shall in general find that they are divergent - just like QED. In order to handle such divergences, we introduce a cut-off Λ, as was done in 4.2.12, for example. These divergences arise when we perform the integration over the internal loop momenta of the diagrams. In fact we only need to consider the "one particle irreducible" Green's functions Γ, which are defined as those diagrams which cannot be cut into two pieces by cutting a single internal line. For example, diagram (a) of Fig. 6.9 is one particle reducible, whilst the two pieces into which it may be cut (shown in (b)) are each one particle irreducible.

Fig. 6.9 One particle reducible and irreducible diagrams.

In fact, diagram (a) is divergent, because of the divergent loop integration in the first diagram of (b); the propagator connecting the two pieces has its momentum fixed by the external momenta, which are finite. We denote by $\Gamma^{(n)}(p, \mu_o, \lambda_o, \Lambda)$ the sum of all the cut off one particle irreducible Green's functions with n external "legs". Thus, both diagrams in (b) of Fig. 6.9 contribute to $\Gamma^{(4)}$; p is a short hand notation to remind us that $\Gamma^{(n)}$ depends upon various external momenta, collectively called "p". The dependence on λ_o and μ_o arises from the vertices and propagators. The Λ-dependence arises from the introduction of the cut off to give integrals definite values. The theory is called "renormalizable", because all of the divergent integrals which arise may be traced back to divergences in $\Gamma^{(2)}$ and $\Gamma^{(4)}$ and may be absorbed into "renormalizations" of quantities appearing in the original Lagrangian, namely mass, coupling constant and the field operator. Specifically, we may absorb the divergences which arise into the <u>constant</u> quantities Z, μ, λ defined as follows

$$\Gamma^{(n)}(p, \mu_o, \lambda_o, \Lambda) = Z^n \bar{\Gamma}^{(n)}(p, \mu, \lambda) , \tag{6.133a}$$

with μ, λ defined by

$$\bar{\Gamma}^{(2)}(p, \mu, \lambda) = i(p^2 - \mu^2) + O(p^4) \tag{6.133b}$$

$$\bar{\Gamma}^{(4)}(p, \mu, \lambda) = - i\lambda + O(p^2) . \tag{6.133c}$$

The "renormalized Green's functions" $\bar{\Gamma}^{(n)}$ are finite functions of p, μ, λ, in the sense that they do not diverge when $\Lambda \to \infty$, and the "bad" Λ-dependence is contained in Z, μ, λ.

Consider, for example, $\Gamma^{(4)}$. This is given, to order λ_o^2, by the sum of the diagrams (a), (b), (c) and (d) of Fig. 6.10.

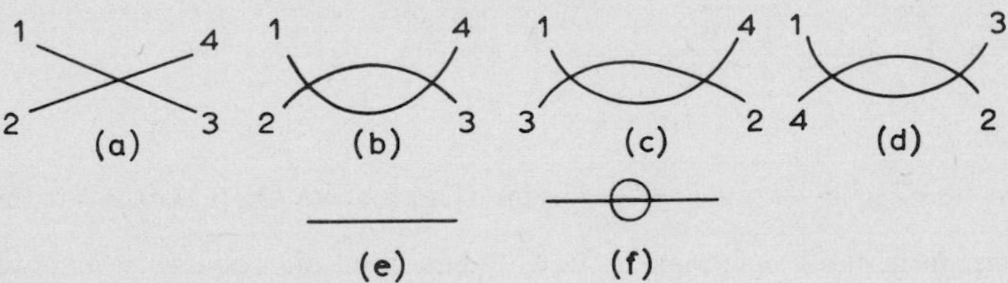

Fig. 6.10 Diagrams contributing to $\Gamma^{(4)}$ and $\Gamma^{(2)}$.

We put all external momenta zero (p = 0), so (b) (c) and (d) are equal, and cut off the divergent integral by introducing a covariant cut off factor $-\Lambda^2(k^2-\Lambda^2)^{-1}$ in the loop integration. Then

$$\Gamma^{(4)}(0,\mu_o,\lambda_o,\Lambda) = -i\lambda_o + 3(\tfrac{1}{2})\lambda_o^2(2\pi)^{-4}\int dk\,(k^2-\mu_o^2)^{-2}(-\Lambda^2)(k^2-\Lambda^2)^{-1} + $$

$$+ O(\lambda_o^3)$$

$$= -i\lambda_o + \frac{3\lambda_o^2}{2(2\pi)^4}\,i\pi^2\left[\ln\frac{\Lambda^2}{\mu_o^2} + \text{finite}\right] + O(\lambda_o^3)\ . \tag{6.134}$$

Similarly, $\Gamma^{(2)}$ is given by the sum of diagrams (e) and (f) so

$$\Gamma^{(2)} = i\,(p^2-\mu_o^2) + O(\lambda_o^2)$$

$$= Z^2\, i\,(p^2-\mu^2) + O(p^4)\ , \tag{6.135}$$

using (6.133a, b). Equating coefficients of p^2 gives

$$Z = 1 + O(\lambda_o^2)\ ,$$

and knowing only this we may calculate λ, using (6.133c),

$$-i\lambda = \bar{\Gamma}^{(4)}(0,\mu,\lambda) = Z^{-4}\Gamma^{(4)}(0,\mu_o,\lambda_o,\Lambda) = \Gamma^{(4)}(0,\mu_o,\lambda_o,\Lambda) + O(\lambda_o^3)\ .$$

So, using (6.134), we find

$$\lambda = \lambda_o - \frac{3\lambda_o^2}{16\pi^2}\ln\frac{\Lambda}{\mu_o} + O(\lambda_o^3)\ . \tag{6.136}$$

Returning to the problem in hand, the Callan-Symanzik equation may be derived[31] from the following trivial property of the $\Gamma^{(n)}$. We noted that the μ_o-dependence arises from the propagators, which have the form $i(k^2 - \mu_o^2)^{-1}$, where k is the momentum carried by the line. Now,

$$\mu_o \frac{\partial}{\partial \mu_o}\left(\frac{i}{k^2 - \mu_o^2}\right) \equiv \frac{i}{k^2 - \mu_o^2}\left(-i2\mu_o^2\right)\frac{i}{k^2 - \mu_o^2} \quad . \tag{6.137}$$

So, if we take $\mu_o \frac{\partial}{\partial \mu_o}$ of any Feynman diagram, we shall obtain a sum of diagrams in which the vertex $(-i2\mu_o^2)$ has been successively inserted in all internal lines of the diagram; notice that the vertex carries zero momentum out of the lines, since the two propagators on the right of (6.137) have the same momenta. In terms of the $\Gamma^{(n)}$, identity (6.137) implies

$$\mu_o \frac{\partial}{\partial \mu_o} \Gamma^{(n)}(p, \mu_o, \lambda_o, \Lambda) = -i\mu_o^2 \Gamma^{(n)}_{\varphi_o^2}(0; p, \mu_o, \lambda_o, \Lambda) , \tag{6.138}$$

where $\Gamma^{(n)}_{\varphi_o^2}$ is the one particle irreducible diagram with the zero momentum insertion of the operator φ_o^2 (which has a vertex 2). In general $\Gamma^{(n)}_{\varphi_o^2}$ will be divergent, and it too must be renormalized

$$\Gamma^{(n)}_{\varphi_o^2}(\Delta; p, \mu_o, \lambda_o, \Lambda) = Z^n Z_1 \bar{\Gamma}^{(n)}_{\varphi^2}(\Delta; p, \mu, \lambda) , \tag{6.139a}$$

where

$$\bar{\Gamma}^{(2)}_{\varphi^2}(\Delta; p, \mu, \lambda) = 2 + O(\Delta^2, \Delta.p, p^2) . \tag{6.139b}$$

Now we rewrite (6.138) in terms of the renormalized Green's function $\bar{\Gamma}$

$$\begin{aligned}\mu_o \frac{\partial}{\partial \mu_o} \Gamma^{(n)}(p, \mu_o, \lambda_o, \Lambda) &= \frac{\partial}{\partial(\ln \mu_o)}\left[Z^n \bar{\Gamma}^{(n)}(p, \mu, \lambda)\right] \\ &= Z^n \left[\frac{\partial(\ln \mu)}{\partial(\ln \mu_o)}\mu\frac{\partial}{\partial \mu} + \frac{\partial \lambda}{\partial(\ln \mu_o)}\frac{\partial}{\partial \lambda} + n\frac{\partial(\ln Z)}{\partial(\ln \mu_o)}\right]\bar{\Gamma}^{(n)}(p, \mu, \lambda) \\ &= -i\mu_o^2 \Gamma^{(n)}_{\varphi_o^2}(0; p, \mu_o, \lambda_o, \Lambda) = -i\mu_o^2 Z^n Z_1 \bar{\Gamma}^{(n)}_{\varphi^2}(0; p, \mu, \lambda) .\end{aligned}$$

Thus

$$\left[\mu\frac{\partial}{\partial \mu} + \beta\frac{\partial}{\partial \lambda} + n\gamma\right]\bar{\Gamma}^{(n)}(p, \mu, \lambda) = -i\mu^2 \delta\, \bar{\Gamma}^{(n)}_{\varphi^2}(0; p, \mu, \lambda) , \tag{6.140a}$$

where

$$\beta \equiv \frac{\partial\lambda}{\partial(\ln\mu_o)} \Bigg/ \frac{\partial(\ln\mu)}{\partial(\ln\mu_o)} \qquad (6.140b)$$

$$\gamma \equiv \frac{\partial(\ln Z)}{\partial(\ln\mu_o)} \Bigg/ \frac{\partial(\ln\mu)}{\partial(\ln\mu_o)} \qquad (6.140c)$$

$$\delta \equiv \left(Z_1\,\mu_o^2/\mu^2\right)\left(\frac{\partial\ln\mu}{\partial(\ln\mu_o)}\right)^{-1} = Z_1 \Bigg/ \frac{\partial\mu^2}{\partial\mu_o^2} \quad . \qquad (6.140d)$$

(6.140) is the Callan-Symanzik equation[30]. Its importance lies in the fact that β, γ, δ are finite, as we shall now show. Taking $n = 2$ and $p = 0$, the boundary conditions (6.133b) and (6.139b) yield

$$\left(\mu\frac{\partial}{\partial\mu} + \beta\frac{\partial}{\partial\lambda} + 2\gamma\right) i(-\mu^2) = -2i\mu^2 + 2\gamma(-i\mu^2)$$

$$= -i\mu^2\,\delta\,2 \quad .$$

Then

$$1 + \gamma = \delta \quad , \qquad (6.141a)$$

and in the same way we can show that

$$\gamma = -\tfrac{1}{2}\,\mu^2\,\delta\,\frac{\partial}{\partial p^2}\,\bar{\Gamma}^{(2)}_{\varphi^2}(0;p,\mu,\lambda)\Bigg|_{p^2=0} \qquad (6.141b)$$

$$\beta + 4\gamma\lambda = \mu^2\,\delta\,\bar{\Gamma}^{(4)}_{\varphi^2}(0;0,\mu,\lambda) \quad . \qquad (6.141c)$$

So far we have kept Λ finite. If we now let $\Lambda\to\infty$, the $\bar{\Gamma}_{\varphi^2}$ are finite, and the above equations are sufficient to show that this implies that β, γ, δ are also finite when $\Lambda\to\infty$. Further, since they are constants, they can only depend upon the renormalized coupling constant λ. In practice they may be calculated as a power series in λ. A similar equation follows for any renormalizable field theory with more fields and/or coupling constants.

It is important to note that β, γ, δ depend upon λ only because of the divergences which forced us to renormalize the theory. <u>If</u> the theory had been finite, that is Γ did not depend upon Λ, then in any renormalization λ and Z would have to be independent of μ_o by dimensional arguments. In that case $\beta = \gamma = 0$ and $\delta = 1$, and the renormalized Green's functions <u>would</u> satisfy an equation precisely analogous to (6.138)

$$\mu \frac{\partial}{\partial \mu} \overline{\Gamma}^{(n)} = - i \mu^2 \overline{\Gamma}^{(n)}_{\varphi^2} . \tag{6.142}$$

This is not true in the theory under consideration, or in any other renormalizable theory, because Z and λ do in fact depend upon μ_o, as is apparent from (6.136). In fact, since (6.135) implies

$$\mu^2 = \mu_o^2 [1 + O(\lambda_o^2)] ,$$

$$\frac{\partial(\ln \mu)}{\partial(\ln \mu_o)} = 1 + O(\lambda_o^2)] .$$

Thus, from the definition (6.140b) of β and (6.136), we find

$$\beta = \frac{3\lambda_o^2}{16\pi^2} + O(\lambda_o^3)$$

$$= \frac{3\lambda^2}{16\pi^2} + O(\lambda^3) . \tag{6.143}$$

So β is indeed non-zero and finite, as promised.

The derivation of the Callan-Symanzik equation which we have presented does not make clear what the physical significance of the equation is. This becomes apparent when we solve it.

Solution in the Deep Euclidean Region

It seems quite evident that at large momenta the inserted Green's function $\overline{\Gamma}^{(n)}_{\varphi^2}$ becomes small compared with the uninserted Green's function $\overline{\Gamma}^{(n)}$; after all, the inserted Green's function has an extra propagator, and at large momenta we expect it to damp the value of $\overline{\Gamma}^{(n)}_{\varphi^2}$. In fact, this is not quite correct. Weinberg has shown[32] that in every order of perturbation theory as $\alpha \to \infty$

$$\overline{\Gamma}^{(n)}(\alpha p) \sim \alpha^{4-n} \text{ (powers of } \ln \alpha) + O(\alpha^{4-n-2}) \tag{6.144a}$$

$$\overline{\Gamma}^{(n)}_{\varphi^2}(\alpha p) \sim \alpha^{4-n-2} \text{ (powers of } \ln \alpha) , \tag{6.144b}$$

provided all momenta p are space-like ($p^2 < 0$). We are concerned with the full Green's function, in which all orders of perturbation theory are included. It is conceivable that the powers of $\ln \alpha$ in the two expressions add up to give a power of α, and that this extra power might make $\bar{\Gamma}^{(n)}_{\varphi^2}$ larger than $\bar{\Gamma}^{(n)}$. Provided this does not happen, $\bar{\Gamma}^{(n)}_{\varphi^2}$ becomes negligible compared with $\bar{\Gamma}^{(n)}$ in the above specified kinematic region, the so-called deep Euclidean region, and in that case the Callan-Symanzik equation is homogeneous and soluble. In this asymptotic region the Green's function satisfies the renormalization group equation[33]

$$\left[\mu\frac{\partial}{\partial\mu} + \beta(\lambda)\frac{\partial}{\partial\lambda} + n\,\gamma(\lambda)\right]\bar{\Gamma}^{(n)}(p,\mu,\lambda) = 0 \ . \tag{6.145}$$

Now, on dimensional grounds alone,

$$\bar{\Gamma}^{(n)}(p,\mu,\lambda) = \mu^{4-n}\,F^{(n)}\left(\frac{p}{\mu},\lambda\right),$$

so

$$\bar{\Gamma}^{(n)}(\alpha p,\mu,\lambda) = \alpha^{4-n}\left(\frac{\mu}{\alpha}\right)^{4-n}F^{(n)}\left(p\frac{\alpha}{\mu},\lambda\right),$$

and by Euler's theorem

$$\left(\mu\frac{\partial}{\partial\mu} + \alpha\frac{\partial}{\partial\alpha}\right)\bar{\Gamma}^{(n)}(\alpha p,\mu,\lambda) = (4-n)\,\bar{\Gamma}^{(n)}(\alpha p,\mu,\lambda) \ .$$

Thus the Callan-Symanzik equation (6.145) may be rewritten as

$$\left[-\alpha\frac{\partial}{\partial\alpha} + \beta(\lambda)\frac{\partial}{\partial\lambda} + 4-n+n\gamma(\lambda)\right]\bar{\Gamma}^{(n)}(\alpha p,\mu,\lambda) = 0 \ . \tag{6.146}$$

This makes it clear that the Callan-Symanzik equation is giving information on the "scaling" behaviour of $\bar{\Gamma}^{(n)}$, i.e. the behaviour if we "rescale" all momenta by a factor α. If β and γ had turned out to be zero, we could immediately solve (6.146)

$$\bar{\Gamma}^{(n)}(\alpha p,\mu,\lambda) = \alpha^{4-n}\,\bar{\Gamma}^{(n)}(p,\mu,\lambda) \ , \tag{6.147}$$

and we see that the scaling behaviour is determined simply by the mass dimension of $\bar{\Gamma}^{(n)}$. Since $\beta,\ \gamma \neq 0$ this is not correct, and we shall see that $\bar{\Gamma}^{(n)}$ has "anomalous (scaling) dimensions".

In fact, (6.146) is a standard partial differential equation, whose solution is given in Piaggio[34], for example. We write (6.146) in the "standard" form, with $\bar{\Gamma}^{(n)}(\alpha p,\mu,\lambda) = f(\alpha,\lambda)$,

$$\alpha\frac{\partial f}{\partial\alpha} - \beta(\lambda)\frac{\partial f}{\partial\lambda} = [4-n+n\gamma(\lambda)]\,f \ ,$$

and seek two independent solutions of the simultaneous equations

$$\frac{d\alpha}{\alpha} = \frac{d\lambda}{-\beta(\lambda)} = \frac{df}{[4-n+n\gamma(\lambda)]\,f} \quad .$$

The first two give one solution, which may be written

$$u \equiv \alpha \exp\left[\int_{\lambda_1}^{\lambda} \frac{dx}{\beta(x)}\right] = c_1 \ ,$$

where c_1 is a constant. This suggests we define a new variable $\bar{\lambda}(\lambda, \alpha)$

$$\alpha \exp\left[\int_{\bar{\lambda}(\lambda,\alpha)}^{\lambda} \frac{dx}{\beta(x)}\right] \equiv 1 \quad , \tag{6.148}$$

which has the property $\bar{\lambda}(\lambda, 1) = \lambda$. Then, in terms of this variable, the above solution is expressible in terms of $\bar{\lambda}$ alone

$$u \equiv \exp\left[\int_{\lambda_1}^{\bar{\lambda}} \frac{dx}{\beta(x)}\right] = c_1 \ .$$

The second two equations give

$$v \equiv f \exp\left[\int_{\lambda_2}^{\lambda} dx \ \frac{4-n+n\gamma(x)}{\beta(x)}\right] = c_2 \ .$$

The general solution of the partial differential equation is then

$$\varphi(u, v) = 0 \ ,$$

or solving for v

$$v = \psi(u) \ ,$$

with φ or ψ arbitrary functions. In our case, if we write f as a function of $\lambda, \bar{\lambda}$,

$$f(\alpha, \lambda) = g(\bar{\lambda}, \lambda) \ ,$$

the general solution is

$$g(\bar{\lambda}, \lambda) \exp\left[\int_{\lambda_2}^{\lambda} dx \ \frac{4-n+n\gamma(x)}{\beta(x)}\right] = \psi\left(\exp\left[\int_{\lambda_1}^{\bar{\lambda}} \frac{dx}{\beta(x)}\right]\right) .$$

So

$$g(\bar{\lambda}, \lambda) = g(\bar{\lambda}, \bar{\lambda}) \exp\left[\int_{\lambda}^{\bar{\lambda}} dx \ \frac{4-n+n\gamma(x)}{\beta(x)}\right] \ .$$

Now, since $\lambda = \bar{\lambda}$ implies $\alpha = 1$,

$$g(\bar\lambda, \bar\lambda) = f(1, \bar\lambda(\lambda,\alpha)) = \bar\Gamma^{(n)}(p,\mu,\bar\lambda(\lambda,\alpha)) \ ,$$

and the general solution is

$$\bar\Gamma^{(n)}(\alpha p,\mu,\lambda) = \bar\Gamma^{(n)}(p,\mu,\bar\lambda(\lambda,\alpha)) \exp\left[\int_\lambda^{\bar\lambda} dx \, \frac{4-n+n\gamma(x)}{\beta(x)}\right]$$

$$= \alpha^{4-n}\,\bar\Gamma^{(n)}(p,\mu,\bar\lambda(\lambda,\alpha))\exp\left\{ n\int_0^{\ln\alpha} dt\,\gamma\left[\bar\lambda(\lambda,e^t)\right]\right\}, \tag{6.149a}$$

where from (6.148)

$$\ln\alpha \equiv \int_\lambda^{\bar\lambda(\lambda,\alpha)} \frac{dx}{\beta(x)} \ . \tag{6.149b}$$

As expected, the α dependence is not confined to the mass dimension α^{4-n}, but occurs also in the combination $\bar\lambda(\lambda,\alpha)$, which we see plays the role of an effective coupling constant. Thus to get any further we must know what happens to $\bar\lambda(\lambda,\alpha)$ as $\alpha\to\infty$.

Now suppose

$$\bar\lambda \to \lambda_F \quad \text{as} \quad \alpha\to\infty \ . \tag{6.150}$$

Then

$$\bar\Gamma^{(n)}(\alpha p,\mu,\lambda) \sim \alpha^{4-n}\,\bar\Gamma^{(n)}(p,\mu,\lambda_F)\,\alpha^{n\gamma(\lambda_F)}\exp\left\{ n\int_0^{\ln\alpha} dt\left(\gamma[\bar\lambda(\lambda,e^t)]-\gamma(\lambda_F)\right)\right\}$$

$$\propto \alpha^{4-n+n\gamma(\lambda_F)} \ , \tag{6.151}$$

provided the integral in the exponential approaches a constant. Thus in this case the "canonical" scaling behaviour given in (6.147) is modified. The Green's function behaves as if each field φ had dimension $1-\gamma(\lambda_F)$. Thus $\gamma(\lambda_F)$ is called the anomalous dimension of the field. But, is the behaviour (6.150) possible ? We see from (6.149b) that this requires

$$\int_\lambda^{\bar\lambda} \frac{dx}{\beta(x)} \to +\infty \quad \text{as} \quad \bar\lambda \to \lambda_F \ , \tag{6.152}$$

which in turn implies

$$\beta(\lambda_F) = 0 \ . \tag{6.153}$$

Thus any possible "fixed points" λ_F must be zeros of $\beta(\lambda)$. Now, one zero of β is guaranteed to exist, since

$$\beta(0) = 0 \ . \tag{6.154}$$

This is clear from what we did earlier. $\lambda = 0$ is the free field theory, and we know that $\beta = 0$ then, as is apparent from (6.138), since we may take $\mu = \mu_0$, $\lambda = \lambda_0$ and Λ is not needed. In fact, $\lambda_F = 0$ is just what we are seeking to explain the success of the naive quark parton model. If $\bar{\lambda} \to 0$ as $\alpha \to \infty$, the high energy behaviour of the field theory is determined by the free field theory, and we could justify our assumption that the constituents of the nucleon were just free quark partons, with no glue binding them.

For the $\lambda\varphi^4$ theory we have been working with, we have shown

$$\beta(\lambda) = \frac{3\lambda^2}{16\pi^2} + O(\lambda^3) \ , \tag{6.143}$$

and as claimed $\lambda = 0$ is a zero of β. So $\beta(\lambda) \to 0+$ as $\lambda \to 0+$. But, since λ must be positive (to give an energy spectrum bounded below), this means that the integral on the left of (6.152) must approach $-\infty$ as $\bar{\lambda} \to 0$. Thus the $\lambda\varphi^4$ becomes a free field theory as $\alpha \to 0$. What we are seeking is an "asymptotically free" theory in which $\bar{\lambda} \to 0$ as $\alpha \to +\infty$. Plainly this requires a (different) field theory for which

$$\beta(x) \to 0- \qquad \lambda \to 0+ \tag{6.155a}$$

$$\beta(x) \to 0+ \qquad \lambda \to 0- \ . \tag{6.155b}$$

6.4.2 ASYMPTOTICALLY FREE GAUGE THEORIES

So far as is known, the <u>only</u> renormalizable field theories satisfying (6.155) are the massless non-abelian gauge theories[35], typical of which is that from which we constructed the Weinberg-Salam model. It must be emphasized that we are referring to the field theory of the gauge vector bosons <u>without</u> the symmetry breaking we used to give them masses; the introduction of scalar fields to make the gauge bosons massive destroys the asymptotic freedom. Thus, in the first place, we are concerned with a field theory described by just the first term of (6.72)

$$\mathcal{L} = -\tfrac{1}{4} F^{\alpha}_{\mu\nu} F^{\alpha\mu\nu} \ , \tag{6.156a}$$

with

$$F^{\alpha}_{\mu\nu} = \partial_{\mu} A^{\alpha}_{\nu} - \partial_{\nu} A^{\alpha}_{\mu} - g\, C^{\alpha\beta\gamma} A^{\beta}_{\mu} A^{\gamma}_{\nu} \quad , \tag{6.156b}$$

and $C^{\alpha\beta\gamma}$ the structure constants of the (non-abelian) strong gauge group G_S. The coupling constant g characterises the strength of the AAA vertex and g^2 the strength of the AAAA vertex, so to calculate $\beta(g)$ in this theory we should need to evaluate the Feynman diagrams for the vertex function $\Gamma^{(3)}$ and self energy $\Gamma^{(2)}$, shown in Fig. 6.11. The wavy lines represent the gauge bosons and the dotted lines are the "Fadeev-Popov ghosts"; these propagate only on internal closed loops and their formal introduction is necessary to quantize singular theories such as those with which we are concerned. As before, these are ultraviolet divergent and are made finite by introducing a cut-off Λ. However, since this is a massless theory, the separation of the Λ-dependence, which is necessary to define the renormalized theory, will require the introduction of a mass μ. We then obtain a Callan-Symanzik equation of the same form as (6.140) with λ replaced by the renormalized coupling constant g. We shall not repeat the derivation of $\beta(g)$. It is clear from Fig. 6.11 that the lowest order contribution is of order g^3, and the

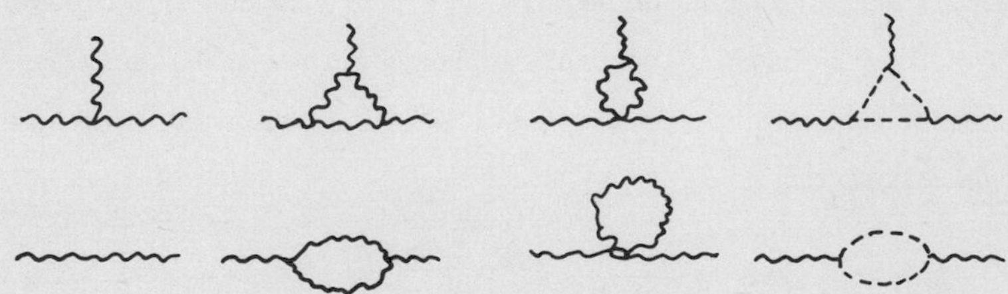

Fig. 6.11 Diagrams contributing to $\Gamma^{(4)}$ and $\Gamma^{(2)}$

result is[35]

$$\beta(g) = -\frac{b g^3}{16\pi^2} + O(g^5) \quad , \tag{6.157a}$$

with

$$b = \frac{11}{3}\, c_2(G_S) \quad , \tag{6.157b}$$

where

$$c_2(G_S)\,\delta^{\alpha\beta} = C^{\alpha\gamma\delta}\,C^{\beta\gamma\delta} \quad . \tag{6.157c}$$

The crucial minus sign in (6.157a) ensures that (6.155) is satisfied. In fact, the negative sign is preserved even if we introduce some (but not too many) fermions into the theory, via the gauge invariant combination given in the third term of (6.72). If we have F multiplets of fermions each in the representation R, then b in (6.157) becomes[35]

$$b = \frac{11}{3}\,c_2(G_S) - \frac{4}{3}\,FT(R) \quad , \tag{6.158a}$$

where

$$T(R)\,\delta^{\alpha\beta} = \mathrm{Tr}\,(R^\alpha R^\beta) \quad . \tag{6.158b}$$

Thus if G_S is SU(3), and the fermions are in the $\underline{3}$ representation, $c_2(G_S) = 3$, $T(R) = \frac{1}{2}$; so $b > 0$, if $F < 16$.

It follows from the definition (6.149b) that $\bar{g}$ satisfies

$$\alpha\frac{\partial}{\partial\alpha}\,\bar{g}(g,\alpha) = \beta[\,\bar{g}(g,\alpha)\,] \quad . \tag{6.159}$$

If g is positive,then, since $\bar{g}(g,1) = g$, we require $\bar{g}$ to decrease as α increases, if the theory is to be asymptotically free. Thus it follows from (6.159) that $\beta(g)$ must be negative. $\bar{g}$ will continue to decrease while β is negative, and $\bar{g}$ will only approach zero if β has no zero between g and the origin. In this case, if β is given by (6.157a), we have from (6.159) that

$$\ln\alpha = \frac{16\pi^2}{2b}\,[\,\bar{g}^{-2} - g^{-2}\,] \quad ,$$

so

$$\frac{\bar{g}}{g} = \left[1 + \frac{g^2}{8\pi^2}\,b\ln\alpha\right]^{-\frac{1}{2}} \quad \text{for } \alpha \gg 1 \quad . \tag{6.160}$$

Thus $\bar{g} \to 0$ as $\alpha \to \infty$, as expected. Now suppose we want to know the scaling behaviour of the Green's function with n external bosons. This is given by (6.149). γ is determined from the two-point function $\Gamma^{(2)}$ and it is clear from Fig. 6.11 that the lowest order contribution is $O(g^2)$

$$\gamma(g) = \frac{dg^2}{16\pi^2} + O(g^4) \quad .$$

Then

$$\int_0^{\ln\alpha} dt\ \gamma[\bar{g}(g,e^t)] = \frac{dg^2}{16\pi^2}\int_0^{\ln\alpha} dt \left[1+\frac{bg^2}{8\pi^2}t\right]^{-1}$$

$$= \ln\left\{\left[1+\frac{bg^2}{8\pi^2}\ln\alpha\right]^{d/2b}\right\}. \quad (6.161)$$

Substituting into (6.149) we find the leading asymptotic behaviour

$$\bar{\Gamma}^{(n)}(\alpha p,\mu,g) \propto \alpha^{4-n}(\ln\alpha)^{nd/2b} . \quad (6.162)$$

So although we have an "asymptotically free theory" , in the sense defined earlier, its scaling behaviour as $\alpha\to\infty$ is not that of a free field theory; instead the free field behaviour α^{4-n} is modified by (calculable) powers of $\ln\alpha$. The source of these logarithms is the fact that $\bar{g}$ vanishes only logarithmically as $\alpha\to\infty$, as is apparent from (6.160). This in turn derives from the proportionality of $\beta(g)$ to g^3, so that $g\beta(g)$ has a double zero in g^2 at $g^2 = 0$. Of course, $\ln\alpha$ is slowly varying, so that this theory can simulate a free field theory over large variations in α . At the present time there is no known field theory which is asymptotically free in the (stronger) sense that its Green's functions possess free field scaling behaviour.

6.4.3 MODELS AND APPLICATIONS

The considerations of the previous section naturally lead one to suspect that the underlying strong interaction field theory must be based on the asymptotically free non-abelian gauge theories. However, to ensure this asymptotic freedom it is necessary that the gauge symmetry is unbroken. As a result the gauge vector bosons will be massless, and we face the immediate problem of explaining why these massless "gluon" fields are not observed in experiment. At present this problem has no satisfying answer, which is why we again caution the reader against too readily accepting the (attractive) speculations in which we are indulging.

Of course, we have been faced with a similar problem in particle physics ever since the advent of SU(3) symmetry. This is most naturally understood by postulating the existence of quarks (p,n,λ), and possible p', if we accept the indications from the weak-electromagnetic gauge group. None of these quarks has ever been seen experimentally, but we can always use the excuse that they are too heavy to be produced by currently available energies, whatever they may be. However, an alternative explanation might be valid. If there is an underlying strong interaction group G_S, as discussed in the last section, then each type of quark would belong to some representation of the group G_S, which might be the "colour" group, for example; then we might have $(p_\alpha, n_\alpha, \lambda_\alpha, p'_\alpha)$ with $\alpha = 1,2,3$ labelling the $\underline{3}$ representation of the colour group SU(3). The ordinary hadrons are presumed to be colour singlets (in general neutral under G_S), and there might be some unexplained dynamical mechanism which prevented the production of colour non-singlet particles in collisions of colour singlet particles[36]; then, since the quarks are colour non-singlets, they could not be produced in ordinary hadronic collisions. This same mechanism would also prevent the production of the colour group gluons, since these too are colour non-singlets. So it might be possible that the colour group G_S is an unbroken symmetry and that the (unobservable) vector gluons have zero mass. At present this is the best "explanation" we have, and with due caution, we proceed to explore some of the dynamical consequences of these theories.

Deep Inelastic Scattering

We saw at the end of 4.2.11 that the hadronic tensor $W_{\mu\nu}$, which determines the deep inelastic structure functions, is the Fourier transform of the product of two current operators at different space-time points. In our simplified analysis we suppress the Lorentz indices on the currents and also their internal symmetry quantum numbers. Thus we consider the product of two scalar currents $J(y)\ J(0)$. We saw also that the kinematic region for deep inelastic scattering is controlled by this product of currents when $y^2 \simeq 0$. The connection with what we have been discussing in this section is made by using Wilson's operator product expansion[37]. This enables us to write

$$J(y)\,J(0) \simeq \sum_n C_n(y)\,O_n(0) \quad \text{for} \quad y^2 \simeq 0 \, , \qquad (6.163)$$

where the C_n are c-numbers and O_n are a complete set of local operators having lowest "twist", i.e. mass dimension minus spin. In the $\lambda\varphi^4$ theory, for example, the operators O_n have twist 2 and are

$$O_n = \varphi \overleftrightarrow{\partial}_{\mu_1} \cdots \overleftrightarrow{\partial}_{\mu_n} \varphi , \tag{6.164}$$

while in theories involving fermion and gauge fields additional operators are needed to make O_n a complete set. In free field theory the above expression derives from the Wick expansion, and the O_n arise from expanding the normal product. On invariance grounds

$$C_n(y) = \mathcal{C}(y^2)\, y^{\mu_1} \cdots y^{\mu_n} . \tag{6.165}$$

In a scale invariant theory, which, aside from the logarithms, is what the asymptotically free gauge theories are, the dimensions of the objects appearing in (6.163) satisfy

$$2\, d_J = d_{C_n} + d_{O_n} . \tag{6.166}$$

In the $\lambda\varphi^4$ theory we might take

$$J(y) = \varphi(y)\, \varphi(y)$$

as our scalar current, so that in mass units

$$d_J = 2 .$$

In the same units (6.164) gives

$$d_{O_n} = n + 2 .$$

Hence (6.166) gives

$$d_{C_n} = 2 - n ,$$

so using (6.165) we have (aside from logarithms)

$$\mathcal{C}_n(y^2) = c_n (y^2)^{-1} . \tag{6.167}$$

Plainly this is singular on the light cone, and we know from free field theory that the correct form is to replace y^2 by $y^2 - i\epsilon y_0$ $(\epsilon > 0)$. So altogether we have[38] for $y^2 \simeq 0$

$$J(y)\, J(0) \simeq \sum_n c_n (y^2 - i\epsilon y_0)^{-1}\, y^{\mu_1} \cdots y^{\mu_n}\, \varphi \overleftrightarrow{\partial}_{\mu_1} \cdots \overleftrightarrow{\partial}_{\mu_n} \varphi , \tag{6.168a}$$

which implies, since $(\alpha \mp i\epsilon)^{-1} = P(\alpha^{-1}) \pm i\pi\delta(\alpha)$, that

$$[J(y), J(0)] \simeq \sum_n 2\pi i c_n \epsilon(y_o)\delta(y^2) y^{\mu_1} \cdots y^{\mu_n} \varphi \overleftrightarrow{\partial}_{\mu_1} \cdots \overleftrightarrow{\partial}_{\mu_n} \varphi \,. \tag{6.168b}$$

Now, we are concerned with the matrix element of (6.168) between spin-averaged single nucleon states $|p\rangle$ of momentum p. Plainly

$$\langle p| \varphi \overleftrightarrow{\partial}_{\mu_1} \cdots \overleftrightarrow{\partial}_{\mu_n} \varphi |p\rangle = a_n p_{\mu_1} \cdots p_{\mu_n} + \cdots ,$$

where $\cdots\cdot$ signifies additional terms containing $g_{\mu_i \mu_j}$; these additional terms yield at least one factor of y^2 when contracted with $y^{\mu_i} y^{\mu_j}$ in (6.168), and so are less singular as $y^2 \to 0$ than the leading term proportional to a_n. Thus the spin averaged matrix element of (6.168) is

$$V(y^2, y.p) \simeq \epsilon(y_o)\, \delta(y^2)\, f(y.p) \,, \tag{6.169a}$$

where

$$f(\lambda) = \sum_n 2\pi i\, a_n c_n \lambda^n \,, \tag{6.169b}$$

and the Fourier transform of this quantity determines the hadronic tensor $W(q^2, \nu)$, as in (4.380),

$$W(q^2, \nu) = 4\pi^2 \int dy\, e^{iqy}\, V(y^2, y.p) \,. \tag{6.170}$$

With the leading behaviour given in (6.169) we may perform the y integration. First we change to the "light-cone variables" introduced in 4.2.11

$$q_\pm = q_o \pm q_3$$

$$y_\pm = y_o \pm y_3 \,.$$

The integration over y_1, y_2 may now be done immediately; defining

$$y_\perp^2 = y_1^2 + y_2^2 \quad \text{and} \quad \delta(y^2) = \delta(y_+ y_- - y_\perp^2) \,,$$

we see that the integral

$$\int dy_1\, dy_2\, \delta(y^2) = \pi \int_0^\infty dy_\perp^2\, \delta(y_+ y_- - y_\perp^2) = \pi\, \theta(y_+ y_-) \,.$$

Hence

$$W(q^2, \nu) = 4\pi^2 \int \tfrac{1}{2} dy_+ dy_-\, e^{\frac{1}{2} i(q_+ y_- + q_- y_+)} \epsilon(y_+ + y_-)\, f(\tfrac{1}{2} m_N (y_+ + y_-))\, \pi\theta(y_+ y_-). \tag{6.171}$$

We have already seen that the Bjorken limit corresponds to taking $q_+ \to \infty$ with q_- fixed, and that this means that the scaling behaviour is controlled by the singularities of the integrand above as $y_- \to 0$. In fact

$$\tfrac{1}{2} i q_+ W(q^2, \nu) = 2\pi^3 \int dy_+ dy_- \left(\partial_- e^{\frac{1}{2} i q_+ y_-} \right) \left(e^{\frac{1}{2} i q_- y_+} \epsilon f \theta \right)$$

$$= -2\pi^3 \int dy_+ dy_- \, e^{\frac{1}{2} i q_+ y_-} \, \partial_- \left(e^{\frac{1}{2} i q_- y_+} \epsilon f \theta \right)$$

$$\simeq -2\pi^3 \int dy_+ dy_- \, e^{\frac{1}{2} i (q_+ y_- + q_- y_+)} \epsilon(y_+ + y_-) f\left(\tfrac{1}{2} m_N (y_+ + y_-) \right) y_+ \delta(y_+ y_-) ,$$

using Gauss' theorem and retaining only the singular derivative of θ

$$\partial_- \theta(y_+ y_-) = y_+ \delta(y_+ y_-) .$$

Performing the y_- integration thus gives

$$q_+ W(q^2, \nu) \simeq 4\pi^3 i \int dy_+ \, e^{\frac{1}{2} i (q_- y_+)} f\left(\tfrac{1}{2} m_N y_+ \right) .$$

Writing $t = m_N y_+ / 2$, and recalling $\nu \simeq \frac{1}{2} q_+$, $x \equiv -q^2/2m_N \nu \simeq -q_-/m_N$, we obtain

$$m_N \nu \, W(q^2, \nu) = 4\pi^3 i \int dt \, e^{-ixt} f(t)$$

$$\equiv F(x) . \tag{6.172}$$

Thus the assumption of scale invariance does lead to Bjorken scaling as observed experimentally and discussed in 4.2.11. Inverting the Fourier transform in (6.172) gives

$$\int_0^1 dx \, e^{ix\lambda} F(x) = 8\pi^4 i f(\lambda) ,$$

so

$$i^n \int_0^1 dx \, x^n F(x) = 8\pi^4 i f^{(n)}(0)$$

$$= -16\pi^5 n! \, a_n c_n , \tag{6.173}$$

using (6.169b). In other words, the "moments" of the structure function $F(x)$ $(\int x^n F(x))$ are proportional to c_n.

Now, the asymptotically free theories are not quite scale invariant, as we have seen. As a result, the c_n in (6.169b) depend upon $\ln y^2$, and the structure function $F(x)$ and its moments depend logarithmically upon q^2. The Callan-Symanzik equation enables us to calculate these logarithmic derivations from Bjorken scaling, as we shall now show. We start from the inserted Green's function $\bar{\Gamma}^{(N)}_{J(q)J(-q)}$ - meaning a renormalized Green's function with N external lines and two insertions of the operator J, one carrying away momentum q and the other bringing in momentum q. Using the Fourier transform of (6.163), we may express this Green's function in terms of other inserted Green's functions

$$\bar{\Gamma}^{(N)}_{J(q)J(-q)} \simeq \sum_n \tilde{C}_n(q)\, \bar{\Gamma}^{(N)}_{O_n} , \tag{6.174}$$

where $\tilde{C}_n$ is the Fourier transform of C_n and $\bar{\Gamma}^{(N)}_{O_n}$ is the Green's function with O_n inserted and carrying zero momentum out of the diagram. All of these inserted Green's functions satisfy Callan-Symanzik equations analogous to (6.140)

$$\left[\mu \frac{\partial}{\partial\mu} + \beta \frac{\partial}{\partial\lambda} + N\gamma + 2\gamma_J\right] \bar{\Gamma}^{(N)}_{JJ} = -i\mu^2\, \delta\, \bar{\Gamma}^{(N)}_{\varphi^2 JJ} \tag{6.175a}$$

$$\left[\mu \frac{\partial}{\partial\mu} + \beta \frac{\partial}{\partial\lambda} + N\gamma + \gamma_n\right] \bar{\Gamma}^{(N)}_{O_n} = -i\mu^2\, \delta\, \bar{\Gamma}^{(N)}_{\varphi^2 O_n} , \tag{6.175b}$$

where γ_J and γ_n are the "anomalous dimensions" of J and O_n calculated, as in (6.140c), from the renormalization constants necessary to define the inserted Green's functions $\bar{\Gamma}^{(N)}_{JJ}$, $\bar{\Gamma}^{(N)}_{O_n}$. Substituting (6.174) into (6.175a), we may derive the equation satisfied by the $\tilde{C}_n$

$$\left(\mu \frac{\partial}{\partial\mu} + \beta \frac{\partial}{\partial\lambda}\right) \tilde{C}_n(q) = (\gamma_n - 2\gamma_J)\, \tilde{C}_n(q) , \tag{6.176}$$

where we have used another consequence of (6.163) that

$$\bar{\Gamma}^{(N)}_{\varphi^2 JJ} = \sum_n \tilde{C}_n(q)\, \bar{\Gamma}^{(N)}_{\varphi^2 O_n} .$$

Thus $\tilde{C}_n$ satisfies a renormalization group equation analogous to (6.145) which may be solved, as before, to give the (anomalous) dependence upon

q^2.

We have performed this analysis in the unrealistic case of scalar currents, no internal quantum numbers and a $\lambda\varphi^4$ field theory. A similar, but more complicated, analysis obtains if we take vector or axial vector currents, include internal quantum numbers (SU(3), charm, colour ....) and use an asymptotically free gauge theory involving gauge fields and fermion fields[39]. Conservation of charge requires that γ_J for the electromagnetic current is identically zero, and because of current algebra this ensures that all $\gamma_J = 0$. Thus $\tilde{C}_n$ is controlled by the anomalous dimension γ_n of the operator O_n. In the more realistic cases referred to above there are several operators O_n^i for each n ; for example, we have boson and fermion operators for each n. In addition, the inserted Green's functions are not multiplicatively renormalized. Instead we have "mixing" for each n.

$$\Gamma^{(N)}_{O_n^i} = Z^N \sum_j Z^{(n)}_{ij} \bar{\Gamma}^{(N)}_{O_n^j} ,$$

and the upshot is that (6.176) becomes a matrix equation for the column vector $\underline{\tilde{C}}_n$ in terms of the anomalous dimension matrix $\underset{\approx}{\gamma}_n$. This may be diagonalized to give a set of equations like (6.176) for each eigenvector $\underline{\tilde{C}}_n^{(\alpha)}$ controlled by the eigenvalue $\gamma_n^{(\alpha)}$ of $\underset{\approx}{\gamma}_n$. On dimensional grounds alone

$$\underline{\tilde{C}}_n(q) = \underline{f}_n\left(\frac{-q^2}{\mu^2}\right)\left(-q^2\right)^{-n} q_{\mu_1}\cdots q_{\mu_n} ,$$

and solving the (diagonalized) renormalization group equations, as in (6.162), gives the q^2 dependence of the eigenvectors

$$\underline{f}_n^{(\alpha)}\left(-\frac{q^2}{\mu^2}, g\right) \propto \left(\ln\frac{-q^2}{\mu^2}\right)^{-d_n^{\alpha}/2b} , \tag{6.177a}$$

where

$$\gamma_n^{(\alpha)} = d_n^{(\alpha)} g^2/16\pi^2 + O(g^3) . \tag{6.177b}$$

So in general $\underline{\tilde{C}}_n$ is a linear combination of (known) powers of $\ln(-q^2)$. We have seen that the structure function moments are proportional (in this case) to the linear combination $\tilde{C}_n^i\, a_n^i$, where a_n^i is determined by the matrix elements of O_n^i between spin averaged single nucleon states. These matrix elements are in general unknown, so the structure function moments are in general given by an unknown linear combination of known powers of $\ln(-q^2)$. As a result it will be extremely difficult to detect these logarithmic deviations from Bjorken scaling in the deep inelastic processes. In particular cases, however, it may be more feasible. For example, the structure function moments corresponding to SU(3) non-singlet combinations of current are controlled by a single power of $\ln(-q^2)$. Also, one linear combination of the $n = 2$ operators is the energy-momentum tensor, which has no anomalous dimensions, and has a known matrix element. So the $n = 2$ moments are also controlled by a single power of $\ln(-q^2)$. Another difficulty is that we do not know at what values of $-q^2$ to expect scaling and its logarithmic deviations to be observed, because we do not know the renormalization mass μ. So even if deviations from Bjorken scaling are observed, we can always say that this is because the scaling region has not yet been reached. Recent raw data from NAL[40] do show some q^2-variation, but it is certainly premature to conclude that deviations from Bjorken scaling (logarithmic or otherwise) have actually been observed; for instance, the observed variation might easily be accounted for by radiative corrections, which are not yet well understood.

Purely Hadronic Weak Interactions

Another application of asymptotic freedom has been in the area of the purely hadronic weak processes[41], which we discussed in Chapter 5. In a unified gauge theory these will certainly be mediated by a W-boson and possibly by the Higgs scalar. Thus the invariant matrix element for the purely hadronic weak process $i \to f$ will be

$$M(i \to f) \sim \alpha \int dx \int dk\, e^{ikx}\, D^{\mu\nu}(k, m_W) < f \,|\, T\left\{ \mathcal{J}_\mu^+(x)\, \mathcal{J}_\nu^-(0) \right\} | \, i > + \cdots \tag{6.178}$$

where $D^{\mu\nu}$ is the W-boson propagator and $\cdots$ indicates the contributions from other possible exchanges. In unified theories the semi-weak coupling constant is proportional to e, which is why we have exhibited the proportionality to α. The W-propagator has a denominator $k^2 - m_W^2$, so the dominant contribution to the integral is when $k^2 \simeq m_W^2$, which is "large". The rapidly oscillating e^{ikx} will then damp contributions from the x integration except when x (not just x^2) is small. So we see that the W-contribution to M is controlled by the "short distance" behaviour of the current product, $T\{\mathcal{J}_\mu^+(x), \mathcal{J}_\nu^-(0)\}$. The short distance expansion[42] is similar to (6.165) but simpler; because x (y in (6.163)) is small, we do not need to retain the whole series, but just the leading contributing terms. The result is that the W-exchange contribution is

$$M_W(i \to f) \sim \alpha\, m_W^{-2} \sum_k c_k \left(\ln \frac{m_W^2}{\mu^2} \right)^{-d_k/2b} \langle f | O_k | i \rangle , \qquad (6.179)$$

where the O_k are the dominant operators and d_k is derived from their anomalous dimensions. These anomalous dimensions depend on the SU(3) quantum numbers of O_k; the SU(3) 27 operators will in general have anomalous dimensions different from the 8 operators. Thus we see that we might be able to understand the observed enhancement of $|\Delta I| = \frac{1}{2}$ transitions relative to the $|\Delta I| = \frac{3}{2}$ ones (just as Wilson conjectured[37]), which was discussed in the previous chapter. In fact it does seem that this is precisely what happens[41]. Taking $m_W \simeq$ 60-100 GeV, $\mu \simeq$ 0.3-1 GeV gives enhancement factors of 5-10 depending on the unified theory used and the strong interaction gauge model. This is too small to explain completely the observed enhancement, which is of the order of 15-20, although it might explain the previously noted fact that these processes seem to have amplitudes of order G rather than $G \sin\theta_c$.

Both of these applications illustrate the power of field theories such as these to explain in a satisfying way hitherto intractable dynamical phenomena. This is why they are so attractive to theorists despite the unpalatable features we have already described.

Chapter 7

CP-VIOLATION

The neutral kaon system has for many years been a source of vital information on the nature of the weak interactions. In 1964 the system provided the first evidence that CP-invariance, which we have hitherto assumed, was not in fact an exact symmetry[1]. Let us see why we are certain that CP-invariance is violated.

We have already observed that the states of relevance in the weak interactions are those superpositions of $|K^o>$ and $|\bar{K}^o>$ which have well defined lifetimes. The two states observed experimentally are denoted $|K_L^o>$, $|K_S^o>$, with L and S signifying that one has a long lifetime and the other a short one. It has for a long time been known that the K_S^o decays primarily to two pion states such as $\pi^+\pi^-$, while K_L^o decays into three pions such as $3\pi^o$. We have also observed (in Chapter 5) that angular momentum conservation and Bose symmetry show that these final states have opposite behaviour under the operation $\mathcal{CP}$

$$\mathcal{CP}|\pi^+\pi^->_{\ell=0} = |\pi^+\pi^->_{\ell=0}$$

$$\mathcal{CP}|\pi^o\pi^o\pi^o>_{\ell=0} = - |\pi^o\pi^o\pi^o>_{\ell=0} \; .$$

Christensen et al[1] showed that a small percentage of K_L^o mesons decay into the state $|\pi^+\pi^->$ besides the predominant modes such as $|3\pi^o>$. In other words K_L^o decays both into a state which is even under CP and into one which is odd, which implies

$$<2\pi|\tilde{\mathcal{H}}|K_L^o> \neq 0 \quad \text{and} \quad <3\pi^o|\tilde{\mathcal{H}}|K_L^o> \neq 0 \quad ,$$

where $\tilde{\mathcal{H}}$ represents the operator responsible for the decay. This shows that <u>some</u> interaction must violate CP-invariance. It <u>might</u> be $\tilde{\mathcal{H}}$ itself.

On the other hand $\tilde{\mathcal{H}}$ might be CP-invariant (and therefore equal to the $\mathcal{H}_W^H$ of Chapter 5); if so,

$$(\mathcal{C}\mathcal{P})\,\tilde{\mathcal{H}}\,(\mathcal{C}\mathcal{P})^{-1} = \tilde{\mathcal{H}} \quad ,$$

and we may be certain that K_L^o is not an eigenstate of $\mathcal{C}\mathcal{P}$. In this case some interaction other than $\tilde{\mathcal{H}}$ would be the source of the CP impurity of the K_L^o state. Thus, although we are sure that CP is not an exact symmetry, we still do not know that the fundamental CP-violating interaction is a weak interaction. We shall see that the observed violation is tantalizingly small (which is why it was not detected earlier), so that, even if the weak interaction is the source of CP-violation, it is believed to be a good approximation to neglect the violation (as we have done) except in the special circumstances that we shall discuss.

Ten years after the experimental discovery of CP-violation there is one "firm favourite" for the role of accepted theory and a number of "other runners" which have not been eliminated. But it is still not clear how this theory is to be grafted on to the main body which we have discussed at length. For this reason I prefer to reverse the approach used hitherto; we shall first discuss the experimental situation and then describe some of the many theories which have been advanced.

7.1 THE NEUTRAL KAON SYSTEM

The neutral kaons K^o and $\bar{K}^o$, having hypercharges $Y = +1$ and -1 respectively, are produced in strong interactions, which conserve hypercharge. However they can only decay via weak interactions which, as we have seen, do not conserve hypercharge. Since their electrical charges are the same, both kaons can decay into the same final state (such as $\pi^+\pi^-$) having zero hypercharge - the K^o by a $\Delta Y = -1$ weak process and the $\bar{K}^o$ by a $\Delta Y = +1$ process. In fact, the kaons are the only case where a particle (stable against strong decays) and its distinct antiparticle can have the same decay products. This is clear, because any candidates must be neutral and have zero baryon number, since charge and baryon number are conserved in all interactions. Thus the only possibilities are the neutral pion, which is its own antiparticle, and the neutral kaon. As a result of this unique feature the K^o state may be connected to the $\bar{K}^o$ state by the weak S-matrix, the first contribution arising in order G^2. A typical contribution is from the $\pi^+\pi^-$ intermediate state

$$K^o \leftrightarrow \pi^+\pi^- \leftrightarrow \bar{K}^o .$$

The (weak) S-matrix may be expanded in powers of G (c.f. (3.92))

$$S = 1 - i\int dx\, \mathcal{H}_W(x) + \frac{(-i)^2}{2!}\int dx\,dy\; T\left\{\mathcal{H}_W(x),\, \mathcal{H}_W(y)\right\} + O(G^3) \quad , \tag{7.1}$$

where $\mathcal{H}_W$ is the full weak Hamiltonian and is proportional to G; the term quadratic in $\mathcal{H}_W$ is responsible for the above mentioned $K^o \leftrightarrow \bar{K}^o$ transition. The (T-matrix) quantity

$$T \equiv \mathcal{H}_W - \tfrac{1}{2} i \int dy\; T\left\{\mathcal{H}_W\, , \mathcal{H}_W(y)\right\} + \cdots \tag{7.2}$$

therefore has the role of <u>effective</u> weak Hamiltonian which is <u>not</u>, however, hermitian. If we restrict our attention only to matrix elements between linear combinations of K^o and $\bar{K}^o$, plainly T is represented by a general 2 x 2 matrix. Including the strong diagonal mass contribution, this leads to a two-channel Wigner-Weisskopf equation describing the time dependence(2) of a Schrödinger picture state which is a coherent superposition of K^o and $\bar{K}^o$

$$i\,\frac{d\Psi}{dt} = (M - \tfrac{1}{2} i \Gamma)\, \Psi \; , \tag{7.3a}$$

where

$$\Psi = \begin{pmatrix} a(t) \\ \bar{a}(t) \end{pmatrix} \tag{7.3b}$$

represents the state

$$|\Psi(t)> = a(t)\,|K^o> + \bar{a}(t)\,|\bar{K}^o> . \tag{7.3c}$$

M and Γ are 2 x 2 hermitian matrices, so $M - \frac{1}{2} i \Gamma$ is a general 2 x 2 matrix. So without loss of generality we may write

$$(M - \tfrac{1}{2} i \Gamma) \equiv a_o I_2 + a_1\sigma_1 + a_2\sigma_2 + a_3\sigma_3 \; , \tag{7.4}$$

where a_i (i = 0,1,2,3) are complex numbers and σ^i (i = 1,2,3) are the 2 x 2 Pauli matrices. We have chosen the phases of the K^o, $\bar{K}^o$ states such that under the operation of charge conjugation $\mathcal{C}$ and parity reversal $\mathcal{P}$:

$$\mathcal{C}\ |K^o> = |\bar{K}^o>\ ,\ \mathcal{C}\,|\bar{K}^o> = |K^o>$$

$$\mathcal{P}\ |K^o> = -\,|K^o>,\ \mathcal{P}\,|\bar{K}^o> = -|\bar{K}^o>\ .$$

Thus

$$\mathcal{C}\mathcal{P}\,|K^o> = -\ |\bar{K}^o> \quad \text{and} \quad \mathcal{C}\mathcal{P}\,|\bar{K}^o> = -\ |K^o>\ . \tag{7.5}$$

It follows that $\mathcal{P}$ is represented by the 2 x 2 matrix

$$\begin{pmatrix} 0 & -1 \\ -1 & 0 \end{pmatrix} \equiv -\ \sigma_1\ . \tag{7.6}$$

Under the anti-linear operation of time reversal $\mathcal{T}$

$$\mathcal{T}\left(a\,|K^o> + \bar{a}\,|\bar{K}^o>\right) = a^*|K^o> + \bar{a}^*\,|\bar{K}^o>\ . \tag{7.7}$$

So $\mathcal{T}$ is represented simply by the operation (K) of complex conjugation. Hence $\mathcal{R} = \mathcal{T}\mathcal{C}\mathcal{P}$ is represented by $-K\sigma_1$. If we assumed TCP-invariance,

$$\mathcal{R}\,S\,\mathcal{R}^{-1} = S^{\dagger}\ . \tag{7.8}$$

The † arises from the anti-linear character of $\mathcal{R}$, which changes the i's in (7.1). Thus TCP-invariance implies

$$(-K\sigma_1)\,(M - \tfrac{1}{2}i\,\Gamma)\,(-K\sigma_1)^{-1} = (M - \tfrac{1}{2}i\,\Gamma)^{\dagger}\ .$$

The left-hand side gives

$$K\,(a_o I_2 + a_1\sigma_1 - a_2\sigma_2 - a_3\sigma_3)\,K^{-1}$$

$$= a_o^*\ I_2 + a_1^*\sigma_1 + a_2^*\ \sigma_2 - a_3^*\,\sigma_3\ ,$$

since σ_2 is purely imaginary. The right hand side is

$$a_o^*\ I_2 + a_1^*\ \sigma_1 + a_2^*\ \sigma_2 + a_3^*\ \sigma_3\ .$$

Thus

$$\text{TCP-invariance} \Rightarrow a_3 = 0\ . \tag{7.9a}$$

In the same way

$$\text{T-invariance} \Rightarrow a_2 = 0\ , \tag{7.9b}$$

$$\text{CP-invariance} \Rightarrow a_2 = a_3 = 0\ . \tag{7.9c}$$

However for the moment we prefer to be quite general. $M - \frac{1}{2}i\,\Gamma$ has two complex normalized eigenvectors Ψ_i $(i = 1, 2)$ with corresponding eigenvalues λ_i

$$(M - \tfrac{1}{2} i \Gamma)\, \Psi_i = \lambda_i\, \Psi_i \qquad \text{(no summation)} , \tag{7.10a}$$

where

$$\Psi_i = \begin{pmatrix} p_i \\ q_i \end{pmatrix} , \quad |p_i|^2 + |q_i|^2 = 1 . \tag{7.10b}$$

We may separate λ_i into real and imaginary parts

$$\lambda_i \equiv m_i - \tfrac{1}{2} i \gamma_i . \tag{7.11}$$

Plainly the eigenvalues determine 4 of the 8 real parameters describing $M - \frac{1}{2} i \Gamma$, while the eigenvectors determine the remaining 4. Explicitly we find

$$\lambda_i = a_o \pm a , \tag{7.12a}$$

where

$$a \equiv (a_1^2 + a_2^2 + a_3^2)^{\frac{1}{2}} . \tag{7.12b}$$

The corresponding eigenvectors are given by (7.10b) with (up to a phase factor)

$$p_i = (a_i - i a_2) \left(|a_1 - i a_2|^2 + |a_3 \mp a|^2 \right)^{-\frac{1}{2}} \tag{7.13a}$$

$$q_i = -(a_3 \mp a) \left(|a_1 - i a_2|^2 + |a_3 \mp a|^2 \right)^{-\frac{1}{2}} . \tag{7.13b}$$

From (7.3) we see that the associated Schrödinger states $\Psi_i(t)$ evolve purely exponentially with time

$$\Psi_i(t) = \Psi_i(0) \exp(-i\lambda_i t) . \tag{7.14}$$

Defining

$$N_i(t) \equiv \Psi_i^\dagger(t)\, \Psi_i(t) ,$$

we see that

$$N_i(t) = N_i(0) \exp(-\gamma_i t) .$$

Thus the normalization of the state Ψ_i cannot be constant if $\gamma_i \neq 0$. But this is just what we expect, since we know that Ψ_i can decay into states outside of the neutral kaon system, such as $\pi \ell \nu$, 2π, etc. As a result, the probability N_i of finding the system in the state Ψ_i must decrease with time, so

$$\gamma_i > 0 \quad , \tag{7.15}$$

and the γ_i are the mean lifetimes of the two eigenstates. By observing the decay of a neutral kaon beam the quantities γ_i and m_i may be measured. However, we see from (7.11) and (7.12) that the determination of these quantities alone is not sufficient to ascertain whether a_2 or a_3 or both are non-zero. But the probability N of finding the system in any state Ψ must also decrease with time. Using (7.3) we see that for any Ψ

$$\begin{aligned} -\frac{d}{dt}(\Psi^\dagger \Psi) &= \Psi^\dagger (iM + \Gamma/2)\Psi + \Psi^\dagger(-iM + \Gamma/2)\Psi \\ &= \Psi \Gamma \Psi \ . \end{aligned} \tag{7.16}$$

It follows that Γ must be positive definite. If we substitute an arbitrary state

$$\begin{aligned} \Psi(t) &= c_1 \Psi_1(t) + c_2 \Psi_2(t) \\ &= c_1 e^{-i\lambda_1 t} \Psi_1 + c_2 e^{-i\lambda_2 t} \Psi_2 \end{aligned}$$

into (7.16), we may deduce

$$(\lambda_2^* - \lambda_1)\Psi_2^\dagger \Psi_1 = i\Psi_2^\dagger \Gamma \Psi_1 \ . \tag{7.17}$$

Then the Schwartz Inequality yields[3]

$$|\lambda_2^* - \lambda_1| \, |\Psi_2^\dagger \Psi_1| \leq \left\{(\Psi_2^\dagger \Gamma \Psi_2)(\Psi_1^\dagger \Gamma \Psi_1)\right\}^{\frac{1}{2}} = (\gamma_1 \gamma_2)^{\frac{1}{2}} \ . \tag{7.18}$$

Now experimentally[4]

$$\gamma_2 \equiv \gamma_S = 1.144 \times 10^{10} \ s^{-1} \tag{7.19a}$$

$$\gamma_1 \equiv \gamma_L = 18.9 \times 10^6 \ s^{-1} \tag{7.19b}$$

$$m_L - m_S = 0.54 \times 10^{10} \ s^{-1} \ . \tag{7.19c}$$

So (7.18) gives a bound on $|\Psi_2^\dagger \Psi_1|$

$$|\Psi_2^\dagger \Psi_1| < 0.06 \ .$$

In fact one can do rather better than this. By applying the Schwartz inequality in each of the separate decay channels (2π, 3π, $\pi\ell\nu$, etc.) which contribute to the right of (7.17), Bell and Steinberger[5] conclude

$$|\Psi_2^\dagger \Psi_1| < 0.006 \ . \qquad (7.20)$$

We may use this inequality to learn something about a_2 and a_3. The observed CP-violation is a small effect (as we shall see shortly), so from (7.9c) a_2 and a_3 must be small. We therefore work to first order in these small quantities. Then

$$\lambda_1 - \lambda_2 = 2a \simeq 2a_1 \qquad (7.21a)$$

and

$$\Psi_2^\dagger \Psi_1 = p_2^* p_1 + q_2^* q_1 \simeq \mathrm{Im}\,\epsilon_2 + i\,\mathrm{Im}\,\epsilon_3 \ , \qquad (7.21b)$$

where

$$a_2 = a_1\epsilon_2 \ , \quad a_3 = a_1\epsilon_3 \ . \qquad (7.21c)$$

Thus from (7.20) we deduce

$$|\mathrm{Im}\,\epsilon_2 + i\,\mathrm{Im}\,\epsilon_3| < 0.006 \ . \qquad (7.22)$$

We may use the strong interaction data on the production of neutral kaons to obtain further information. The ratio of the cross sections for $p\bar{p} \to K_1^o K^- \pi^+ \pi^o$ and $p\bar{p} \to K_1^o K^+ \pi^- \pi^o$ gives a measure(6) of $|q_2/p_2|$

$$|q_2/p_2|^2 = 1 \pm 0.03 \ .$$

Using the above approximation this gives

$$|\mathrm{Im}\,\epsilon_2 - \mathrm{Re}\,\epsilon_3| < 0.015 \ . \qquad (7.23)$$

Thus combining (7.22) and (7.23) we obtain a bound on $|\epsilon_3| = |a_3/a_1|$. From (7.19) we see that $\lambda_1 - \lambda_2$ is known and therefore (from (7.21a)) a_1 is also known. Combining all of these gives(7)

$$|a_3| < 0.008\,\gamma_S = 1.2 \times 10^{-16}\, m_K \ . \qquad (7.24)$$

This looks like pretty good evidence that TCP is indeed an exact symmetry of nature. However, the dimensionless measure of any TCP-violation depends upon the quantity with which we scale a_3; this is determined by the nature of the interaction which violates the symmetry. If the violation occurs in the strong interaction, of which m_K is a characteristic quantity, then (7.24) indicates that TCP is conserved to about one part in 10^{16}. On the other hand, if the violation were to reside in the $\Delta Y = 0$ weak interaction, we should conclude that TCP is conserved to one part in 10^{10} or so. However, if the violation is weak but with $|\Delta Y| \neq 0$, then a_3 must be a second order weak effect, of which γ_S is characteristic. In this case we should conclude that the upper limit on TCP-violation is only of the order of 1%, which is of the same order as the known CP-violation, as we shall see.

There are of course strong theoretical reasons for believing in TCP-invariance, which can be proved for a local Lorentz invariant field theory[8]. We shall therefore assume it henceforth; the further analysis of the neutral kaon system is thereby considerably simplified.

With this assumption, $a_3 = 0$ and we see from (7.12) and (7.13) that (changing the phase) we may take

$$\Psi_{\substack{1\\2}} = (1 + |r|^2)^{-\frac{1}{2}} \begin{pmatrix} 1 \\ \pm\, r \end{pmatrix}. \tag{7.25}$$

In the notation of (7.12) and (7.13),

$$r = a\,(a_1 - i\,a_2)^{-1} .$$

So,

$$2\,a_o = \lambda_1 + \lambda_2 \tag{7.26a}$$

$$2\,a_1 = (\lambda_1 - \lambda_2)\,\tfrac{1}{2}\,(r + r^{-1}) \tag{7.26b}$$

$$2\,i\,a_2 = (\lambda_1 - \lambda_2)\,\tfrac{1}{2}\,(r - r^{-1}) \tag{7.26c}$$

$$a_3 = 0 . \tag{7.26d}$$

Then

$$\Psi_2^\dagger\,\Psi_1 = \frac{1 - |r|^2}{1 + |r|^2} , \tag{7.27}$$

and using (7.20) we see that

$$|\,1 - |r|^2\,| < 0.012 . \tag{7.28}$$

Thus r is a number having modulus close to unity. In fact, using the notation of (7.21),

$$r \simeq 1 + i\,\epsilon_2 , \tag{7.29a}$$

and from (7.22), since $a_3 = 0$,

$$|\operatorname{Im} \epsilon_2| < 0.006 \ . \qquad (7.29b)$$

We are now in a position to analyse the processes in which CP-violation has actually been observed.

7.2 EXPERIMENTAL DATA ON CP-VIOLATION

7.2.1 TWO-PION DECAY MODES OF THE NEUTRAL KAONS

It is apparent from (7.19) that one of the neutral kaon modes Ψ_1 has a much longer lifetime than the other Ψ_2. We shall henceforth denote the modes by Ψ_L and Ψ_S corresponding to the long-lived K_L and short-lived K_S mesons. Now suppose we have a beam of neutral kaons (K^o and $\overline{K}^o$) in flight having been produced at some target by strong interactions. The state of the system, in our picture, is a linear superposition of K^o and $\overline{K}^o$, or equivalently of K_L and K_S. If we know the velocity (energy) of the beam, we may set up experimental apparatus sufficiently far from the production point as to ensure the virtual absence of any K_S component; the K_L lifetime is 600 times that of K_S, so, if we set up at a distance of 300 K_S lifetimes down beam, then the beam is essentially 100% pure K_L. This is what Christensen et al[1] did in their famous experiment. They observed, of course, the $K_L \to 3\pi$ decays discussed in Chapter 5. But they also saw a small number of 2π events - much larger than could be explained by the odd K_S still surviving in the beam. The only conclusion is that the K_L decays via both the two-pion and three-pion modes, and we have already observed that this implies a violation of CP-invariance.

We write the invariant amplitude for the decay of a K^o into a two-pion state with isospin $I (= 0, 2)$ in the form

$$M(K^o \to 2\pi, I) \equiv a(I) \exp[i\delta_o(I)] \ , \qquad (7.30)$$

where $\delta_o(I)$ in the s-wave $\pi\pi$ phase shift introduced in 5.1.1. Then TCP-invariance requires

$$M(\overline{K}^o \to 2\pi, I) = -a(I)^* \exp[i\delta_o(I)] \ . \qquad (7.31)$$

Thus from (7.25) we may write down the amplitudes for $K_{L,S} \to 2\pi$

$$M(K_{\substack{L\\S}} \to 2\pi, I) = (1+|r|^2)^{-\frac{1}{2}} [a(I) \mp r\, a(I)^*] \exp[i\delta_o(I)] \ . \tag{7.32}$$

Now CP-invariance implies

$$r = \pm 1 \quad \text{and} \quad a(I) = a(I)^* \ . \tag{7.33}$$

The first follows from (7.9b) and (7.26c), and the second from (7.30) and (7.31) using (7.5). Thus only one of the modes can decay into two pions, if CP is exact. Since we know experimentally that the K_S^o certainly does and the K_L^o usually not, we conclude

$$r \simeq 1 \ . \tag{7.34}$$

This is the justification of our previous approximation (7.29). The quantities actually measured by the experimentalists are

$$\eta^{+-} \equiv \frac{M(K_L \to \pi^+\pi^-)}{M(K_S \to \pi^+\pi^-)} \tag{7.35a}$$

$$\eta^{oo} \equiv \frac{M(K_L \to \pi^o\pi^o)}{M(K_S \to \pi^o\pi^o)} \ . \tag{7.35b}$$

Both of these vanish if there is exact CP-invariance. It is a straightforward matter to express the $\pi^+\pi^-$ and $\pi^o\pi^o$ states in terms of the $I = 0,2$ states. Only the relative phase of $a(2)$ and $a(0)$ is observable, so without loss of generality we may take $a(0)$ to be real and positive. Then we may write

$$\eta^{+-} = (\epsilon + \epsilon')(1 + \omega/\sqrt{2})^{-1} \tag{7.36a}$$

$$\eta^{oo} = (\epsilon - 2\epsilon')(1 - \sqrt{2}\,\omega)^{-1} \ , \tag{7.36b}$$

where

$$\epsilon \equiv (1-r)(1+r)^{-1} \tag{7.36c}$$

$$\epsilon' \equiv \frac{1}{\sqrt{2}} [a(2) - r\, a(2)^*](1+r)^{-1} a(0)^{-1} e^{i\Delta} \tag{7.36d}$$

$$\omega \equiv [a(2) + r\, a(2)^*](1+r)^{-1} a(0)^{-1} e^{i\Delta} \tag{7.36e}$$

$$\Delta \equiv \delta_o(2) - \delta_o(0) \ . \tag{7.36f}$$

We see from (7.33) and (7.34) that ϵ and ϵ' vanish if CP-invariance is exact, so, if either is found to be non-zero, CP-invariance is violated. From the definitions it is apparent that ϵ is characteristic purely of the decaying states, while ϵ' depends both on the decaying state (via r) and on the nature of the $K \to 2\pi$ interaction amplitude $a(I)$. This illustrates the point made at the beginning of this chapter. If the weak interaction is CP-invariant $a(I) = a(I)^*$, but we would still have CP-violation if $r \neq 1$. On the other hand, if the weak interaction is not CP-invariant, then $r \neq 1$ anyway. The parameter ω reduces to the value (5.31c) when CP is exact; since it is proportional to the $I = 2$ $K \to 2\pi$ amplitudes, it is a measure of the extent to which the $|\Delta I| = \frac{1}{2}$ rule is violated in the 2π decays of K_S^o. Equation (5.34) indicates that $|\omega| \sim 0.05$. Thus to fairly good approximation we may neglect it and take

$$\eta^{+-} \simeq \epsilon + \epsilon' \tag{7.37a}$$

$$\eta^{oo} \simeq \epsilon - 2\epsilon' . \tag{7.37b}$$

Now we present the data. We write

$$\eta^{+-} = |\eta^{+-}| e^{i\varphi^{+-}} , \quad \eta^{oo} = |\eta^{oo}| e^{i\varphi^{oo}} ,$$

and the latest measurements give[(4,10)]

$$|\eta^{+-}| = (2.17 \pm 0.07)\, 10^{-3} \tag{7.38a}$$

$$\varphi^{+-} = (45.1 \pm 1.3)^o \tag{7.38b}$$

$$|\eta^{oo}| = (2.25 \pm 0.09)\, 10^{-3} \tag{7.38c}$$

$$\varphi^{oo} = (49.1 \pm 13.2)^o . \tag{7.38d}$$

Nobody looking at these data could resist conjecturing that $\eta^{+-} = \eta^{oo}$, so that $\epsilon' = 0$. In fact the latest fit gives

$$|\eta^{oo}/\eta^{+-}| = 1.008 \pm 0.041 \tag{7.39a}$$

and

$$|\epsilon'/\epsilon| < 0.02 . \tag{7.39b}$$

Of course, the data have not always been so clear cut. The remarkable thing is that the "clear favourite" theory, referred to at the beginning, predicted all of this (and more) shortly after the original experiment which established the existence of CP-violation.

7.2.2 LEPTONIC DECAY MODES OF THE NEUTRAL KAONS

The neutral kaon system also has semileptonic decays which can be studied for CP-violating effects. We assume that the leptons enter the weak Hamiltonian only in the combination L_λ of Chapter 3. Then the hadrons must enter via a "current" whose $|\Delta Y| = 1$ pieces we denote by $\mathscr{J}_\lambda^\pm$. Using only TCP-invariance as expressed in (4.482), the hadronic matrix elements may be written in the form

$$< \pi^+ | \mathscr{J}_\lambda^+ | \bar{K}^o > = g_+ (q^2)(K+\pi)_\lambda + g_-(q^2) q_\lambda = - < \pi^- | \mathscr{J}_\lambda^- | K^o >^*$$

$$< \pi^+ | \mathscr{J}_\lambda^+ | K^o > = h_+ (q^2)(K+\pi)_\lambda + h_-(q^2) q_\lambda = - < \pi^- | \mathscr{J}_\lambda^- | \bar{K}^o >^* .$$

The transitions in the second equation can only proceed if $\mathscr{J}_\lambda^\pm$ has a piece with quantum numbers satisfying $\Delta Y = -\Delta Q$. We may combine these amplitudes to give those for K_L. The q_λ term gives m_ℓ when contracted with the leptonic current, so we neglect it; obviously this is an excellent approximation for electronic modes at least. We also ignore the q^2 dependence of g_+ and h_+. Then

$$\frac{\Gamma(K_L \to \pi^- \ell^+ \nu_\ell)}{\Gamma(K_L \to \pi^+ \ell^- \bar{\nu}_\ell)} = \frac{|g_+^* + r h_+^*|^2}{|h_+ + r g_+|^2} = \frac{|1 + r^* x|^2}{|x + r|^2} ,$$

where

$$x \equiv \frac{h_+}{g_+}$$

measures the $\Delta Y = -\Delta Q$ amplitude relative to that with $\Delta Y = +\Delta Q$. Since we know ϵ is small, we may work to first order, and then

$$\delta_L(\ell) \equiv \frac{[\Gamma(\ell^+) - \Gamma(\ell^-)]}{[\Gamma(\ell^+) + \Gamma(\ell^-)]} = 2 \operatorname{Re} \epsilon \; \frac{1 - |x|^2}{|1 + x|^2} ,$$

Thus a measurement of δ_L enables us to calculate $\operatorname{Re}\epsilon$, provided x is known independently. The data are as follows[4]

$$\operatorname{Re} x = 0.000 \pm 0.022 \qquad (7.40a)$$

$$\operatorname{Im} x = 0.012 \pm 0.030 \qquad (7.40b)$$

$$\delta_L = (3.4 \pm 0.1)\, 10^{-3} . \qquad (7.40c)$$

Together these imply

$$\mathrm{Re}\,\epsilon = (1.72 \pm 0.10)\,10^{-3} \quad . \tag{7.40d}$$

7.3 MODELS OF CP-VIOLATION

It is clear from the results presented in the last section that CP-violation is a well established experimental fact. Thus from (7.9c) a_2 and/or a_3 must be non-zero. If we assume TCP-invariance, $a_3 = 0$ (as in (7.9a)), so a_2 must be non-zero. In which case (7.9b) shows that T-invariance is also violated. In fact, a detailed analysis of K^o decays yields the conclusion that T-invariance is violated even without the assumption of TCP-invariance. We shall therefore use the two terms interchangably. The magnitude of the observed effect is characterized dimensionlessly by

$$|\epsilon| \sim 10^{-3} \quad , \tag{7.41}$$

or from (7.26)

$$|a_2/a_1| \sim 10^{-3} \quad . \tag{7.42}$$

Now from (7.12) and (7.19)

$$|a_1| \sim \gamma_S \sim 10^{-5}\ \mathrm{eV} \quad ,$$

so

$$|a_2| \sim 10^{-8}\ \mathrm{eV}$$

is the energy characteristic of the CP-violating interaction. This explains why the effect is negligible in many other processes. The task of the theorist (after the event) has been to come up with models of the CP-violating interaction which "predict" (7.41) and (hopefully) make other falsifiable predictions. We start with the present "clear favourite".

7.3.1 SUPERWEAK THEORY[11]

The off-diagonal terms of $M - \frac{1}{2}i\Gamma$, to which a_2 contributes, are those connecting the K^o and $\overline{K}^o$ states. So one possibility of explaining the asymmetry a_2 is that the fundamental T-(or CP-) violating interaction $\mathcal{H}^T$ has selection rule $|\Delta Y| = 2$, and that the non-zero value of a_2 arises as a first-order effect in the characteristic coupling constant F of $\mathcal{H}^T$. The 'ordinary' off-diagonal terms a_1 arise for the first time in second order in the weak coupling constant G, since $\mathcal{H}_W$ has $|\Delta Y| \leq 1$. So, if we use m_N to give a dimensionless measure of F and G, the above model will give

$$|a_2/a_1| \sim (F m_N^2)/(G m_N^2)^2 .$$

Thus from (7.42) this requires

$$F \sim 10^{-8} G , \tag{7.43}$$

since $G m_N^2 \sim 10^{-5}$. Since F is eight orders of magnitude smaller than G, Wolfenstein's theory is called "superweak". The fundamental CP-violating interaction is <u>not</u> the ordinary weak interaction which gives rise to the $K \to 2\pi$ transition, so

$$a(2) = a(2)^* .$$

Consequently (7.36d, e) imply

$$\epsilon' = \frac{1}{\sqrt{2}} \epsilon \omega . \tag{7.44}$$

Substituting into (7.36 a, b) we deduce

$$\eta^{+-} = \epsilon = \eta^{oo} . \tag{7.45}$$

Further, working in first order in ϵ, (7.26c) gives

$$2 i a_2 = (\lambda_L - \lambda_S) \tfrac{1}{2}(r^2 - 1)/r \simeq -(\lambda_L - \lambda_S) 2\epsilon .$$

But, since a_2 arises as a <u>first</u> order effect in $\mathcal{H}^T$ in this model, it is <u>real</u>. Thus, since $\epsilon(\lambda_L - \lambda_S)$ is purely imaginary,

$$\mathrm{Re}\, \epsilon\, (m_L - m_S) + \tfrac{1}{2}\, \mathrm{Im}\, \epsilon\, (\gamma_L - \gamma_S) = 0 .$$

So the phase φ_ϵ of ϵ satisfies

$$\tan \varphi_\epsilon = \frac{2(m_L - m_S)}{\gamma_S - \gamma_L} , \tag{7.46a}$$

and using the data (7.19) this gives (modulo π)

$$\varphi_\epsilon = (43.8 \pm 0.02)^o . \tag{7.46b}$$

(7.45) shows that this is also the phase of η^{+-} and η^{oo}. Thus in addition to (7.45) the superweak theory predicts

$$\varphi^{+-} = \varphi^{oo} = (43.8 \pm 0.02)^o .$$

Using the experimental value of $|\eta^{+-}|$ given in (7.38a) this gives

$$\text{Re}\,\epsilon = 1.57 \pm 0.05 \ , \tag{7.47}$$

which agrees well with value (7.40d) found in the $K_{\ell 3}$ decays.

The superweak theory successfully explains three of the four real numbers (7.38) characterizing the observed CP-violation, but the overall magnitude is not predicted. The obvious attraction of this theory is that it was advanced before really definitive data was available, and that it has stood the test of time. Further, since the new interaction is so fantastically weak, it has the additional feature (seductive for theorists) that it does not require any drastic overhaul of the rest of the theory.

We can see how this arises another way. The Feynman diagram giving rise to the CP-violating $K_L \to 2\pi$ is shown in Fig. 7.1.

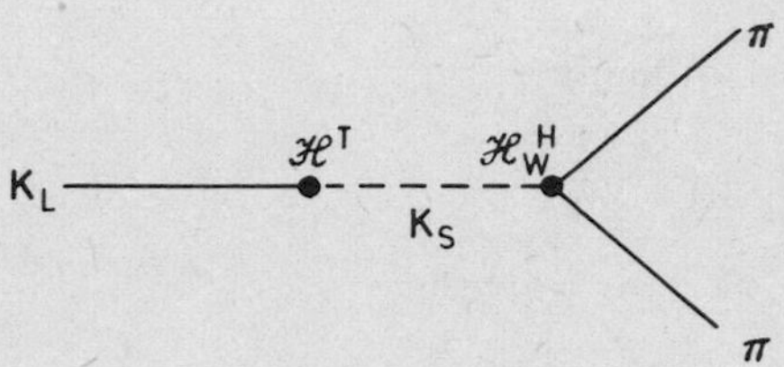

Fig. 7.1 Feynman diagram for $K_L \to 2\pi$ in Superweak theory.

The superweak interaction $\mathcal{H}^T$ gives a minute $K_L - K_S$ vertex whose effect is magnified by the minute energy denominator, which occurs because the $K_L - K_S$ mass difference is a second order weak effect. We may use this picture to fix the magnitude of the T-violating matrix element. Evidently,

$$\begin{aligned} M(K_L \to 2\pi) &= M(K_S \to 2\pi)\,\frac{1}{m_L^2 - m_S^2} < K_S | \mathcal{H}^T | K_L > \\ &= 2.10^{-3}\ M(K_S \to 2\pi)\ , \end{aligned}$$

using the data (7.38). Thus

$$\begin{aligned} < K_S | \mathcal{H}^T | K_L > &\simeq 2.10^{-3}\ (2m_K\,\Delta m) \\ &\simeq 4.10^{-17}\ m_K^2\ , \end{aligned} \tag{7.48}$$

since

$$\frac{\Delta m}{m_K} \equiv \frac{m_L - m_S}{m_K} = 10^{-14} .$$

We have already alluded to the uniqueness of the neutral kaon system in having this mixing feature which makes possible the small energy denominator. Thus the superweak theory makes the additional negative prediction that CP-violation will be immensely difficult to detect outside of the neutral kaon system.

One area where extremely accurate measurements have been and continue to be made is in the attempts to detect an electric dipole moment for the neutron. The presence of a non-zero dipole moment requires P- and T-violation. In the superweak theory a dipole moment could arise via the Feynman diagrams shown in Fig. 7.2,

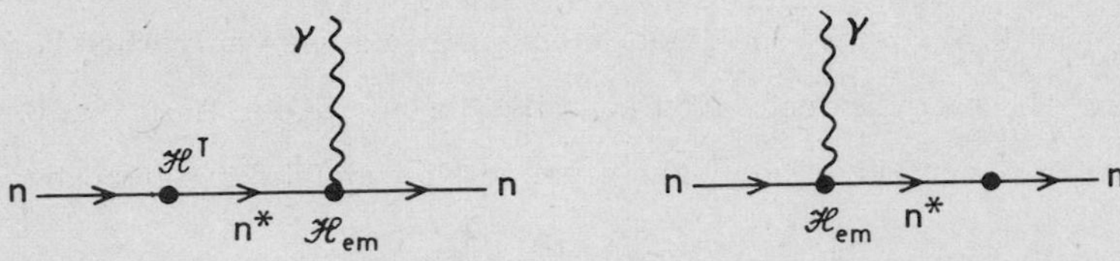

Fig. 7.2 Feynman diagrams for the electric dipole moment of the neutron.

and we assume that the electromagnetic field interacts via its normal C-, P- and T-conserving interaction. The electric dipole moment D_n of the neutron is then given by[12]

$$D_n \simeq \langle n|\mathcal{D}|n^*\rangle (m^{*2} - m_n^2)^{-1} \langle n^*|\mathcal{H}^T|n\rangle + \langle n|\mathcal{H}^T|n^*\rangle (m^{*2} - m_n^2)^{-1} \langle n^*|\mathcal{D}|n\rangle , \tag{7.49}$$

where $\mathcal{D}$ is the electric dipole moment operator. The intermediate states n^* have opposite parity to the neutron, and the T-violating property of $\mathcal{H}^T$ is required so that the two terms do not cancel. We estimate

$$\langle n|\mathcal{D}|n^*\rangle \sim e\, m_n^{-1} (m_n + m^*) \tag{7.50a}$$

$$m^* - m_n \sim m_K \tag{7.50b}$$

$$\langle n^*|\mathcal{H}^T|n\rangle \sim \langle K_S|\mathcal{H}^T|K_L\rangle \ . \tag{7.50c}$$

The first two are justified if n^* is identified with the 1550 S_{11} state[13]. Combining (7.49), (7.50) and (7.48) gives

$$D_n \sim 2.10^{-17}\ e\,m_n^{-1} \sim 10^{-30}\ e\ \text{cm.}\ ,$$

and to be safe Wolfenstein concludes

$$D_n < 10^{-29}\ e\ \text{cm.}\ . \qquad (7.51)$$

Of course, all of the above <u>assumes</u> that the superweak Hamiltonian $\mathcal{H}^T$ possesses a <u>parity-violating</u> piece with quantum numbers $\Delta Y = 0$ which couples with the same strength as the piece responsible for $K_L \to 2\pi$. This latter piece, it will be recalled, has $|\Delta Y| = 2$ and is <u>parity-conserving</u>, as is apparent from (7.48). Thus there is a whole class of superweak theories which includes, in addition to the above discussed version, models in which $\mathcal{H}^T$ is either parity-conserving or has $|\Delta Y| \neq 0$. For these the prediction for D_n will be five or six orders of magnitude lower than (7.51), since the dipole moment then arises only when both $\mathcal{H}^T$ and $\mathcal{H}_W^H$ are present. But even the "large" upper limit (7.51) is six orders of magnitude below the present experimental upper bound[14]

$$|D_n| < 1.0 \times 10^{-23}\ e\ \text{cm.} \qquad (7.52)$$

7.3.2 MILLIWEAK THEORIES

The milliweak theories are so called because they are designed so that the CP-violating effects <u>generally</u> show up as a 10^{-3} correction to the regular CP-conserving weak interaction. Thus in these theories $\mathcal{H}^T$ has a piece with $|\Delta Y| = 1$ (and/or 3), and the non-zero value of a_2 arises from interference between $\mathcal{H}^T$ and the ordinary $\mathcal{H}_W^H$. Consequently

$$|a_2/a_1| \sim F\,m_N^2\ G\,m_N^2/(G\,m_N^2)^2\ ,$$

and (7.42) requires

$$F \sim 10^{-3}\,G\ .$$

Hence the term "milliweak"[15]. Of course, these theories have to get the neutral kaon decays "right", and in this area their predictions usually coincide with the superweak predictions. This is achieved by making the piece of $\mathcal{H}^T$ responsible for $K_L \to 2\pi$ have pure $I = \frac{1}{2}$. This gives a(2) and a(0) a small phase difference ξ, without changing their magnitudes. Then

$$\epsilon' \simeq \frac{1}{\sqrt{2}}\,\omega(\epsilon + i\xi) \quad ,$$

and from (7.36a, b) we find

$$\eta^{+-} \simeq \epsilon \simeq \eta^{oo} \quad ,$$

since $|\omega|$ and $|\xi|$ are small. Thus milliweak theories can approximate the superweak prediction (7.45). And particular milliweak theories - the "isoconjugate" theories[16] - can reproduce (7.45) exactly. This is achieved by arranging that the (parity-violating) piece of $\mathcal{H}^T$, which contributes to $K \to 2\pi$, has both $I = \frac{1}{2}$ and $I = \frac{3}{2}$ pieces in such a way that a(2) and a(0) are rotated equally, without changing their magnitudes. Then ξ is zero and (7.44) and (7.45) follow.

To get the remaining superweak predictions we have to go back to (7.17) and evaluate the right-hand side

$$\Psi_2^\dagger \, \Gamma \, \Psi_1 = \sum_n M(K_S \to n)^* \, M(K_L \to n) \quad . \qquad (7.53)$$

This follows from (7.2), since Γ is non-zero only by virtue of the second order term. The sum in (7.53) is over all possible states n. Of these the two and three-pion states are likely to be dominant, since these are the principal decay modes of the neutral kaons. Experimentally CP-violation in the three-pion modes is found to be small, so using (7.53) and (7.17) we have

$$(\lambda_S^* - \lambda_L)(\Psi_S^\dagger \, \Psi_L) \simeq i[\,\eta^{+-}\,\Gamma(K_S \to \pi^+\pi^-) + \eta^{oo}\,\Gamma(K_S \to \pi^o\pi^o)\,]$$

$$= i\,\epsilon\,\gamma_S \quad ,$$

since $\eta^{+-} = \eta^{oo} = \epsilon$ by construction. Using (7.27) and (7.36c) we have

$$\Psi_S^\dagger \, \Psi_L \simeq 2\,\mathrm{Re}\,\epsilon \quad ,$$

so

$$[\,(m_S - m_L) + \tfrac{1}{2} i (\gamma_S + \gamma_L)\,]\, 2\,\mathrm{Re}\,\epsilon \simeq i\,(\mathrm{Re}\,\epsilon + i\,\mathrm{Im}\,\epsilon)\;\gamma_S \quad .$$

Hence

$$\tan \varphi_\epsilon \simeq \frac{2(m_L - m_S)}{\gamma_S + \gamma_L} .$$

But $\gamma_S \gg \gamma_L$, so this gives a value of φ_ϵ essentially the same as (7.46a). Thus, by the use of this extra dynamical assumption concerning the saturation of the Bell-Steinberger sum rule, we see how the milliweak theories can simulate the superweak theory[17] within the neutral kaon system.

The difference between the two classes arises outside of the neutral kaon system. In general the milliweak theories will predict CP-violation at the 10^{-3} level in other systems, which the superweak theory does not. Unfortunately the experimental searches for CP-violation rarely reach the level of accuracy necessary for the detection of such a small effect. A recent experimental search for T-violating effects in the beta decay of ^{19}Ne failed to show any evidence for these at the 10^{-3} level[18]. Although there is no reason why the milliweak Hamiltonian should have a $\Delta Y = 0$ piece, this must be taken as (fragile) evidence against this class of theories.

The milliweak theories are also likely to give larger values for the neutron's dipole moment than the superweak limit (7.51). A rough estimate would be

$$D_n \sim 10^{-3} \,(G m_N^2/4\pi)\,(e/m_N) \sim 10^{-23} \text{ e cm.} ,$$

where $10^{-3}\,G$ is the milliweak coupling constant. A more realistic and model-independent estimate was made by Barton and White[13]. They assumed that the selection rules of $\mathcal{H}^T$ were the same as those for $\mathcal{H}_W^H\,(\Delta Y = 0)$ in which, we recall, $I = 1$ occurs only when multiplied by $\sin^2 \theta_c$, see (5.134). In the present case a T- and P-violating pion nucleon vertex $\bar{N}N\pi_3$ $(I = 1)$ can be formed arising from the $\cos^2 \theta_c$ part of the Hamiltonian. Thus their result should be enhanced by a factor of $\cot^2 \theta_c$, but damped by a factor of m_π/m_N, because only the neutral pion is coupled[19]. Thus overall we expect a dipole moment about three times larger than their estimate which would be

$$|D_n| \sim 1.5 \times 10^{-24} \text{ e cm.}$$

This is one order of magnitude smaller than the present experimental bound (7.52). It will obviously be most interesting when the accuracy of the experiments can be increased. However, even if D_n remains consistent with zero, the milliweak theory will not have been falsified. One simply has to arrange that $\mathcal{H}^T$ cannot contribute directly to D_n. For example, if $\mathcal{H}^T$ has only $|\Delta Y| = 1$ then it can contribute to D_n only in conjunction with another $|\Delta Y| = 1$ weak interaction. As a result the predicted value will be suppressed to the superweak bound. Thus the milliweak theories are difficult to falsify. However, the extent to which they are accepted will depend on the contortions which are required to accommodate the data.

7.3.3 ELECTROMAGNETIC CP-VIOLATION

If the fundamental CP-violating Hamiltonian has $\Delta Y = 0$, it can contribute to the $K^o - \overline{K}^o$ transition only when accompanied by an ordinary second order weak interaction. Thus

$$|a_2/a_1| \sim F\,(G\,m_N^2)^2 / (G\,m_N^2)^2 ,$$

so

$$F \sim 10^{-3} \sim \alpha/\pi \quad ,$$

where $\alpha = \frac{1}{137}$ is the fine structure constant. This observation led to the suggestion [20] that CP-violation was electromagnetic in origin, and further that it could be a maximal effect in these interactions. Thus the observed CP-violation in $K \to 2\pi$ might result from the T-violating radiative corrections to the ordinary CP-conserving amplitude. Since the limits on P-violation in electromagnetic processes are very small, it was suggested that the electromagnetic current contains a piece K_λ, which is even under C and T, as well as the normal part j_λ, which is odd under C and T [20]. The calculation of radiative corrections to nonleptonic processes cannot be done reliably, so only order of magnitude arguments are available to explain the observed values of η^{+-} and η^{oo}. Plainly the existence of K_λ should lead to observable C- and T-violation in electromagnetic decays. However no evidence of any such violation has been observed. Of course, the prediction of C-violation in any particular process can be avoided by judicious choice of the assumed isospin properties of K_λ.

One might expect the neutron's dipole moment to be the acid test for this class of models. If the electromagnetic interaction is T-violating and P-conserving, we need also the ordinary T-conserving and P-violating weak interaction to generate D_n. Then roughly

$$|D_n| \sim (G\, m_N^2/4\pi)\,(e/m_N) \sim 10^{-20}\, e\, cm.$$

However, the observed effect may be suppressed, because of the soft photon involved in defining the dipole moment. With large suppression the dipole moment may arise only when a virtual photon is exchanged; this provides an additional $\alpha/\pi \sim 10^{-3}$. A realistic calculation of the dipole moment arising with electromagnetic CP-violation yields[(21)]

$$|D_n| \sim 4 \times 10^{-23}\, e\, cm.$$

It <u>is</u> possible to get lower values, as the result depends upon a "mixing angle". However the above is the characteristic scale - lower values require an explanation of the mixing angle. For this reason the present experimental limit (7. 52) is significant, but not conclusive, evidence against electromagnetic CP-violation.

Appendix

FIERZ IDENTITIES

The Fierz identities derive from the observation that the direct product of any two matrices $A_{ij}\ B_{k\ell}$ may be decomposed into a sum of other direct products :

$$A_{ij}\ B_{k\ell} = \sum_r C^{(r)}_{i\ell}\ D^{(r)}_{kj} \quad . \tag{A.1}$$

This follows from the trivial identity

$$A_{ij}\ B_{k\ell} = \sum_{p,q} (A_{ip}\ \delta_{q\ell})\ (B_{kq}\ \delta_{pj}) \quad .$$

In practice they are used when A, B are particular combinations of γ-matrices. The behaviour (2.58a) of these under Lorentz transformations restricts the possible combinations which appear on the right of (A.1). For example, if $A = B = I_4$, (2.58a) requires that

$$\begin{aligned}(I)_{ij}\ (I)_{k\ell} = {} & a_S\ (I)_{i\ell}\ (I)_{kj} + a_P\ (\gamma_5)_{i\ell}\ (\gamma_5)_{kj} \\ & + a_V\ (\gamma^\alpha)_{i\ell}\ (\gamma_\alpha)_{kj} + a_A\ (\gamma^\alpha\gamma_5)_{i\ell}\ (\gamma_\alpha\gamma_5)_{kj} \\ & + a_T\ (\sigma^{\alpha\beta})_{i\ell}\ (\sigma_{\alpha\beta})_{kj} \quad .\end{aligned} \tag{A.2}$$

In writing down (A.2) we have used, in addition, the property that the left-hand side transforms as a scalar, rather than a pseudoscalar, under the space-inversion transformation, see (2.89); thus (I) (γ_5) terms, for example, do not appear on the right-hand side. To find the five constants a_S, a_P, a_V, a_A, a_T, we multiply the equation successively by $(I)_{jk}$, $(\gamma_5)_{jk}$, $(\gamma_\rho)_{jk}$, $(\gamma_\rho\gamma_5)_{jk}$, $(\sigma_{\rho\sigma})_{jk}$ and use the trace properties given in section 2.2.8. This yields $a_S = a_P = a_V = -\ a_V = 2\ a_T = \frac{1}{4}$. Thus in an obvious notation we have

$$I \otimes I = \tfrac{1}{4}\{ I \times I + \gamma_5 \times \gamma_5 + \gamma^\alpha \times \gamma_\alpha - \gamma^\alpha \gamma_5 \times \gamma_\alpha \gamma_5 + \tfrac{1}{2}\sigma^{\alpha\beta} \times \sigma_{\alpha\beta} \}. \tag{A.3}$$

This is the fundamental identity from which all others may be deduced. For example, if we multiply (A.2) or (A.3) by $(\gamma_\rho a)_{pi}\ (\gamma^\rho a)_{\ell q}$, the left-hand side is

$$(\gamma_\rho a) \otimes (\gamma^\rho a), \qquad \text{where} \qquad a = \tfrac{1}{2}(1-\gamma_5)$$

$$a' = \tfrac{1}{2}(1+\gamma_5) \quad .$$

The first term on the right of (A.3) is then

$$a_S(\gamma_\rho a I \gamma^\rho a) \times I = a_S(\gamma_\rho \gamma^\rho a' a) \times I = 0 \quad .$$

In the same way a_P and a_T do not contribute. The a_V term is

$$a_V(\gamma_\rho a \gamma^\alpha \gamma^\rho a) \times \gamma_\alpha = a_V(\gamma_\rho \gamma^\alpha \gamma^\rho a) \times \gamma_\alpha$$

$$= -2\, a_V(\gamma^\alpha a) \times \gamma_\alpha \quad .$$

Similarly a_A contributes

$$a_A(\gamma_\rho a \gamma^\alpha \gamma_5 \gamma^\rho a) \times (\gamma_\alpha \gamma_5) = -2\, a_A(\gamma^\alpha a) \times (\gamma_\alpha \gamma_5) \quad ,$$

and finally we have

$$(\gamma_\rho a) \otimes (\gamma^\rho a) = -(\gamma^\alpha a) \times (\gamma_\alpha a) \quad . \qquad \text{(Q.E.D.)}$$

In the same way it follows that

$$(\gamma_\rho a') \otimes (\gamma^\rho a') = -(\gamma^\alpha a') \times (\gamma_\alpha a')$$

$$(\gamma_\rho a) \otimes (\gamma^\rho a') = 2\, a' \times a$$

$$(\gamma_\rho a') \otimes (\gamma^\rho a) = 2\, a \times a' \quad .$$

Of course, with sufficient energy one can compute all 15 identities giving the decomposition of the general product $\Gamma \times \gamma$, where

$$\Gamma = 1, \gamma_5, \gamma_\rho, \gamma_\rho \gamma_5, \sigma_{\rho\sigma} \qquad \text{and} \qquad \gamma = 1, \gamma_5, \gamma_\lambda, \gamma_\lambda \gamma_5, \sigma_{\lambda\mu} \quad .$$

We leave this as an exercise for the energetic reader who seeks practice in manipulating the γ-matrices.

REFERENCES for Preface

1. R.P. Feynman and Gell-Mann, Phys. Rev. 109, 143 (1959).
2. E. Fermi, Nuovo Cim. 11, 1 (1934); Z.f. Phys. 88, 161 (1934).
3. R.E. Marshak, Riazuddin and C.P. Ryan, Theory of Weak Interactions in Particle Physics (Wiley-Interscience, 1969); E.D. Commins, Weak Interactions (McGraw-Hill, 1973).
4. L.B. Okun, Weak Interactions of Elementary Particles (Pergamon, 1965).
5. J.D. Bjorken and S.D. Drell, Relativistic Quantum Mechanics (McGraw-Hill, 1965); Relativistic Quantum Fields (McGraw-Hill, 1965).
6. M. Gell-Mann and Y. Ne'eman, The Eightfold Way (Benjamin, 1964); P. Carruthers, Introduction to Unitary Symmetry (Wiley, 1966).

REFERENCES for Chapter 2

1. P.A.M. Dirac, Proc. Roy. Soc. A117, 610 (1928); A118, 351 (1928); Principles of Quantum Mechanics, 4th edition (Oxford, 1958).
2. J.D. Bjorken and S.D. Drell, Relativistic Quantum Fields, pp. 112, 116 (McGraw-Hill, 1965).
3. G. Luders, Dansk. Mat. Fys. Medd. 28, 5 (1954). See also W. Pauli, Neils Bohr and the Development of Physics, ed. W. Pauli, p. 30 (Pergamon, 1955).
4. E. Fermi, Nuovo Cim. 11, 1 (1934); Z. Phys. 88, 161 (1934).
5. N. Takaoka and G. Ogata, Z. Natur. 21A, 84 (1966), E.K. Gerling, Yu. A. Shukolyukov and G. Sh. Ashkimadze, Yad. Fiz. 6, 311 (1967); Engl. transl: Sov. J. Nuc. Phys. 6, 226 (1968). T. Kirsten et al., Phys. Rev. Letts. 20, 1300 (1968).
6. H. Primakoff and S.P. Rosen, Rep. Prog. Phys. 22, 121 (1959); Proc. Phys. Soc. 78, 464 (1961); Phys. Rev. 184, 1925 (1969).
7. K. Borer et al., Phys. Lett. 29B, 614 (1969).
8. G. Danby et al., Phys. Rev. Lett. 9, 36 (1962); J.L. Bienlein et al., Phys. Lett. 13, 80 (1964).
9. S.J. Barish et al., Phys. Rev. Lett. 33, 448 (1974).

REFERENCES for Chapter 3

1. R.P. Feynman and M. Gell-Mann, Phys. Rev. 109, 143 (1958).
2. T. Kinoshita and A. Sirlin, Phys. Rev. 113, 1652 (1959).
3. C. Jarlskog, Nucl. Phys. 75, 659 (1966); D. Fryberger, Phys. Rev. 166, 1379 (1968); S.E. Derenzo, Phys. Rev. 181, 1854 (1969).
4. M. Fierz, Z. Phys. 104, 553 (1937); R.H. Good, Rev. Mod. Phys. 27, 187 (1955).
5. This will be true if we impose T-invariance upon the more general matrix element (3.58). The proof of this is left as an exercise, but remember $\mathcal{T}$ is an anti-linear operator.
6. M. Roos and A. Sirlin, Nucl.Phys. B29, 296 (1971).
7. H. Gurr, F. Reines and H. Sobel, Phys. Rev. Lett. 28, 1406 (1972); Proc. Trieste Conf. (1973).
8. M.A. Ruderman, Weak Interactions and Low Energy Nuclear Physics, ed. J.S. Bell, p.111 (CERN Geneva, 1969).
9. R.B. Stothers, Phys. Rev. Lett. 24, 538 (1970).
10. F.J. Hasert et al., Phys. Lett. 46B, 121 (1973).
11. H. Yukawa, Proc. Phys. Math. Soc., Japan, 17, 48 (1935); J. Schwinger, Ann. Phys., N.Y., 2, 407 (1957).
12. See e.g. N.N. Bogoliubov and D.V. Shirkov, Introduction to the Theory of Quantized Fields, p.144 (Interscience, 1959).
13. T.D. Lee, Phys. Rev. 128, 899 (1962).
14. R.A. Shaffer, Phys. Rev. 128, 1452 (1962); D. Bailin, Phys. Rev. 135, B166 (1964).
15. S.E. Derenzo, Phys. Rev. 181, 1854 (1969).
16. A.C.T. Wu, C.P. Yang, K. Fuchel and S. Heller, Phys. Rev. Lett. 12, 57 (1964).
17. G. Bernadini et al., Phys. Lett. 13, 86 (1966); M.M. Block et al., Phys. Lett. 12, 281 (1964).

REFERENCES for Chapter 4

1. M. Gell-Mann, Phys. Rev. 125, 1067 (1962); M. Gell-Mann and Y. Ne'eman, The Eightfold Way (Benjamin, 1964).

2. S. Okubo, Lectures on Unitary Symmetry, University of Rochester report, unpublished; P. Carruthers, Introduction to Unitary Symmetry (Wiley, 1966); R.E. Behrends, J. Dreitlein, C. Fronsdal and B.W. Lee, Rev. Mod. Phys. 34, 1 (1962); J.J. de Swart, Rev. Mod. Phys. 35, 916 (1963).

3. E.U. Condon and G.H. Shortley, Theory of Atomic Spectra (Cambridge, 1935).

4. M. Gell-Mann and M. Levy, Nuovo Cim. 16, 705 (1960).

5. We follow the treatment presented by Jackiw in S.B. Treiman, R. Jackiw and D.J. Gross, Lectures on Current Algebra and its Applications, p.101 (Princeton, 1972).

6. N. Cabibbo, Phys. Rev. Lett. 10, 531 (1963).

7. S.S. Gerstein and Y.B. Zeldovitch, Soviet Phys. J.E.T.P. 2, 576 (1956); R.P. Feynman and M. Gell-Mann, Phys. Rev. 109, 193 (1958).

8. J. Schwinger, Phys. Rev. Lett. 3, 296 (1959); P. Federbush and K. Johnson, Phys. Rev. 120, 1926 (1960); S. Okubo, Nuovo Cim. 44A, 1015 (1966); Ann. Phys. N.Y., 38, 377 (1966).

9. M. Gell-Mann, Physics 1, 63 (1964).

10. Particle Data Group, Rev. Mod. Phys. 45, No. 2, Part II Supp. (1973).

11. E. Di Capua et al., Phys. Rev. 133, B1333 (1964).

12. S.M. Berman, Phys. Rev. Lett. 1, 468 (1958); T. Kinoshita, Phys. Rev. Lett. 2, 477 (1959).

13. J.E. Augustin et al., Phys. Rev. Lett. 20, 129 (1968); Phys. Lett. 28B, 588 (1969).

14. N.P. Chang, Phys. Rev. 131, 1272 (1963).

15. P. Depommier et al., Nucl. Phys. 134, 189 (1968).

16. M.L. Goldberger and S.B. Treiman, Phys. Rev. 110, 1178 and 1478 (1958).

17. S.L. Adler, Phys. Rev. Lett. 14, 1051 (1965); Phys. Rev. 140, B.736 (1965); W.I. Weisberger, Phys. Rev. Lett. 14, 1047 (1965), Phys. Rev. 143, 1302 (1966).

18. S. Fubini, G. Furian and C. Rossetti, Nuovo Cim. 40, 1171 (1965).

19. R.P. Feynman, unpublished (1966).

20. S.L. Adler and R.F. Dashen, Current Algebras (Benjamin, 1968).

21. H. Lehman, K. Symanzik and W. Zimmerman, Nuovo Cim. 1, 1425 (1955); 6, 319 (1957).

22. M.L. Goldberger, Phys. Rev. 99, 979 (1965); R. Karplus and M. Ruderman, Phys. Rev. 98, 971 (1955).

23. Yu. I. Pomeranchuk, Sov. Phys. J.E.T.P. 7, 499 (1958).

24. A. Kropf and H. Paul, Z. Phys. 267, 129 (1974).

25. A. Sirlin, Phys. Rev. Lett. 18, 1224 (1967); Phys. Rev. 164, 1767 (1967).

26. J.C. Hardy, H. Schmeing, J.S. Geiger and R.L. Graham, Nucl. Phys. A223, 157 (1974).

27. C.J. Christensen et al., Phys. Rev. D5, 1628 (1972).

28. M. Gell-Mann, Phys. Rev. 111, 362 (1958).

29. R.T. Shann, Nuovo Cim. 5A, 591 (1971).

30. F.D. Calaprice, E.D. Commins and D. Girvin, Phys. Rev. D9, 519 (1974).

31. K. Kubodera, J. Delorme and M. Rho, Nucl. Phys. B66, 253 (1973); D.H. Wilkinson, Phys. Lett. 48B, 169 (1974).

32. H. Pagels, Phys. Rev. 179, 1337 (1969); R.A. Coleman and J.W. Moffatt, Phys. Rev. 186, 1635 (1969); H. Pagels and A. Zepeda, Phys. Rev. D5, 3262 (1972); H.J. Braathen, Nucl. Phys. B44, 93 (1972); H.F. Jones and M.D. Scadron, Phys. Rev. D11, 174 (1975).

33. M.R. Goldman, Nucl. Phys. B49, 621 (1972); D. Phil thesis, University of Sussex, unpublished.

34. A.A. Quaranta et al., Phys. Rev. 177, 2118 (1969).

35. A. Halpern, Phys. Rev. Lett. 13, 660 (1964); W.R. Wessel and P. Phillipson, Phys. Rev. Lett. 13, 23 (1964).

36. E.J. Bleser et al., Phys. Rev. Lett. 8, 228 (1962).

37. J.E. Rothberg et al., Phys. Rev. 132, 2664 (1963).

38. R. Hildebrand, Phys. Rev. Lett. 8, 34 (1962).

39. E. Bertolini et al., Proc. Int. Conf. on High Energy Physics, ed. J. Prentki, p. 421 (CERN, Geneva, 1962).

40. R.J. Blin-Stoyle, Fundamental Interactions and the Nucleus (North-Holland, 1972).

41. M. Gell-Mann and A. Pais, Phys. Rev. 97, 1387 (1955).

42. L.B. Okun, Weak Interactions of Elementary Particles, p.113, (Pergamon, 1965).

43. See e.g. J.K. Kim, Phys. Rev. Lett. 19, 1079 (1967).

44. J. Dreitlein and H. Primakoff, Phys. Rev. 125, 1671 (1962).

45. S. Coleman and S.L. Glashow, Phys. Rev. Lett. 6, 423 (1961).

46. T.D. Lee and C.N. Yang, Phys. Rev. 126, 2239 (1962); A. Pais, Phys. Rev. Lett. 9, 117 (1962).

47. M. Gourdin and A. Martin, CERN-TH 261 (1962); M. Gourdin and G. Charpak, Cargese Lectures, ed. M. Levy (Benjamin, 1963).

48. J. Lovseth, Phys. Lett. 5, 200 (1963).

49. S.L. Adler, Nuovo Cim. 30, 1020 (1963); E 32, 309 (1964).

50. N. Dombey, Rev. Mod. Phys. 41, 236 (1969); N. Dombey and B.J. Read, Nucl. Phys. B60, 65, (1973).

51. T.D. Lee, Lecture Notes of Int. School of Physics 'Enrico Fermi', ed. T.D. Lee, p.142 (Academic Press, 1966).

52. See e.g. S.L. Adler, Ann. Phys. N.Y. 50, 189 (1968).

53. C.H. Llewellyn Smith and A. Pais, Phys. Rev. Lett. 28, 865, E1356 (1972).

54. S.L. Adler, Phys. Rev. 135, B963 (1964).

55. C.H. Llewellyn Smith, Phys. Reps. 3C, 261 (1972).

56. S.V. Bonetti et al., Lett. Nuovo Cim. 2, 817 (1964).

57. S.L. Adler, Phys. Rev. 140, B736 (1965).

58. S.L. Adler, Phys. Rev. 143, B1144 (1966).

59. Lee and Yang, ref. (46).

60. M.G. Doncel and E. de Rafael, Nuovo Cim. 4A, 363 (1971).

61. J.D. Bjorken, Phys. Rev. 179, 1547 (1969).

62. J.D. Bjorken and E.A. Paschos, Phys. Rev. D1, 3151 (1970); S.D. Drell, D.J. Levy and T.M. Yan, Phys. Rev. Lett. 22, 744 (1969); Phys. Rev. 187, 2159 (1969); D1, 1035, 1617 (1970); Llewellyn Smith ref. (55); S.D. Drell and T.M. Yan, Ann. Phys. N.Y. 66, 555 (1971); R.P. Feynman, Phys. Rev. Lett. 23, 1415 (1969) and High Energy Collisions (Gordon and Breach, 1969). P.V. Landshoff, J.C. Polkinghorne and R.D. Short, Nucl. Phys. B28, 325 (1971).

63. C.G. Callan and D.J. Gross, Phys. Rev. Lett. 22, 156 (1969); D.J. Gross and C.H. Llewellyn Smith, Nucl. Phys. B14, 337 (1969).

64. T. Eichten et al., Phys. Lett. 46B, 274 (1973).

65. R.C. Barish et al., Proc. XVII Int. Conf. on High Energy Physics (Rutherford Lab., Chilton, Berks, 1974).

66. C.H. Llewellyn Smith, Nucl. Phys. B17, 277 (1970).

67. E. Bloom et al., SLAC-PUB 815, 907 (1970); G. Miller et al., Phys. Rev. D5, 519 (1970).

68. C.H. Llewellyn Smith, Phys. Rev. D4, 2392 (1971).

69. R.A. Brandt, Phys. Rev. Lett. 23, 1260 (1969); H. Leutwyler and J. Stern, Nucl. Phys. B20, 77 (1970); R.A. Brandt and G. Preparata, Phys. Rev. Lett. 25, 541 (1970); Nucl. Phys. B27, 541 (1971). See also, D.J. Gross, Lectures on Current Algebra and its Applications, S.B. Treiman, R Jackiw and D.J. Gross, p. 329 (Princeton, 1972).

70. R.E. Behrends, R.J. Finkelstein and A. Sirlin, Phys. Rev. 101, 866 (1956); S.M. Berman, Phys. Rev. 112, 267 (1958).

71. T. Kinoshita and A. Sirlin, Phys. Rev. 113, 1652 (1959).

72. T.D. Lee, Phys. Rev. 128, 899 (1962); D. Bailin, Phys. Rev. 135 B166 (1964); Nuovo Cim. 40A, 822 (1965).

73. A. Bailin and D. Bailin, Nucl. Phys. B17, 317 (1970).

74. J.D. Bjorken, Phys. Rev. 148, 1467 (1966); E 160, 1582 (1967); E.S. Abers, R.E. Norton and D. Dicus, Phys. Rev. Lett. 18, 676 (1967); E.S. Abers, D.A. Dicus, R.E. Norton and H.R. Quinn, Phys. Rev. 167, 1461 (1968).

75. A. Sirlin, Phys. Rev. Lett. 19, 877 (1967).

76. E.S. Ginsberg, Phys. Rev. 142, 1035 (1966); 162, 1570 (1967); 171, 1675 (1968).

77. C.G. Callan and S.B. Treiman, Phys. Rev. Lett. 16, 153 (1966); V.S. Mathur, S. Okubo and L.K. Pandit, Phys. Rev. Lett. 16, 371 (1966).

78. M Gell-Mann, Caltech report CTSL-20, unpublished (1961); S. Okubo, Prog. Theor. Phys. 27, 949 (1962).

79. M. Ademollo and R. Gatto, Phys. Rev. Lett. 13, 264 (1964); P. Langacker and H. Pagels, Phys. Rev. Lett. 30, 630 (1973).

80. G. Donaldson et al., Phys. Rev. Lett. 31, 337 (1973); D. Hitlin, Proc. of Summer Inst. in Particle Physics, SLAC Report 167, 2, 167 (1973).

81. L.M. Chounet, J.M. Gaillard and M.K. Gaillard, Phys. Reps. 4C, 199 (1972); S. Wojcicki, Proc. of Summer Inst. of Particle Physics, SLAC Report 167, 2, 35 (1973).

82. D. Amati, C. Bouchiat and J. Nuyts, Phys. Lett. 19, 59 (1965); C.A. Levinson and I.J. Muzinich, Phys. Rev. Lett. 15, 715 (1965); L.K. Pandit and J. Schechter, Phys. Lett. 19, 56 (1965); Weisberger ref. (17).

83. W. Willis and J. Thompson, Advances in Particle Physics, eds. R.L. Cool and R.E. Marshak, 2, p. 295 (Interscience, 1968).

84. K. Kleinknecht, Proc. XVII Int. Conf. on High Energy Physics, (Rutherford Lab., Chilton, Berks, 1974).

85. R.E. Marshak, Riazuddin and C.P. Ryan, Theory of Weak Interactions in Particle Physics, p. 411 (Wiley-Interscience, 1969).

86. G. Ebel, H. Pilkuhn and F. Steiner, Nucl. Phys. B17, 1 (1970).

87. M. Roos, Phys. Lett. 36B, 130 (1971).

88. M. Roos, Nucl. Phys. B77, 420 (1974).

89. Adler, ref. (49); A. Pais, Ann. Phys. 63, 311 (1971); E 69, 604 (1972).

90. T.D. Lee and C.N. Yang, Phys. Rev. 119, 1410 (1961).

REFERENCES for Chapter 5

1. P. McNamee and F. Chilton, Rev. Mod. Phys. 36, 1005 (1964).

2. Particle Data Group, Phys. Lett. 50B, 1 (1974).

3. N. Cabibbo, Phys. Rev. Lett. 12, 62 (1964); M. Gell-Mann, Phys. Rev. Lett. 12, 155 (1964); K. Itabashi, Phys. Rev. 136, B221 (1964); D. Bailin, Nuovo Cim. 38, 1342 (1965); S.P. Rosen, S. Pakvasa and E.C.G. Sudarshan, Phys. Rev. 146, 1118 (1966)

4. C. Bouchiat and Ph. Meyer, Phys. Lett. 22, 198 (1966); M. Suzuki, Phys. Rev. 144, 1154 (1966).

5. S. Weinberg, Phys. Rev. Lett. 4, 87; E 585 (1960).

6. R.H. Dalitz, Phil. Mag. 44, 1068 (1953); Phys. Rev. 94, 1046 (1954).

7. S.K. Bose and S.N. Biswas, Phys. Rev. Lett. 16, 346 (1966); C. Bouchiat and Ph. Meyer, ref. (4), Phys. Lett. 25B, 282 (1967); C.G. Callan and S.B. Treiman, Phys. Rev. Lett. 16, 153 (1966); D.K. Elias and J.C. Taylor, Nuovo Cim. 44A, 518 (1966), E 48A, 814 (1967); J. Weyers, L.L. Foldy and D.R. Speiser, Phys. Rev. Lett. 17, 1062 (1966); Suzuki, ref. (4); H.T. Nieh, Phys. Rev. Lett. 20, 82 (1968).

8. Bouchiat and Meyer, ref. (7).

9. M. Suzuki, Phys. Rev. 137, B1602 (1965); Bailin, ref. (3); Rosen et al., ref. (3).

10. B.W. Lee, Phys. Rev. Lett. 12, 83 (1964); H. Sugawara, Prog. Theor. Phys. 31, 213 (1964); M. Gell-Mann, Phys. Rev. Lett. 12, 155 (1964); S. Okubo, Phys. Rev. Lett. 8, 362 (1964).

11. M. Suzuki, Phys. Rev. Lett. 15, 986 (1965); H. Sugawara, Phys. Rev. Lett. 15, 870; E 997 (1965).

12. Y. Hara, Y. Nambu and J. Schechter, Phys. Rev. Lett. 16, 380 (1966); S. Badier and C. Bouchiat, Phys. Lett. 20, 529 (1966); L.S. Brown and C.M. Sommerfield, Phys. Rev. Lett. 16, 751 (1966).

13. Fayyazuddin and Riazuddin, Nuovo Cim. 47A, 222 (1967).

14. See e.g. J.K. Kim, Phys. Rev. Lett. 19, 1079 (1967).

15. Brown and Sommerfield, ref. (12).

16. Callan and Treiman, ref. (7); M.K. Gaillard, Phys. Lett. 20, 533 (1966); R. Gatto, L. Maiani and G. Preparata, Nuovo Cim. 41, 622 (1966); Y. Nambu and Y. Hara, Phys. Rev. Lett. 16, 855 (1966); Riazuddin and K.T. Mahanthappa, Phys. Rev. 147, 972 (1966).

17. M-Y. Han and Y. Nambu, Phys. Rev. 139, B1006 (1965); D10, 674 (1974); O.W. Greenberg, Phys. Rev. Lett. 13, 598 (1964); A. Tavkheldize, Seminar on High Energy Physics and Elementary Particles, p. 763 (IAEA, Vienna, 1965); M. Gell-Mann, Act. Phys. Aust. Supp. 9, 733 (1972); Proc. XVI Int. Conf. on High Energy Physics, eds. J.D. Jackson and A. Roberts, 4, p.333 (NAL, Batavia, Ill., 1972); H. Fritzsch, M. Gell-Mann and H. Leutwyler, Phys. Lett. 47B, 365 (1973).

18. J. C. Pati and C. H. Woo, Phys. Rev. D3, 2920 (1971).

19. G. Barton, Nuovo Cim. 19, 512 (1961); E.M. Henley, Ann. Rev. Nucl. Sci. 19, 367 (1969).

20. D. Bailin, Rep. Prog. Phys. 34, 491 (1971).

21. E. Fischbach and K. Trabert, Phys. Rev. 174, 1843 (1968).

22. R.J. Blin-Stoyle, Phys. Rev. 118, 1605 (1960); F.C. Michel, Phys. Rev. 133, B329 (1964).

23. E. Fischbach, D. Tadic and K. Trabert, Phys. Rev. 186, 1688 (1969); E. Fischbach and D. Tadic, Phys. Rep. 6C, 123 (1973).

24. P. Olesen and J.S. Rao, Phys. Lett. 29B, 233 (1964).

25. B.H.J. McKellar, Phys. Lett. 26B, 107 (1967); E. Fischbach, Phys. Rev. 170, 1398 (1968); Fischbach and Trabert, ref. (21).

26. D.H. Wilkinson, Phys. Rev. 109, 1603 (1958).

27. R.J. Blin-Stoyle, Fundamental Interactions and the Nucleus, Chs. 8, 9 (North-Holland, 1973).

28. M. Gari, Phys. Rep. 6C, 317 (1973),

29. F.C. Michel, Phys. Rev. 133, B329 (1964); M. Gari and H. Kummel, Phys. Rev. Lett. 23, 26 (1969); M. Gari, Phys. Lett. 31B, 627 (1970).

30. H. Hattig, K. Hünchen, P. Roth and H. Waffler, Nucl. Phys. A 137, 144 (1969); H. Waffler et al., to be published.

31. G.S. Danilov, Phys. Lett. 18, 40 (1965).

REFERENCES for Chapter 6

1. W. Heisenberg, Ann. Phys. Lpz. 32, 20 (1938).

2. P.W. Higgs, Phys. Lett. 12, 132 (1964); Phys. Rev. 145, 1156 (1966); T.W.B. Kibble, Phys. Rev. 155, 554 (1967).

3. B.W. Lee, Phys. Rev. D5, 823 (1972).

4. S. Weinberg, Phys. Rev. Lett. 19, 1264 (1967); A. Salam, Elementary Particle Theory, ed. N. Svartholm (Almquist and Wiksell, Stockholm, 1968).

5. G. 't Hooft, Nucl. Phys. B35, 167 (1971). See also B.W. Lee and J. Zinn-Justin, Phys. Rev. D5, 3121, 3137, 3155 (1972).

6. G.M. Kendall, University of Sussex, M.Sc. thesis, unpublished (1968).

7. F.J. Hasert et al., Phys. Lett. 46B, 121 (1973).

8. H.S. Gurr, F. Reines and H.W. Sobel, Phys. Rev. Lett. 28, 1406 (1972).

9. J. Godine and A. Hankey, Phys. Rev. D6, 3301 (1972); A. Love, Lett. Nuovo Cim. 5, 113 (1972); V.K. Cung, A.K. Mann and E.A. Paschos, Phys. Lett. B41, 355 (1972); D.A. Dicus, Phys. Rev. D8, 890 (1973); R. Budny, Phys. Lett. B45, 345 (1973).

10. D.A. Ross, Nucl. Phys. B51, 116 (1973).

11. See e.g. S. Weinberg, Rev. Mod. Phys. 46, 255 (1974).

12. H. Georgi and S.L. Glashow, Phys. Rev. Lett. 28, 1494 (1970).

13. S.L. Glashow, J. Iliopoulos and L. Maiani, Phys. Rev. D2, 1285 (1970); S. Weinberg, Phys. Rev. Lett. 27, 1689 (1971).

14. W. Angerson, Nucl. Phys. B69, 493 (1974); A. Sirlin, Phys. Rev. Lett. 32, 966 (1974).

15. C.H. Albright, Phys. Rev. D8, 3162 (1973); L.M. Seghal, Nucl. Phys. B65, 141 (1973); S.L. Glashow, Proc. Erice Summer School on Particle Physics (1973); R.B. Palmer, Phys. Lett. 46B, 240 (1973).

16. F.J. Hasert et al., Phys. Lett. 46B, 138 (1973).

17. A. Benvenuti et al., Phys. Rev. Lett. 32, 800 (1974).

18. A. Love, D.V. Nanopoulos and G.G. Ross, Nucl. Phys. B49, 513 (1972).

19. S.L. Adler, Phys. Rev. 177, 2426 (1969); J.S. Bell and R. Jackiw, Nuovo Cim. 60A, 47 (1969).

20. W. Bardeen, Phys. Rev. 184, 1848 (1969); J. Wess and B. Zumino, Phys. Lett. 37B, 95 (1971).

21. D.G. Sutherland, Nucl. Phys. B2, 433 (1967); M. Veltman, Proc. Roy. Soc. A301, 107 (1967).

22. M. Gell-Mann, Proc. XVI Int. Conf. on High Energy Physics, eds. J.D. Jackson and A. Roberts, 4, p.333, (NAL, Batavia, Ill., 1972).

23. A.D. Dolgov, L.B. Okun and V.I. Zakharov, Phys. Lett. 47B, 258 (1973).

24. M.Y. Han and Y. Nambu, Phys. Rev. 139, B1006 (1965).

25. C.H. Llewellyn Smith, Phys. Lett. 46B, 233 (1973); See also J.M. Cornwall, D.N. Levin and G. Tiktopoulos, Phys. Rev. Lett. 30, 1268; E 31, 572 (1973).

26. R.N. Mohapatra, Phys. Rev. D6, 2023 (1972).

27. A. Pais, Phys. Rev. D8, 625 (1973).

28. T.D. Lee, Phys. Rev. D8, 1226 (1973).

29. H. Georgi and A. Pais, Phys. Rev. D10, 1246 (1974).

30. C.G. Callan, Phys. Rev. D2, 1541 (1970); K. Symanzik, Comm. Math. Phys. 18, 227 (1970); 23, 49 (1971).

31. C.G. Callan, Erice Summer School Notes (1973).

32. S. Weinberg, Phys. Rev. 118, 838 (1960).

33. M. Gell-Mann and F.E. Low, Phys. Rev. 95, 1300 (1954); N.N. Bogoliubov and D.V. Shirkov, Intro. to the Theory of Quantized Fields (Interscience, 1959).

34. H.T.H. Piaggio, Elementary Treatise on Differential Equations and their Applications, p.133 (Bell, 1950).

35. H.D. Pollitzer, Phys. Rev. Lett. 26, 1346 (1973); D.J. Gross and F. Wilczek, Phys. Rev. Lett. 26, 1343 (1973).

36. W. Bardeen, H. Fritzsch and M. Gell-Mann, Proc. of Topical Meeting on Conformal Invariance in Hadron Physics (Frascati, 1972).

37. H. Fritzsch and M. Gell-Mann, Proc. XVI Int. Conf. on High Energy Physics, eds. J.D. Jackson and A. Roberts, 2, 135 (NAL, Batavia, Ill., 1972).

38. K.G. Wilson, Phys. Rev. 179, 1499 (1969).

39. R.A. Brandt and G. Preparata, Nucl. Phys. B27, 541 (1971).

40. H. Georgi and H.D. Pollitzer, Phys. Rev. D9, 416 (1974); D.J. Gross and F. Wilczek, Phys. Rev. D9, 980 (1974); D. Bailin, A. Love and D.V. Nanopoulos, Lett. Nuovo Cim. 9, 501 (1974).

41. D.J. Fox et al., Phys. Rev. Lett. 33, 1504 (1974).

42. G. Altarelli and L. Maiani, Phys. Lett. 52B, 351 (1974); M.K. Gaillard and B.W. Lee, Phys. Rev. Lett. 33, 108 (1974).

43. A. Likte et al., Phys. Rev. Lett. 30, 1189 (1973); G. Tarnopolsky et al., Phys. Rev. Lett. 32, 432 (1974).

44. J.D. Bjorken, Phys. Rev. 148, 1467 (1966).

REFERENCES for Chapter 7

1. J.H. Christenson, J.W. Cronin, V.L. Fitch and R. Turlay, Phys. Rev. Lett. 13, 138 (1964); Phys. Rev. 140, B74 (1965).

2. See e.g. P.K. Kabir, The CP Puzzle (Academic Press, 1968), or R.E. Marshak, Riazuddin and C.P. Ryan, Theory of Weak Interactions in Particle Physics, p. 610 (Wiley-Interscience, 1969).

3. T.D. Lee and L. Wolfenstein, Phys. Rev. 138, B1490 (1965).

4. Particle Data Group, Phys. Lett. 50B, 1 (1974).

5. J.S. Bell and J. Steinberger, Proc. Oxford Int. Conf. on Elementary Particles, eds. R.G. Moorhouse, A.E. Taylor and T.R. Walsh, p. 195 (RHEL, Didcot, 1965).

6. P.L. Crawford, Phys. Rev. Lett. 15, 1045 (1965).

7. Kabir, ref. (2), p. 110.

8. G. Luders, Dansk, Mat. Phys. Medd. 28, 5 (1954); W. Pauli, Neils Bohr and the Development of Physics, ed. W. Pauli, p. 30 (Pergamon, 1955).

9. T.T. Wu and C.N. Yang, Phys. Rev. Lett. 13, 180 (1974); L.B. Okun and C. Rubbia, Proc. Heidelberg Int. Conf. on Elementary Particles, ed. H. Filthuth, p. 301 (North-Holland, 1968).

10. See K. Kleinknecht, Proc. XVII Int. Conf. on High Energy Physics (Rutherford Lab., Chilton, Berks, 1974).

11. L. Wolfenstein, Phys. Rev. Lett. 13, 180 (1964).

12. L. Wolfenstein, Nucl. Phys. B77, 373 (1974).

13. G. Barton and E. D. White, Phys. Rev. 184, 1660 (1969).

14. W. B. Dress, P. D. Miller and N. F. Ramsey, Phys. Rev. D7, 3147 (1973).

15. L. Wolfenstein, Theory and Phenomenology in Particle Physics, p. 218 (Academic Press, 1969); M. A. B. Beg, Ann. Phys. N.Y. 52, 577 (1969).

16. R. N. Mohapatra and J. C. Pati, Phys. Rev. D8, 2317 (1973).

17. B. Stech, IIeme Conf. Int. sur les Particules Elementaire, Aix-en-Provence 1973, J. Phys. Coll. C 1, Supp. no. 10, Tome 34, 71 (1973).

18. D. Girvin, University of California thesis, unpublished (1972); F. P. Calaprice, E. D. Commins and D. Girvin, Phys. Rev. D9, 519 (1974).

19. D. Bailin, Rep. Prog. Phys. 34, 491 (1971).

20. S. Barshay, Phys. Lett. 17, 78 (1965); J. Bernstein, G. Feinberg and T. D. Lee, Phys. Rev. 139, B1650 (1965); T. D. Lee, Phys. Rev. 140, 959 (1965); F. Salzman and G. Salzman, Phys. Lett. 15, 91 (1965); L. B. Okun, Phys. Lett. 23, 595 (1968).

21. D. J. Broadhurst, Nucl. Phys. B20, 603 (1970)

INDEX